THE

OLD VEGETABLE NEUROTICS

Hemlock, Opium, Belladonna and Henbane

THEIR

PHYSIOLOGICAL ACTION AND THERAPEUTICAL USE
ALONE AND IN COMBINATION

BEING

THE GULSTONIAN LECTURES OF 1868

EXTENDED AND INCLUDING A

COMPLETE EXAMINATION OF THE ACTIVE CONSTITUENTS OF OPIUM

BY

JOHN HARLEY, M.D. Lond. F.R.C.P. F.L.S.

HONORARY FELLOW OF KING'S COLLEGE, LONDON; LATE ASSISTANT
PHYSICIAN TO KING'S COLLEGE HOSPITAL; ASSISTANT
PHYSICIAN TO THE LONDON FEVER HOSPITAL

London
MACMILLAN AND CO.
1869

PREFACE.

FACTA PRÆSTANTIORA VERBIS.

THE LECTURES which form the basis of this work were delivered at the Royal College of Physicians in the spring of last year. Since this time I have, as leisure has permitted me, enlarged my observations, and extended them to the active constituents of Opium, each of which has been examined in detail, alone and in combination with Belladonna. The question of the antagonism of Opium and Belladonna, which has assumed a very serious and important phase, has been fully examined.

My object throughout has been to ascertain, clearly and definitely, the action of the drugs employed on the healthy body in medicinal doses, from the smallest to the largest; to deduce simple practical conclusions from the facts observed; and then to apply the drug to the relief of the particular conditions to which its action appeared suited.

Observations on the lower animals have been as carefully noted as those on man, and my labours have in this respect been amply rewarded; for I find, generally, that for every variation in the effects of a particular drug on man, we may expect to see its exact counterpart in some one or other of the lower animals. It appears, indeed, that the effects which result from a given action depend, primarily, on the specific development of the nervous system; and, secondarily, on individual peculiarity. Many and varied as the effects of the action of a particular medicine often are, they constitute but one connected series, the members of which are reciprocally complementary. It follows that carefully-observed experiments on the animal series will elicit the whole of the phenomena which may result from the action of the same drug in different individuals of the human species. And in order, therefore, to obtain a *complete* view of these, we must subject the medicine to the analytical action of that variety of nervous system which characterises the different species of animals.

Much time and labour has been expended in laying before the reader observations followed throughout from the beginning to the end of the action of a given dose; and my first endeavour has been to present to the mind, not only the particular effects in the order of their sequence, but also those more general features which convey, as well as imperfect words may do, impressions of the actual condition of the individual subjected to the influence of the drug.

Tedious detail and some repetition may have resulted from this endeavour, but such defects are trivial in comparison of insufficiency in the evidence adduced.

The study of the action of medicines is more difficult than any other, inasmuch as the conditions are more complex. To gain any knowledge in this direction, the disturbing causes must be eliminated, and the question otherwise reduced to its simplest terms.

To recognise and eliminate the disturbing agencies

which modify a given effect, is a process much more difficult than to deduce conclusions from any number of simple facts. The main difficulties, indeed, of therapeutical enquiry lie in correct observation: there are errors to be eliminated, or variations to be accounted for, in the drug itself; while the peculiarity of nervous constitution or idiosyncrasy, the conditions of health and disease, and every variety of external agencies, all interpose disturbances, the value of which in each case it is difficult to estimate.

In questions relating to the action of medicines, medical men often betray a want of intelligence as great as that of those who have no special knowledge of the subject. We ask each other to believe statements dependent on the slenderest proofs, and sometimes without proof at all, or even when it points in the opposite direction. Impatient to observe, we satisfy ourselves with general impressions, which, taking the place of logical inferences from carefully-observed facts, ultimately become deeply-rooted convictions. These to other minds are unsatisfactory, and engender at first difficulty, then doubt, indifference, and scepticism. Medicine is proportionately degraded, and the treatment of disease is changed according to fashion. In the present day, the use of secret nostrums is openly sanctioned and adopted. Patients are allowed to drug themselves to death with anodynes and narcotics. The profession includes numbers of men who, if they have faith in their practice, evince an ignorance discreditable to an Anglo-Saxon Leech, and who, if they have not, are the basest of charlatans.

Medicine appeals to her honest followers, to maintain the dignity of science and the purity of truth; and if we would answer her appeal, we must avoid hasty conclusions, and patiently study those slighter variations and less obtrusive phenomena, of which the more general and obvious effects are compounded.

Our first impressions on entering the wilderness of therapeutical enquiry must indeed be discouraging, and the prospect of reducing anything to order, at first sight, hopeless. A lifetime will seem too short to effect any change, and we shall be inclined to turn back. But let us shut out the desert and the jungle from our view, and turn to the nearest object. Let us clear away the suffocating undergrowth from about it, denude it of the tangled climbers that conceal its trunk, and the moss which covers its branches. Let us lop off the parasites that deform it, and the foreign branches, it may be, which some previous hand has engrafted, and, thus isolated and reduced to its natural simplicity, let us choose it as the special object of our study and care. Life may be long enough to know this single individual; and if we each one effect so much, what is now an uncultivated wild, with scarcely one well-ordered patch to rest the eyes upon, will soon show signs of culture; and with continued labour become in future generations a fair garden —a health resort—where, with simple directions, we may send our patients to cull the good gifts which a Beneficent Hand has planted and purposed for the relief of 'the thousand ills that flesh is heir to.'

78 Upper Berkeley Street,
Portman Square, W.
January 1860.

CONTENTS.

CONIUM MACULATUM.

CHAPTER I.

THE history of HEMLOCK has been often written, and those who wish for references to this subject may find them in a work 'De Cicuta,' by J. Viventius, published at Naples in the year 1777.

It is evident that the ancient Greek physicians possessed a tolerably accurate knowledge of the physiological action of this plant in poisonous doses, but it does not appear that they had so clear a perception of its effects in medicinal doses. The subsequent history of the plant presents us with a dreary view of the progress of medical knowledge. So late as the year 1845, a well ascertained and carefully observed case of poisoning by hemlock disclosed early truths respecting its toxical action which had lain dormant for many centuries and become encrusted with error. A glimmering of the ancient light has indeed occasionally appeared to those who from time to time have sought to determine the properties of the plant, as will be shown hereafter; but the general state of our knowledge may be summed up in the words of two of our most eminent therapeutists. Eleven years ago Pereira wrote:—'In the present state of uncertainty with respect to the real physiological operation of hemlock it is obviously impossible to lay down indications or contra-indications for its use, which can be much relied on.'[1] And Dr. Christison closes his admirable investigations upon the active principle of the plant with the following remarks: 'I wish I could have added to these observations on the

[1] Elem. Mat. Medica, vol. ii. part ii. p. 202.

poisonous effects of conia and hemlock some account of their physiological action in small doses. This branch of the enquiry into their action I have not yet been able to investigate; it cannot be pursued with any accuracy by experiments on the lower animals. The phenomena must be ascertained in the human subject chiefly, which I have not hitherto been able to accomplish. On this head it may merely be observed that if any physicians or physiologists would acquire definite information as to the physiological effects of hemlock in small or medicinal doses they must begin the enquiry anew. Little importance can be attached to anything already done in this field, as I have no doubt whatever that by far the greater portion of the preparations of hemlock hitherto employed have been of very little energy, and, in the doses commonly used, are absolutely inert.'[1]

Since these words were written, little or nothing has been done in the direction which they indicate; and they are as generally applicable at the present time as they were when they were first uttered. Reference to the most modern and approved treatises on the actions of medicines shows how little, apart from its bare toxical effects, is known of the physiological action of hemlock; and a critical examination of the preparations most recently provided for our use, and of their doses, prove, as I shall hereafter have occasion to show, that, as a medicine, we are practically ignorant of the virtues of this most valuable plant. I do not here speak of the toxical effects of hemlock, which have been well illustrated in one or two fatal cases that have occurred in the last few years. Experiments, too, on its toxical effects on the lower animals are most exhaustive and complete. Leaving then, with very few exceptions, the *corpus vile* out of consideration, I shall devote myself almost exclusively to a consideration of the subtle agency of this plant in man, and, generally speaking, in the able-bodied.

Having assured myself of the effects of hemlock in my own person, I have thus been able to fully appreciate its operation in others, and to proceed much more boldly than I could have done without such personal knowledge.

[1] Edinburgh Philosophical Trans., vol. xiii. p. 398.

Physiological Action of Hemlock.—The *first* effect of hemlock is a depression of the motor function; and its *last* is the complete obliteration of all muscular movement derived from the cerebro-spinal motor tract.

1. After taking three drachms of the *succus conii* of the 'British Pharmacopœia,' I set out *walking*; and three-quarters of an hour after the dose, I felt a heavy clogging sensation in my heels. There was a distinct impairment of motor power. I felt, so to speak, that 'the go' was taken out of me. It was not that I felt fatigued just then, but it seemed as if a drag was suddenly put upon me, and that it would have been impossible to walk fast, if urged to do so. After walking about a mile up-hill, this sensation was very decided; and on putting a foot on the scraper at the door of the hospital, the other leg was shaky, and felt almost too weak to support me. My movements appeared clumsy to myself, and it seemed necessary that I should make an effort to control them. At the same time, there was a sluggishness of the adaptation of the eye. My vision was good for fixed objects; but, when an uneven object was put in motion before the eyes, there was a haze and dimness of vision, producing a feeling of giddiness. The pulse and pupils were unaffected. These were the whole of the effects; and, after continuing for an hour, they rapidly disappeared, and left me in the possession of my usual vigour.

2. If a strong, active individual take five or six drachms of the succus on getting up in the morning, and start off for a long *walk*, he will be overtaken in the course of half or three-quarters of an hour with a feeling of general tiredness, and a special weakness of the knees, as if he had been regularly tired out by walking all day to the full extent of his powers. If he be unusually active and strong, he will not, perhaps, yield to the inclination to rest, but will proceed slowly on his way, feeling a strange lightness and powerlessness of the legs, with a tendency to drop forward on his knees. This will be associated with some giddiness, and a feeling of heaviness over the eyes. At first, the feeling of languor will be most oppressive, but it will soon become more tolerable; and, if he should continue his journey for an hour, he will find

that the feeling of fatigue has by this time nearly passed off. In the course of another hour, he will be as active as ever.

3. The following were the effects produced in my own person during a *period of rest*, and they contrast well with the foregoing. Three-quarters of an hour after taking five drachms and a half of the succus conii, on raising my eyes from the object upon which they had been fixed to a more distant one, the vision was confused, and a feeling of giddiness suddenly came over me. That these symptoms were due to impairment of power in the muscular apparatus employed in the adaptation of the eye, was obvious to me; for, so long as my eyes were fixed on a given object, the giddiness disappeared, and the definition and capacity of vision for the minutest objects were unimpaired. But the instant that I directed the eyes to another object, all was haze and confusion, and I felt giddy; and, in order to recover my vision and dismiss the sense of giddiness, it was necessary to lay hold upon some object, as it were, with my eyes, and rest them securely upon it. It was clear to me that the adjusting muscular apparatus of the eye was enfeebled, and its contractions so sluggishly performed that they could no longer keep pace with the more active movements of the external muscles of the eyeball. Within ten minutes of the appearance of this disorder of vision, a general muscular lethargy affected me, and the eyelids felt as heavy as if they were oppressed with the deepest drowsiness. The pupils were considerably dilated. I sat down to note these observations; but, being afraid to maintain this posture lest the rapidly increasing muscular lethargy should get the better of me, I rose up again and tried to shake it off. An hour and a quarter after taking the dose, I first felt decided weakness in my legs. The giddiness and diminution of motor power continued to increase for the next fifteen minutes. An hour and a half after taking the dose, these effects attained their maximum; and at this time I was cold, pale, and tottering. The pulse, which had been emotionally excited by the sudden accession of the foregoing symptoms, was now 68, quite regular, and of undiminished force and volume. The legs felt as if they would soon be too weak to support me.

There was a positive diminution of voluntary power in every part of the muscular system, and this nearly amounted to complete paralysis as far as the hamstring and levator palpebræ muscles were concerned. At one time, the greatest exertion was required to elevate the eyelids. The mind remained perfectly clear and calm, and the brain active throughout; but the body seemed heavy, and well-nigh asleep. After continuing for about half an hour at their maximum, the symptoms began rapidly to decline, and within three hours and a half of taking the dose they had totally disappeared. As a proof of this, I may mention that, for the hour following the disappearance of the conium symptoms, I was engaged in writing letters; I then walked briskly a distance of three miles, and finished the day's work by drawing a microscopic object.

In order to illustrate the further effects of conium, I select the following examples:—

4. I gave to a man aged 57, of powerful muscular development, the succus conii, in doses increased from three drachms to an ounce. The medicine produced no appreciable effect until the quantity was increased to six drachms. This dose was followed, twenty minutes after taking it, by sudden giddiness, and so much weakness of the legs as rendered the patient incapable of walking; and he was obliged to lie down. There was an aching pain across the eyebrows, and mistiness of vision. He could hardly raise the eyelids, which seemed pressed down with a heavy weight; and he was disposed to fall off to sleep. After twenty minutes he got up and walked a mile; but the legs were so weak that they could hardly support him; the knees tended to fall forwards; and his gait was tottering. An hour and a half after the dose, the effects had almost entirely passed off, and he felt as if nothing unusual had happened to him. On another occasion (see p. 17), after taking an ounce, these symptoms were repeated; but the accession of the giddiness and weakness was so rapid, that he would have fallen, but that he caught hold of a support. All the symptoms were intensified, and continued a little longer than when only six drachms were taken.

5. A delicate young woman, of inactive habits, took four drachms of the succus. Twenty minutes afterwards, and while attending to her usual duties, she experienced nausea and giddiness. She dropped an inkstand which she was holding in her hands, and was unable to walk; and she was placed in the recumbent posture. These symptoms came on with alarming swiftness, and the pulse went up to 120 from emotional excitement; but in a few minutes the heart regained its usual quietude, and she remained perfectly comfortable and calm, but without power to move the arms or legs. An hour after taking the medicine, there was nearly complete muscular paralysis; the eyelids were closed, the pupils widely dilated; and the mind clear, calm, and active, and she expressed herself quite comfortable. She tried perseveringly to raise the eyelids when I requested her to do so, but she was quite unable to separate their margins. The pulse and respiration were normal; the surface warm. At the end of an hour, these symptoms passed off; and after three hours she had completely recovered her activity, and resumed her duties. The next day she complained of slight wearisome pain in the muscles of the legs.

Such are the general and constant effects produced by hemlock when administered in full medicinal doses. When the dose falls short of producing any of the above-mentioned symptoms while the individual is in a state of ordinary activity, then we have absolutely no indications of the action of hemlock.

In my own person, I have sometimes thought that, in the absence of any appreciable symptoms, a feeling of dulness and depression might be attributed to its use; but, from my observations in others, I am rather inclined to attribute this to the anxiety about the expected operation of the medicine. When the eyes and legs are both at rest, then a feeling of languor will be the only effect of a dose of hemlock just sufficient, when these organs are in active play, to produce slight and transient giddiness and weakness of the legs. The operation of hemlock is uniform and invariable in man; and the same remark doubtless applies to the lower animals also. All animals, I believe, may be brought within its influence, and its effects will be in proportion to the degree of development of their motor apparatus.

It has been doubted whether the horse is affected by hemlock; but this doubt is solved by the following experiments made by Mr. Frederick Mavor, of Park Street, and myself, upon a thoroughbred two-year-old colt. At intervals of a week, we gave the succus conii of the Pharmacopœia, in doses of six, eight, twelve, and sixteen ounces, by the mouth. No effects followed any but the last dose, which is equivalent to a pound of the fresh leaves. He continued lively, and eating his food as usual, until twenty-five minutes after the dose, when he was observed to remain standing stockstill, with the ears fallen, and the head and neck pendent, the upper eyelids swollen and drooping, so as to nearly cover the eyes. He presented at the same time a dull, heavy, tumble-down appearance. The pulse, pupils, and tongue were unchanged. An experienced veterinary surgeon happened to come in at the time; and observing the general expression of languor exhibited by the animal, and the swollen and nearly closed eyelids, said, 'That horse has the influenza, so prevalent just now.' Five minutes afterwards, the animal dropped upon his knees, and in recovering his position, nearly tumbled down. After a little stumbling, he regained his legs, and continued for the next twenty minutes in the same state of dulness and perfect quietude as before, excepting that now and then, a fore or hind leg gave way, and he was obliged to exert himself to regain his equilibrium. At the end of this time he was walked out. After a little stumbling, he went along slowly and languidly, with the ears down, the head and neck depressed, and the eyelids half closed, swaying a little as he went. Two hours afterwards, the effect had entirely passed off, and he was as active and lively as before the dose. The head and ears were erect; the swelling and drooping of the upper lids—from the paralysis of the levator palpebræ and orbicularis—had disappeared; the pulse was of good volume and power, and unchanged. No excreta were voided during the action of the medicine.

I proceed now to consider in detail the action of conium upon the nervous system. The earliest indications of the operation of the medicine are invariably those that arise from depression of the motor function of the *third pair of nerves.* They are: giddiness; the sensation of a heavy weight depressing the eyelids, or actual ptosis; a dull lazy or fixed expressionless stare like that of a drunken person; dilatation of the pupils. After moderate doses, the interference of vision is only such as results in haziness, as if a thin film of transparent vapour were floating between the eye and the object; the effect being identical with that observed on looking through a medium of unequal density, such as the mixture of hot and cold air enveloping a highly heated stove. It occurs independently of any dilatation of the pupil, and is compatible with good definition for fixed objects. It is due to imperfect adjustment of the refracting media of the eye from partial paralysis of the ciliary branches of the third nerve. It is thróugh these minute branches that the individual first becomes conscious of the effect of hemlock, and if he should be reading at the time, he will suddenly find the occupation fatiguing, and, very soon afterwards, it may be impossible; and he will be glad to close the eyes to relieve himself of the symptom, and as the muscular lethargy begins to be felt, content to lie perfectly still as if asleep.

In full doses the depressing influence involves the other branches of the nerve and the lazy movements of the eyeball, or dull fixed and occasionally divergent stare, indicate the partially paralysed condition of the external *muscles of the eyeball*; while more or less drooping of the upper lids express a similar condition of the levator palpebræ.

Double vision, from inability to maintain the convergence of the optic axes, excepting as a very evanescent effect, is a comparatively rare result of the action of hemlock. I have only observed it in a few persons. In one of these, a delicate invalid confined by weakness and ovarian disease chiefly to the recumbent position, fl. ʒij. of the succus conii produced full effects accompanied by double vision. This was a constant symptom—it came on half an hour after taking the medicine and lasted twenty minutes. After having taken the hemlock for six months, she told me as often as I happened to see her during the operation of the medicine, that she saw each object in the room double, that my eyes were also doubled, and that she felt as if she were squinting.

Dilatation of the pupil occurs usually after only very large doses, and then it is often but slight, and only observable in a subdued light—the excitement of strong light overcoming the tendency to dilate, just as the exertion of a strong will strengthens for a time an enfeebled limb.

The absence of any preponderating action of the muscles supplied by the *fourth and sixth pairs of nerves* shows that they are equally affected with the third pair.

A proportionate diminution of power is also observed in the muscles supplied by the motor branches of the *fifth and seventh pairs.* The contractile power of the *m. orbicularis* in particular is distinctly weakened. This was observed in the horse; in the mouse (see below); and also in case 16, &c.

Upon the *eighth pair* the action of conium is not very apparent in a state of health, but in the spasmodic affections arising from irritation of this nerve its influence is very decided as will be shown hereafter (p. 49).

As to the *hypoglossal nerve,* I have never observed any decided loss of voluntary power in the tongue during the action of conium, unless, as in chorea, some derangement of this centre pre-existed. In the disease instanced, I have certainly observed the speech to become more halting and defective when the action of the medicine has been at its maximum, than before the dose.

In the absence of irritation, functional or lesional, it is equally difficult to recognise any particular influence upon the *spinal cord.* Hemlock affects the motor function of this part of the nervous system last of all, and short of a poisonous dose it does not interfere with its motor activity, or reflex function, as it is called, in any appreciable degree. This will be apparent from the following instructive experiment.

At 5.30 P.M., August 23, I injected ♏xv. of the succus conii beneath the skin of a full-grown, active, male mouse. Seven minutes afterwards he began to stumble; at the tenth minute he tumbled over several times while sitting, kangaroo fashion, upon his hind legs cleaning his fur. Up to the twentieth minute the little animal continued tolerably active and self-possessed, getting up as if nothing was the matter, as often as, in sitting or walking, he happened to roll over upon his side. He now gathered himself together in the usual couching posture, and resting the nose upon the table became very still and dull, with the eyes partly closed; respiration 160—the normal rate—and regular. When disturbed he was unable to run, and on attempting to walk he rolled over on the side. Without any visible change or movement the little animal now passed into a state of complete paralysis, in which he remained until 8.15 P.M.—two hours and three quarters after the dose was given. During the whole of this time he lay motionless in the position in which he was placed, with the eyes nearly closed; perfectly flaccid, and exhibited no indication of sensibility when dangled by an ear, or a toe, or by the tip of the tail. As he lay upon the side the only indications of life throughout this period were the following:—distinct and regular respiratory movements of the sides of the chest abdomen, decreasing during the two hours from 160 to 135, and interrupted by one or two full swelling inspirations during the minute. On gently passing the point of a pencil along the half-closed margins of the eyelids so as to touch them, a sluggish contraction, so faint as to be scarcely perceptible, occurred; but on separating the lids and touching the cornea no contractile action of the m. orbicularis was observable. On rolling the rump portion of the tail gently between the thumb and finger a reflex movement, consisting of sudden backward jerk, of all four legs was simultaneously excited; the vibrissæ were at the same time momentarily agitated. At 8.15 P.M. the effects of the hemlock began to subside, as indicated by a little increase in the depth of the inspirations, which were 135, and on irritating the tail as above mentioned, the head, body, and legs were simultaneously jerked backwards, the general movement being such as to throw the body backwards nearly an inch. The eyes were at the same time opened and the vibrissæ strongly worked. Shortly afterwards there was a slight movement of the fore paws, apparently of a voluntary nature—the first performed since the paralysis came on. But there was no further movement until 9 P. M., when on disturbing him as he lay upon his side, he struggled forwards a little, but did not succeed in getting upon his legs. This, however, was accomplished when I again disturbed him at 9.50 P.M., and he then drew himself together, opened the eyes and began to look about him and sniff for food. Shortly afterwards he crawled a few paces and meeting with a bit of bread and butter he licked off the latter, and then sat upon his haunches and cleaned his fore paws. At 11 P.M. he was in his usual condition (r. 140). The next day the little animal was as lively and active as ever (r. 160), and has continued so up to this present day, August 28. The secretory functions were uninterfered with, and the appetite reappeared with the return of power to eat. The motor centres within the brain slept soundly for about three hours, and then slowly awoke, but the spinal marrow acted as a faithful sentinel throughout.[1]

When, however, there is a morbid excitability of the reflex function of the spinal cord, the influence of conium in subduing it is powerful and direct. Illustrations of this are furnished at page 50.

Conium then in a state of health and in the fullest medicinal doses that we can venture to give, exerts its power chiefly, if not exclusively, upon the motor centres within the cranium. And of these the *corpora striata* of course are the parts principally affected. This appears to be indicated by the extreme rapidity with which the paralysing influence radiates through the body. So sudden and powerful is its action in full doses, that the patient, if he be standing at the time of its accession, has scarcely time to throw out the

[1] If the dose be a poisonous one the spinal cord sooner or later succumbs to its sedative influence, the respiratory movements grow slower and shallower, and die away with the life of the animal. Two fatal cases in man are reported, one in *Edinb. Month. Journ. Med. Sc.* 1845; the other in the *Gazette des Hôpitaux*, Nov. 1847.

Matthiolus, *Comment. on Dioscorides*, says 'that in Italy asses fall into such a profound sleep after eating hemlock that the ignorant country-people take them for dead, and begin to skin them, and that when in the middle of their work, the poor asses wake up.' It may be true that they were only asleep—the effect by the hemlock having nearly passed off at the time, but it is just as likely that some strong reflex movements of the hinder quarters were interpreted into a 'waking-up.'

arms and lay hold upon some support to prevent himself from falling. And in lesser doses the sudden depression of muscular power is such that the patient lets the child she is carrying in her arms, or the heavy object she is holding in her hand, fall to the ground. Again, many patients experience when the action of hemlock is at its height, a dull aching pain across the brows, over the roofs of the orbits, and at the back of the eyeballs; sensations manifestly referable to the corpus striatum itself.

Excepting then the reflex action of the cord, the whole motor function of an individual under the full influence of conium is actually asleep; and this is the simplest view that we can take of the physiological action of hemlock. It is to the corpora striata, to the smaller centres of motion, and to the whole of the motor tract, precisely what opium is to the brain of a person readily influenced by its hypnotic action; and just as opium tranquillises and refreshes the over-excited and weary brain, so does conium soothe and strengthen the unduly excited and exhausted centres of motor activity.

At first sight, we should be apt to regard conium as a depressor of the muscular vigour; but this, I am convinced from repeated observations, would be a very erroneous view of its action; and I am prepared to say that, in repressing and removing irritative excitement of the motor centres, conium is a *tonic* to these parts of the nervous system, in cases which require its use.

The influence of conium appears to be in proportion, not to the *muscular strength* of the individual, but to his *motor activity*. This important conclusion is derived from the following facts.

1. The operation of hemlock in the same individual varies in degree according to his motor activity. A dose of conium, which in the ordinary condition of the patient shall be just sufficient to produce the peculiar effects of the plant in a mild degree, will, during the exhaustion following a profuse seminal discharge for example, operate much more decidedly and intensely. Again, the effect will be found to vary, in proportion as the activity of the patient varies. Thus in those whose bodily vigour gradually declines as the day

wears away, a dose which will be followed by no appreciable effect in the early morning, will produce decided symptoms in the evening, and *vice versâ.*

2. Those leading a sedentary inactive life are more readily affected by conium than those of active habits. A delicate person of active habits will therefore bear a larger dose of hemlock than one possessing abundance of strength with but little energy.

3. An active restless child will often take, with scarcely any appreciable effect, a dose of conium sufficient to paralyse an adult of indolent habits; and such as would reduce a powerful muscular man to a tottering condition, and force him to assume the recumbent position, and retain it for a quarter of an hour or more. (See cases 1, 9, and 16.)

4. The same rule applies to children themselves; to produce a given effect, a dull inactive child requires only half the quantity that a lively, active one does.

The general inference from these observations is that the dose of conium must be proportioned, not to the muscular strength of the individual, but to the degree of his motor activity, functions which the operation of conium indicate to be so far distinct from each other. Indeed, it appears that by means of conium we may comparatively measure the bodily activity of the individual.

Upon the cerebrum hemlock is powerless. I have induced its full physiological action again and again hundreds of times, in at least a hundred different individuals of all ages, and have never been able to recognise the least *narcotic*, nor directly *hypnotic*, effects.

If sleep followed complete repose of the muscular system as a necessary consequence, then there would be no more powerful or direct hypnotic than hemlock. But other conditions over which this drug has no control are required to procure sleep. In reducing the motor centres to a state of perfect repose hemlock powerfully predisposes the brain for sleep—brings it within its reach, so to speak, but there leaves it. If indeed the imagination should happen to be dull, or the ideas, as in childhood, few, simple, and superficial, the brain may accept the invitation, and partake of an influence, which in its results, is identical with that which previously affected the motor centres only.

Under the influence of an effectual dose, a child often presents the aspect of sleep. Suddenly compelled by the overspreading muscular lethargy to relinquish his amusements, he lays his head upon the carpet, the lids are seen to droop over the dull listless eyes, and disinclined—probably from a degree of torpidity of the hypoglossal centre—for speech, the child is content to lie tranquil. To a superficial observer he *appears* to be asleep, and if the effects of the conium be prolonged he may really become so. But such an event will rarely happen to an adult. His mind will continue during the whole action of the hemlock as calm and active as was Socrates' when he said to his friend, 'Crito, we owe Æsculapius a cock; pay the debt, and do not forget it.' And he will tell us that he feels a strong desire to keep the eyes closed and remain quiet and undisturbed.

In poisonous doses the eyes will be completely fixed and the pupils dilated; articulation and deglutition impossible; expression and all other power of motion gone, and yet, while there is every appearance of the most profound coma, the perceptive faculties and reasoning powers may be as acute as ever. Such a condition I believe is not an uncommon one in other states than that produced by hemlock.

Like the cerebrum, the *sensory part of the nervous system* is altogether unaffected directly by the action of conium. Its anodyne power in certain diseases (see p. 54) may be fairly attributed to muscular relaxation in the diseased parts rather than to a direct influence upon the sensory nerves, conium being to the muscular apparatus involved what the knife, for example, is to irritable ulcer of the sphincter muscle of the bowel. Of its anodyne influence upon the facial branches of the *fifth nerve,* I have been unable hitherto to obtain other than very doubtful evidence; but in the painful dentition of children, I know of no anodyne so valuable as hemlock. Here indeed we fail to recognise painful compression of sensitive nerves by muscular contraction, but we have, as a consequence of a slight irritation of the dental branches of the fifth nerve, a general irritability of the motor centres, often so excessive as to result in convulsions. Probably in such a condition, there is a perpetual reaction of the motory and sensory centres one upon the other, converting a peripheral irritation into a centric pain. I think, therefore, that we may trace the anodyne influence which hemlock undoubtedly exercises over the sensory branches of the fifth nerve, to the power that it possesses of calming the general irritability of the motor centres, rather than to a direct and exceptional action upon the sensitive nerve itself.

During the action of hemlock a sensation of muscular weariness almost amounting to aching is not uncommon, particularly if bodily exertion be continued the while. Other muscular sensations may occasionally be present. In a case of partial *paraplegia* in which I employed conium as a means of quieting the reflex jerkings of the limbs, the patient, in addition to the usual symptoms of its action, always experienced a deep-seated intermittent tingling sensation pervading the limbs to their extremities. It was different in character and more intense than the vibrations produced by the mildest current of the magneto-electric machine, with the sensations produced by which he was quite familiar. It came on with a gentle creeping action every few minutes, and lasted only a few seconds at a time. Invariably associated with the muscular weakness it passed off with it after continuing for about thirty minutes. What I have said of the influence of conium upon the cerebrum and centres of sensation, is equally applicable to the *sympathetic nervous system.* Conium has no influence upon the circulating organs, upon the secretions, or excretions, or directly upon nutrition. Its influence upon the sexual function (see p. 50) is traceable to a direct action upon the spinal cord.

The dilatation of the pupils arises from a depression of the centre of the third nerve, and not from a stimulant action of the sympathetic.

Excepting as a transient emotional effect in nervous individuals upon the sudden accession of the symptoms after a first dose of hemlock, the heart and blood vessels are absolutely unaffected by its operation. I have carefully determined this in persons of all ages—in the weakly infant not three months old; in the strong and the debilitated; and in those who have intermittent action of the heart.

The sudden accession of the symptoms after a first dose of hemlock will in a nervous person sometimes excite the action of the heart for a few minutes, or induce a slight feeling of nausea, or a cold sweat, or even an action of the bowels; but these emotional effects are very rare, and when they occur are totally independent of the action of hemlock.

As to its *effects upon nutrition*, I have administered conium for months—in one case for more than six months (see p. 24)—and in such doses as have *daily* produced its peculiar physiological action, and the result has invariably been an improvement in the general nutrition and vigour of the body. This is true even when the drug is taken in poisonous doses; thus John Hunter's patient 'took one ounce of extract of hemlock in the course of the day for some time, and increased the dose to two and a half ounces. It produced indistinct vision and blindness, loss of voice, falling of the lower jaw, and a temporary palsy of the extremities, and once or twice a loss of sensation, and notwithstanding he was almost every night in a state as it were of complete intoxication from the hemlock, his health did not suffer, but on the contrary kept pace with the improvement in the ulcers' for which the medicine was prescribed. Ultimately the patient killed himself by taking in the course of the morning an ounce and a quarter of the extract.[1] It is doubtless by removing sources of central irritation that conium acts thus indirectly in improving the nutrition of the body.

Having now completed my observations on the physiological effects of hemlock in medicinal doses and short of producing toxical effects, I cannot conclude without advancing an inquiry respecting its primary action upon the nervous centres from whence the effects above-recorded issue. I take it for granted that conia represents the whole of the active principle of hemlock. But whether it exercises its powerful and subtle influence as conia or as some product of

[1] John Hunter's works, by James F. Palmer, vol. ii. p. 379.

its decomposition is doubtful. Unlike several other poisonous agents, conia does not pass through the system unchanged; and it is probable that it exercises its power at the instant of its decomposition. I have assured myself that conia as such does not pass out of the system. It cannot be detected in either the breath, the sweat, the fæces, or the urine.

I have carefully examined the urine voided immediately before and from 1 to 2¼ hours after the administration of large doses of conium many times, and find that this excretion is in no way altered by the powerful action of the medicine in the system.

The following is a typical case:—John R., aged 57, a powerful smith, affected with muscular tremor of the left hand. He took a meal of tea, bread and butter, with a little fish, at 5.30 P.M., and emptied the bladder at 7.30 P.M. At 7.45 he emptied the bladder a second time, and took fl. ℥j. of the succus conii. Half an hour after, he was suddenly seized with giddiness and weakness of the legs, and was unable to stand or walk for a quarter of an hour. During the next quarter of an hour he continued very tottering, but could walk slowly with assistance. Pulse 72, unchanged. Pupils unchanged. For the next hour he remained quiet, feeling great heaviness of the eyelids, and a tendency to sleep. At the end of this time the giddiness and weakness of the legs had entirely passed off. The pulse throughout was 72, and of good volume and power. The pupils which at the commencement measured the ⅛th of an inch in diameter were now dilated to nearly 1/7th. Two and a quarter hours after taking the medicine, and having had nothing whatever in the interim, the patient emptied the bladder a third time, voiding fl. ℥ivss. of urine. The urine passed on taking the dose (urine A), and that voided 2¼ hours afterwards (urine B) were examined. Both specimens had precisely the same physical characters—a full sherry colour, clear, naturally acid, and a specific gravity of 1023·2.

1000 gr. measures URINE A contained:		1000 gr. measures URINE B contained:	
Chlorine	5·5 grs.	Chlorine	5·66 grs.
Urea	16·0 grs.	Urea	16·6 grs.
Phosphates and sulphates .	10·2 grs.	Phosphates and sulphates .	10·5 grs.
Uric acid, normal amount.		Uric acid, normal amount.	

Urine B was examined for conia, but not a trace of it could be detected.

In some cases, the solid constituents were more abundant in the urine passed before the action of medicine; in others, more abundant in that passed after the operation of the drug; the variations in all cases having reference to the time of ingestion of the last meal, and being altogether independent of the action of the conium.

In examining the urine for conia I have adopted two processes. First, after precipitating the chlorine and urea by solutions of nitrate of silver and proto-nitrate of mercury, the clear filtrate was placed in a flat dish and allowed to evaporate spontaneously in a warm room. The crystalline matter and light brown syrupy residue were then supersaturated with caustic potash, and washed with æther. The æther was separated and allowed to distil from a water bath.

When the caustic potash was added, a strong odour of conia, and one that bore the test of comparison side by side with an aqueous solution of conia, or extract of conia mixed with potash, was evolved; and I concluded for a time that I had found conia in the urine. This, however, was a mistake; for on repeating the process on two mixtures—the one composed of equal quantities of the urines of three patients passed immediately before administering conium to them, and the other composed of the same quantities of their urine secreted during the operation of an aggregate dose of fl. ℥xvij. of succus conii—I found that *both* evolved a strong and decided conia odour, and side by side it was impossible to distinguish the least difference between them. From both mixtures the æther extracted a minute film of oily matter free from any trace of conia. I mention this fallacy in order to show that in examining the animal fluids or tissues for conia, we must bear in mind that the addition of caustic potash to them will often develope an odour indistinguishable from conia, and that nothing short of the isolation of the principle itself should satisfy us (see page 81). It is for this reason that I cannot attach any value to Mr. Judd's observation, that the blood, thecal fluid, portions of the cerebrum, of the spinal cord, and of the muscles of cats poisoned by conium, all yielded evidence—odoriferous evidence—of the presence of conia.[1]

The second process which I adopted for the detection of conia was as follows:—About one part of a saturated solution of oxalic acid was added to four parts of the warm urine, and the mixture allowed to evaporate spontaneously. The brown sirupy residue was separated from the crystalline matter, and both separately treated with excess of caustic potash, and washed with æther.

Of those two modes of determining the presence of conia in the urine I prefer the latter.

In order to assure myself that the conia was not decomposed or dissipated in these endeavours to isolate it, I conducted comparative experiments by adding ʒij. or ʒiij. of succus conii to similar quantities of warm urine; and subjecting the mixture to the processes above described, I did not fail in any case to obtain a small quantity of conia. In one experiment I exposed a mixture of ℥viij. of urine, ʒiij. of succus conii, and ℥ij. of a cold saturated solution of oxalic acid, in a flat dish, to the air for a month. At the end of this time I placed the dusty and moistish residue upon a filter, and washed it with ʒiv. of water. On supersaturating the dark brown filtrate with potash an acrid odour of conia was evolved, and by means of æther I separated a minute drop of impure conia from the mixture.

Means of Resuscitation in cases of Poisoning by Hemlock.—Without indulging in speculations as to the manner in which conium acts upon the nerve centres, I will make a few observations on the subject of resuscitation from the poisonous influence of the plant. I have said that the effect of conium is inversely as the motor activity of the individual. I might have said with equal truth that it is inversely as the activity of the respiratory function, or, in other words, the more active the breathing the greater the resistance to the action of hemlock. One thing is clear, that a given dose of conium produces in the same individual, and, cæteris paribus, a greater effect in a state of inactivity, than when,

[1] Trans. Royal Medico-bot. Soc. of London, 1834–1837, pp. 143, 146.

by forced muscular exercise, he tries to ward off the influence of the drug. Now in poisoning dogs by prussic acid, I have as a matter of experience found it necessary to keep the mouth closed for a few seconds after pouring a drachm or two of the fluid into the pharynx. As often as animals have been able in their convulsions to get the mouth open, and inspire good volumes of air, they have recovered. There is something in common, I think, between the ultimate action of hydrocyanic acid and conium, the chief difference being that the former attacks at once the motor centres of organic life, while the latter expends the chief part of its power upon the motor centres of animal life, and only slowly and at a late period of its action paralyses the movements of organic life.

From these and the general considerations which the phenomena of hemlock poisoning suggest, it appears that the whole of our endeavours must be directed to excite and maintain the reflex action of the spinal cord, and of the medulla especially—in a word, to increase the quantity of oxygen taken into the lungs. The application of electricity to the upper part of the spine and chest walls simultaneously, will be the most hopeful means. More indirectly we may appeal to the vagus through the fifth nerve by applying ammonia to the nostrils, and sprinkling cold water on the face. The lower part of the spinal cord may receive an impression by the introduction of brandy or ammonia, well diluted in hot water, into the rectum. If deglutition be still possible, the peripheral branches of the vagus may be stimulated by the introduction of the same stimulants into the stomach. Brandy should be used sparingly, for we must bear in mind that large doses, in addition to an inebriant action, naturally depress the motor function of the craniospinal axis, and so assist, instead of restraining, the action of hemlock. Some persons will tell us in general terms that the symptoms produced by conium resemble those of alcoholic intoxication. This is true in a very limited sense, and those who know what the effect of brandy really is, will assure us that the effects of hemlock are peculiar and distinct, and such as they never experienced before.

That alcohol, in moderate doses even, does not counteract in the least degree the effects of hemlock, I have taken care to ascertain in female patients very susceptible of the action of both agents. Still, as a local stimulant to the stomach, and as a diffusible stimulant to the circulation, this approved antidote of the ancients may be of service when given in limited quantities.

Of some other questions relating to the Use of Hemlock. *Tobacco.*—I have observed that persons who use tobacco freely usually require a large dose of conium to produce its physiological effects. And the reverse, that those cannot tolerate tobacco who are readily influenced by comparatively small doses of hemlock. This is what we should have concluded theoretically from a knowledge of the separate action of these two agents. Great care, therefore, is required in the simultaneous use of them.

Hydrocyanic Acid.—Large doses of conium should be carefully employed in conjunction with full doses of hydrocyanic acid.

Alkalies and Acids.—The effects of conium juice are neither increased nor diminished by the copious addition of alkalies or of either of the mineral acids.

The effects of conium, when combined with henbane, belladonna, and opium, respectively, will be duly considered in the succeeding chapter.

The Medicinal Use of Conium.—Having now completed my account of the physiological action of hemlock, I come now to practical inquiries respecting its value as a therapeutical agent.

The use of inert preparations of conium has rendered, with very few exceptions, all previous statements concerning the medicinal value of this plant almost worthless; and it is necessary, I believe, to begin this inquiry anew. The following examples are given as a small contribution to this inquiry. It is all that I can offer at present. In perusing these cases I would ask the reader to bear in mind these two necessary facts: *First*—That they are not picked examples, showing success to the exclusion of cases in which no benefit was obtained. They represent the whole of the cases in which I have used hemlock as a therapeutical agent. *Secondly*—The treatment has consisted in the administration of hemlock alone, and I have rigidly avoided altering the condition of the patient by any other means, medicinal or *hygienic*. If in any case additional treatment has been required, this will always be mentioned. I must preface my remarks on this topic by the following conclusion. It is the result of a good deal of close observation, and I am thoroughly convinced of its truth:—That hemlock given in doses which fall far short of producing its proper physiological action is useless in the treatment of diseases for which it is adapted.

The mode of action of the neurotic poisons is an unsolved problem, involving as it does a complete knowledge of the nature of nerve force—of its development, its affinities, and its mode of radiation or conduction throughout the body. The influence of hemlock in particular is truly marvellous. We have some clue to the action of its opposite or complement strychnia, for we not only see it repeated in some of the commonest morbid states of the body, but we can induce it by electrical or other irritation of the motor portion of the nervous system. But the action of hemlock is quite beyond our comprehension. The power of the drug is so direct and simple, and at the same time so exclusive, that it resolves, as if by electrolysis, nervous action into its chief constituents; and, leaving untouched the one portion, suddenly and sometimes almost with the rapidity of lightning, depresses or altogether suspends the other.

Deprived of all voluntary movement, the patient resembles one who has been suddenly stricken with ordinary palsy; but, an hour hence, his power of motion will return, and no trace of this powerful influence will remain. Such is the simplicity and certainty of the action of hemlock that I believe there are morbid conditions of the nervous system in which it may be used to analyse the diseased action and point out its cause (see p. 29 et seq.).

Nor can disease be said to furnish us with any clue to the action of hemlock. Functional paralysis is more, I believe, of a name than a reality. But even granting such a condition, the paralytic influence is only partial, and the recovery therefrom slow, even when progressive. The paralysis of hemlock is, as I have said before, a sleep—a sleep from which there is a speedy and complete awakening. Hemlock is the complete realisation of our ideas of a medicine. Over natural function no medicine has so strictly limited and transient an influence. Upon the morbid conditions to which its action is suited, no drug exercises a more powerful influence or leaves a more enduring impression. But to act beneficially—and I cannot too strongly insist upon the fact—it must be given in doses sufficient to produce an impression. To give hemlock in doses that fail to produce an appreciable effect upon the motor system is to give repeatedly the hundredth of a grain of morphia to one dying for want of sleep, or a grain of quinine to cure an ague fit.

In selecting hemlock as a remedy in the treatment of nervous diseases, we must dissociate from it all notions of a deliriant, hypnotic, or convulsive action; all ideas of a sedative power over the heart, or of any influence upon the secretions. We must be guided by that simple view of its physiological action which I have now so fully stated, and then the only question to be proposed will be, Is there irritation, direct or reflex, of the motor centres? If there be, conium is the appropriate and hopeful remedy.

In treating of the medicinal use of hemlock, I will take first the convulsive diseases of children. Hemlock is essentially a children's medicine. It has often appeared to me that in the early development of the body, motor activity, from a variety of exciting causes, is liable to a disproportionate development, and that thus a tendency to convulsive disease often arises. Anyhow, whenever that tendency exists, it may be readily subdued by the judicious use of conium.

I. **Undue excitement of the motor centres** occurring at or near the time of dentition, and producing general irritability of the system with strong tendency to convulsion, and in many cases resulting in actual convulsions. I have treated eleven such cases, and they have presented every variety and degree of irritation. All have recovered under the use of hemlock, and the slighter cases with great rapidity. The following are examples of two of the severer cases :—

1. Stephen S., aged 1¾ years, came under my care March 1867. His mother gave the following history :—During the time she was carrying the child, she was attending the death-bed of a near relative, and was weak, nervous, and depressed. On one occasion she was much startled by the unexpected striking of a large bell overhead as she passed under a gateway. At his birth the child was emaciated, but he made flesh rapidly, and at the end of a month was quite plump. For the first four months he screamed almost constantly. When only *three weeks* old he was seized with violent convulsions, recurring at intervals for several days in succession. These fits were of an epileptic character, each lasting three or four hours, and attended with unconsciousness. The convulsions chiefly affected the left limbs. Between the *second and third month* he had a second series of fits. A *month afterwards*, he had whooping cough badly and the fits were again renewed. Since then he has been liable to frequent attacks of **Laryngismus stridulus**, and a recurrence of the fits every three or four weeks. Dentition began at the eighth month, and at this time the fits were very violent; on one occasion he was unconscious for five hours, and leeches were applied to the temples. From this time to the present, the fits have increased in frequency and severity, and the attacks of laryngismus stridulus are induced by the least excitement. As yet only five teeth (incisors) have appeared, but there are indications of the eruption of a sixth. The child has never walked, the left leg cannot be straightened on account of contraction of the hamstring muscles; the left arm is nearly useless and drawn backwards; there is internal *strabismus* of the right eye, congenital, but much increased since birth. The patient is a florid, healthy-looking, remarkably large and powerful child. He is noisy, restless, and unmanageable beyond description; his temper is so violent that, upon the slightest provocation, he becomes furious, screaming incessantly, and in his struggles striking his head against objects in the way, alike indifferent to restraint and

to pain. Has startings in his sleep, and frequently wakes up with a scream. His appetite is voracious, and thirst insatiable. On putting my finger into the mouth to examine his gums, I immediately brought on a violent attack of spasmodic coughing interrupted by prolonged stridulous inspiration, the face became congested, the tongue was protruded, swollen, and of a dark colour, and he struggled desperately. The poor mother was well-nigh worn out with her unremitting attentions to the child, and she stated that he is wickedly disposed, and that he strikes her upon the slightest provocation. The treatment of this case extended over a year, and consisted in the administration of hemlock alone. Beginning with ♏xv. of the succus, I gradually increased the dose to ʒivss. taken once every day. The beneficial effect of the drug was decided and immediate. During the *first five weeks* the patient had only one fit, and this was associated with the cutting of the sixth tooth. At the end of this time the laryngismus stridulus had finally disappeared; the left arm was no longer drawn backwards and he had begun to use the hand; the left leg could be farther extended. From this time his progress was rapid and uniform; and with the exception of occasional twitchings and grinding of the teeth during sleep, and a single fit to be mentioned hereafter, all convulsive movements ceased. At the *sixth month* of treatment he was able to walk with the help of a hand; five more teeth had appeared, making twelve in all; he was much more quiet and tractable, and was improving in intelligence. Dentition progressed favourably, and at the end of the *ninth month* all the temporary teeth were cut but two; he walked well with the help of a hand, and used the left limbs almost as well as the right; the strabismus was less; and he began now to articulate a few words; he was constantly on the move, and still very rough and boisterous, and his appetite for food and drink remained as insatiable as ever; the nocturnal twitchings had ceased. A *fortnight afterwards* he was able to walk alone, and now began to run about all day long. *Two weeks later on* he awoke in the night with a start, and his mother apprehending that he was about to have a fit poured cold water upon his head.

In the evening of the next day he had a fit which lasted twenty minutes. This was his last convulsive attack; it weakened the left leg for a few days, but under the influence of increased doses of the medicine he speedily regained his former improved condition, cut the remaining two teeth without difficulty, and has continued well up to the present time.

The effect of the medicine upon this young patient was very marked. Twenty minutes after taking it the eyes had a heavy, stiff, dreamy appearance, and he seemed often incapable of raising the lids. He would then stop suddenly in the middle of his play, lay his head upon the carpet, and remain perfectly quiet and motionless for a period varying from half an hour to two hours. On one occasion he lay still and apparently torpid with the eyes closed for four hours, when his mother being somewhat alarmed roused him. This condition was induced by an increased dose of ʒ ij. The effect of the hemlock upon the contracted muscles was very decided. The left leg, which before the dose was slightly bent, and moved stiffly in walking, would become released during the action of the medicine, and as he walked, the heel could be fairly brought upon the ground.

At the time that he began to run about without assistance he was taking ʒ ij.; of the succus daily, and it produced the usual effects of arresting him in his play, causing the eyes to look dull and heavy, and inducing a state of repose lasting from half an hour to an hour and a half. But five or six days afterwards his mother, who had watched him with the most anxious care, observed that the medicine had all at once lost its effect. On mentioning this to me I increased the dose to ʒ ijss.; but after the first two or three doses no effect followed, and nine days afterwards the last fit occurred. I then increased the dose rapidly, and soon found it necessary to give the patient more than double the quantity that he was taking at the time he began to walk without assistance, in order to produce equal physiological effects—viz. ʒ ivss.

This is a very remarkable and interesting observation. It would appear therefrom, that simultaneously with the acquirement of the power requisite for running and freely exercising the body in the erect posture, a rapid development of motor force takes place.

If we can trust this single observation—and I think we may safely use conium for our measure—it appears that the motor force may be more than doubled during the short period of a fortnight. Nor is the above a very uncommon case, for although the child ran about from morning till night as soon as he could walk alone, many children attain an equal degree of activity in the same space of time.

2. John C., aged $2\frac{3}{4}$ years, a pallid slightly-developed child, had always enjoyed good health up to the *seventh month*, when he suffered from an attack of measles, and the irritation attending the eruption of two incisor teeth. These left him very weak, and at the *ninth month* he was attacked with fits which have recurred at variable intervals ever since. These attacks were preceded by a violent start and screaming, then followed rolling of the eyes, loss of consciousness, and strong contractions of the flexor muscles, lasting usually twenty minutes. At first the fits occurred every day, sometimes he had four fits in the day, and occasionally as many as ten or thirteen in the week.

During the last five months the character of the fits has somewhat changed. There is now no loss of consciousness, the attack is preceded by a cry of pain, and accompanied by a feeling of dread. The trunk and limbs are affected with tetanic spasm, the body being arched backwards and slightly curved to the left, the left arm pulled backwards and the right forwards over the body to the left side, the mouth drawn to the left, the legs rigidly extended with a slight curve backwards, and the feet strongly flexed. This state continues, sometimes for only five or ten minutes, sometimes with short interruptions for several hours, and the little patient cries out with pain, frequently exclaiming, 'aches so bad, make me better, make me better.' He came under my care April 3, 1867. During the whole of the time he was waiting in the out-patients' department at the Hospital, he was in an alternate state of cramp and partial relaxation, and when his mother brought him into the prescribing room he was fully extended in her lap, the thumbs were drawn into the palms of the hands, and firmly grasped by the fingers; the toes firmly flexed, the feet rigidly arched backwards, and the muscles of the calves firmly contracted. He complained of aching pains chiefly in the feet and legs, and manifested great dread lest anyone should touch his feet in passing. The head was rather large but quite cool, the fontanelle firmly closed. A slight internal strabismus (congenital) affected both eyes. The functions were naturally performed, and the child was remarkably intelligent and patient. Dentition was completed three months previously, and two months ago he began to walk. Soon, however, the abovementioned attacks supervened and prevented all further attempts, and for the last few weeks he has been unable to set a foot upon the ground or even to bear the pressure of shoes. The cramps distress him by day and take away all appetite for food, and interfere with sleep by night, and the poor little patient is much exhausted. The mother stated that three of her other children had suffered from convulsions during dentition, and one had died in consequence, at the age of seventeen months. Under the influence of conium an improvement was observable in a few days; the contractions of the fingers and toes relaxed, and with the diminution of muscular tension the pain decreased. He was rapidly improving when his mother was called to her father's deathbed, and the treatment was suspended for a fortnight. At the end of this time he had a violent tetanic attack with opisthotonos, lasting 1½ hour, and he was lapsing into his former condition of continuous cramp. Upon resuming the hemlock in increased doses he rapidly improved; the muscles of the calf relaxed, the foot could be extended at right angles with the leg, and he was able to wear boots, and began to walk again. The improvement continued, and within three months from the time I first saw him he was able to walk well. The attacks had totally disappeared, and he was greatly improved in health and strength. The hemlock was now suspended. Since this time he has not manifested the slightest tendency to spasm, and has become a healthy and active child.

The action of conium upon the spinal cord in such cases as these may be likened to the influence of opium upon the overstrained and exhausted brain.

II. **Epilepsy.**—In a disease which may arise from central or peripheral irritation, from emotion or actual injury of the nervous matter, or from disorder of almost any of the functions, there must often be much difficulty in ascertaining its primary cause. In this affection conium, I have reason to believe, will be of use in two ways: first, as a means of separating the disease into two distinct classes—those cases in which the attacks are due to morbid excitement of the motor centres, and those in which the primary irritation is located elsewhere; and second, as the appropriate remedy in the former class of cases. Conium can do no harm in epilepsy, whatever be its cause, and in the particular class of cases indicated I have seen much benefit result from its use. To speak confidently of the advantage of any given medicine in the treatment of epilepsy, requires a very extensive and prolonged experience. I have not used conium in more than twenty cases, and the majority of these were hospital patients, and remained only a month or two under treatment, a period much too limited for satisfactory observation. All of these cases improved excepting those in which the disease sprung from peripheral disorder of sensation, from irregularity of the menstrual function, or from emotion, and in some of these the remedy was fairly tried.

In the cases where the disease arose from sexual abuse, or from the irritation of dentition, the improvement was decided and rapid. I must content myself with furnishing a single instance.

3. Rose H., aged 10 years, a strong, active and well-developed girl of dark complexion. From the age of four years she had suffered from severe epileptic fits, and from this time to the present, she has rarely been free from fits for two consecutive days. The attacks vary in intensity, the graver last from twenty to forty minutes, and she is alternately 'drawn up like a ball,' and suddenly extended; the slightest fits last 5 minutes, are also attended with loss of

consciousness, falling, foaming at the mouth, and strong jerking of the arms and legs. Some days she has as many as three or four of the strong fits, sleeping for about two hours in the intervals between them. The common average for a long time preceding the date she came under treatment, was from twelve to fifteen strong fits during the week. The child is nearly reduced to a state of idiotcy, she has a very rough, impetuous, boisterous manner, and requires a constant attendant. The repetition of a few short hymns by heart represents the whole of her acquirements. She cannot be trusted to execute the most trifling errand. The functions are naturally performed, the appetite is large, and there is generally great thirst. Her sisters and brothers are older, and all healthy and free from any apparent tendency to nervous disease. Her parents are likewise healthy, but the mother is addicted to intemperance.

Treatment and Results.—At first I employed henbane in daily doses of from ʒij. to ʒiv. of the tincture or succus, but finding that little or no benefit resulted, I substituted conium after the second week. I began with fl. ʒj. of the succus morning and evening, and gradually increased the dose during the next six months to fl. ʒvj. and occasionally ʒvij. twice a day. From the first there was a decided improvement. During the first month (November) she had only twenty strong fits and eight slight ones, exclusive of a succession of violent attacks throughout the whole of one day. During the next two months she experienced only sixteen attacks altogether, and four of these were slight. She now had intervals of two and three weeks' respite, and throughout the three months following, the fits diminished considerably in severity, the worst lasting not longer than fifteen minutes. They numbered twenty-nine; and the majority of these attacks occurred on the morning preceding her journey to town to see me, or on the first or second day after her return into the country. So long as she was kept free from unusual excitement, she evinced but a slight tendency to a fit once every few weeks. Pari passu with progressive diminution in the severity of the attacks, her memory and mental condition improved, and her noisy bustling and often violent manner gave place to a calmer and more rational state. The conium had a very marked influence upon her. Twenty minutes after taking the larger doses, she complained either that she could not see, or that the objects of her vision were tinged with all varieties of colour. Sometimes the eyelids drooped a little. For one or two hours after the accession of these symptoms, she remained perfectly quiet, or only occasionally and with slow movement changed her place. Under my own observation—and I always watched the effect of an increased dose for at least two hours—I never observed any other effect than the most complete quietude, amounting almost to apathy. In this state she would stand by my side perfectly still and silent for twenty minutes at a time.

III. **Convulsive Action of the Muscles.**— The two following cases serve to link together as in a continuous chain, epilepsy on the one hand and chorea and muscular tremor on the other. In these also, the beneficial influence of conium is apparent.

4. John M., aged 57, a spare, grey-haired man, was attacked, while using a tailor's heavy pressing iron, with violent convulsive action of the triceps and flexor and extensor muscles of the right arm. The limb hung by his side three-fourths extended, and the muscles above mentioned were powerfully contracted from forty to sixty times a minute, causing, as often, slight abduction of the arm from the side, abduction and extension of the thumb, the index and the second fingers, and flexion of the little finger.

The triceps appeared to be most affected, and its attachments were clearly mapped out every time the muscle was affected with spasm. At the same instant the muscular masses lying on the inner and outer sides of the forearm became very rigid, the fingers were twitched, and my own fingers were jerked off the pulse. The whole limb was at the same time slightly abducted by a faint spasm of the deltoid.

Once or twice the action of the flexors has predominated, the forearm has been bent at right angles to the arm, and the whole limb then shaken with convulsions so violent, that he could only partially restrain them by grasping the wrist in the other hand and holding it with all his force.

During the last four years he has had two such attacks as that I have described. They came on suddenly, and, after lasting about a fortnight, gradually intermitted and declined. During each attack, the convulsive action, extending upwards, has involved the muscles of the shoulder, and the sides of the chest, neck, and face, causing much twitching of the right side of the mouth and some faltering of speech. After continuing twenty minutes the violent spasm has ceased, and the arm has then dropped powerless to his side, and so void of sensation for two hours afterwards that he could not feel the thrust of a pin.

On two occasions, the convulsions, after extending to the face, have become general, and attended with loss of consciousness, and the fit has lasted for twenty minutes.

The present attack came on about thirty-six hours before I saw him. At first there were considerable intervals between the spasms, but they very soon became more severe and continuous and rendered sleep impossible.

Treatment and Results.—I gave him fl. ʒiv. succus conii at once, and kept him under observation for an hour. At the end of this time the spasms were weaker, and only twenty-four a minute. No other appreciable effects followed during this time, but on crossing the square on his way home he was suddenly taken with a swimming sensation, and was obliged to take hold of the railings for a few minutes. The giddiness soon passed off, but he felt so fatigued that he was glad to get into his house, which was close by, and throw himself upon a couch, and remain there for the next half hour.

The twitchings, somewhat subdued in intensity, continued throughout the night, and on putting his hands into cold water on rising (at 8.30 A.M.) the right hand was suddenly clenched and flexed at right angles to the forearm, which was simultaneously bent at an acute angle with the upper arm. After remaining fixed in this rigid state of spasm for two or three minutes, the whole limb was powerfully convulsed, and he feared that he was going to have one of the general fits above described, as on former occasions. This, however, did not occur, and the attack after lasting five minutes suddenly subsided, and the limb returned to its previous state of continuous twitchings. At 11 A.M. the same morning I gave him a second dose of fl.ʒv. of succus conii, and 1½ hour afterwards the spasms were reduced from twenty to fourteen a minute: the brachialis anticus was completely tranquillised; the triceps nearly so, and the spasms were almost entirely confined to the muscular masses arising from the condyles of the humerus. The power of the spasm was at the same time reduced to one half, at least.

There was no further improvement for the next twenty hours, and he did not get more than five hours' sleep that night. After a dose of ʒiv. of the succus next morning he slept tranquilly for two hours, and on awaking found a marked improvement—the spasms no longer involved the triceps. He slept at intervals, for an hour, several times during the day. ʒij. more of the medicine was taken at bedtime, and he had a fair night's repose. Fl.ʒv. taken the following morning was followed after an hour by very decided hemlock symptoms—viz., sudden giddiness, with great heaviness over the eyes, followed by weakness of the legs and tottering gait. After continuing for three-quarters of an hour these effects entirely passed off, and at the end of this time the weakened and less frequent spasms were subdued altogether *visibly*, but on grasping the wrist slight tensions of the flexor tendons could still be *felt*. He had now regained almost perfect command over the fingers, and said he felt that firm strength was returning to the hand. He could pick up a pin, and grasp almost as firmly as with the sound hand. On walking away from my house the last remnant of the twitchings entirely and finally disappeared. He resumed his work two days afterwards, but I continued the conium, morning and night, for several days more. He remained perfectly well three months, when the spasms suddenly returned in a mild form. They disappeared again after three full doses of conium, and the patient has remained well ever since.

5. *Spasm of the sterno-mastoid and platysma muscles of both*

sides; occasional spasm of the diaphragm and muscles of the larynx; hereditary tendency to insanity; susceptibility to the action of hemlock.

Mary A. J., aged 16, a well developed girl, of nervous temperament and excessively inactive habits, and very silent, shy, and reserved manners. Unless actually driven away from her books she will sit still and read the whole of the day. She appeared devoid of all vivacity, and spoke to me in a whisper. The grandmother was twenty-three years in a lunatic asylum, and a cousin aged twenty-three is now in similar confinement. The patient herself has never before suffered from any nervous disorder; the menstrual function was regularly established.

The day before she came under my care she was seized with crampy pain under the right blade bone accompanied by short sobbing inspirations, and rigid spasms of both sterno-mastoid muscles, and occasionally of the platysma, of both sides. The twitchings and short sighing or sobbing continued all day long. She slept well during the night, but on awaking the spasms returned, and, involving at intervals the muscles of the larynx, caused alarming choking sensations. At the time I saw her, the choking spasms had ceased, but the sterno-mastoids and platysmas were in a state of alternate contraction and relaxation, the spasms occurring pretty regularly, and numbering from thirty to forty a minute. The contractions were very strong, and bringing out the sterno-mastoids into rigid relief, showed the depressions between the sternal and clavicular origins of these muscles. The shoulders were occasionally shrugged at the same time by spasmodic action of the trapezius and pectoralis minor. A short simultaneous sob marked the implication of the diaphragm.

Treatment and Results.—I administered fl. ʒiij. succi conii at once; its physiological effects were declared within fifty minutes. During the second hour of the action of the medicine she was occupied in walking home, a distance of two miles. She complained of giddiness and weakness of the knees all the way, and two or three times her knees bent under her, and she had some difficulty in supporting herself. At the end of the first hour the spasms were reduced to two in twenty minutes. For the next three days she took ℨij. of the succus conii in the morning, and ℨj. at night. On the second day only an occasional twitch of the affected muscles was observed, and this finally ceased before night. She remained under observation for several days longer, and there was no renewed tendency to spasm. ℨij. of the succus subsequently produced great heaviness of the eyelids, giddiness, and staggering. Yielding to the feeling of excessive languor, she would lie down and remain in a state of torpor for one or two hours. The influence of conium in reference to bodily activity in this case is very striking, and accords with the case referred to at page 8. In both fl. ℨij. of the succus produced full physiological action. They contrast strongly with cases 1, 3, 9, and 16.

IV. **Chorea.**—A consideration of the pathology of this disease involves the question of the existence of a co-ordinating centre of nerve force. If we assume the existence of such a power, independent of will, sensation, and motion, then we may at first sight as appropriately attribute the disorderly and involuntary movements of chorea to a derangement of the co-ordinating function, as to a morbid excitation of the motor centres. But it appears to me that we have no satisfactory evidence of the existence of a co-ordinating power independent of the will. If it were a real individual function we should often see it manifested in proportion as the other functions were depressed. It would stand forth prominent and sometimes alone as the others fall back. Do we observe this? If chorea does not furnish the demonstration I do not know where to look for it, certainly not in those cases of incomplete paraplegia in which a man walks by the aid of his eyes. But let it be granted that chorea is an instance of disorder of the co-ordinating centres, and we have a case in which the function of co-ordination is excited, and that of voluntary motion depressed. It should follow that in proportion as the latter function is still further depressed, the choreic movements would be increased. But the reverse of this is actually the case. Conium, as the following cases will show, represses and ultimately removes choreic movements, and it has been proved to possess a simple and exclusive action upon the motor centres. Conium therefore disproves, so far, the existence of a co-ordinating function, and points us to the true cause of chorea—an undue excitability of the motor centres, which throw off impressions so rapidly that the will is puzzled to control them.

The following are instances of the use of conium in chorea. I have taken the cases indiscriminately, just as they have come before me, and they are the *only* ones in which I have employed hemlock. I have regarded the disease simply as a primary disorder of the nervous centres, and, avoiding the use of all other remedies, medicinal and hygienic, even to the exclusion of a simple aperient and a shower of water, have relied upon hemlock alone. In the history of these cases I have felt it necessary to describe the nervous symptoms somewhat minutely, in order that the effect of the drug may be the more readily traced and precisely determined.

6. *A third attack of Chorea immediately preceding the menstrual nisus.*

Fanny M., aged 13 years, a slender but healthy-looking girl. Had had three attacks of chorea, each fresh one exceeding the previous attack in severity. The first occurred at nine and a half years, the second at eleven, and the third and present attack appeared about a fortnight before she came under my care on January 17, 1867. She had always been free from other nervous diseases, and from rheumatism; the heart was quite healthy; all the functions were naturally performed, and there was no known cause for the disease.

At the above-mentioned date the choreic movements were severe, and not only abridged the hours of repose, but often awoke the patient from sleep. The general health had suffered in consequence. The involuntary movements were almost exclusively confined to the left side; the leg was weak and uncertain, and she walked lamely; the wrist and hand were in constant and rapid motion, and as the muscles of the upper arm and chest were involved, the whole member was snatched here and there, and writhed about in various directions. She was unable to pick up such an object as a penholder, and when a larger body was placed within her grasp she could not retain it for many seconds. The side of the neck, the face, and the tongue were frequently contorted by irregular movements.

Treatment and Result.—I prescribed ʒj. of the succus conii every morning, and increased the dose on the *third day* to ʒjss., giving it on alternate days, in order to judge of its influence on the disease. According to the mother's statement, there was a slight but decided diminution of the choreic movements on the conium days. On the *sixteenth day* there was a manifest improvement, and the patient slept better. As the medicine did not produce any of its peculiar effects, I now increased the dose to fl. ʒij. every second, and, after a time, third day. Slight giddiness and dimness of vision followed this increased dose. On the *twenty-sixth day* she was able to pick up small objects, and, with fixed attention, to retain them in her hand; to keep the affected arm, when horizontally extended, nearly steady. The fingers, however, continued in constant play. She steadily improved, and on the *thirty-fourth day* was able to dress herself. On the *sixtieth day* she was quite well, excepting an occasional slight twitch of the angles of the mouth; she was able to thread and use her needle; her sleep was sound and undisturbed, and her general health and appearance much improved. Shortly afterwards, the last trace of chorea disappeared, and the hemlock was discontinued. During the short period she was under treatment, the patient underwent a rapid development, and from being spare, pale, and childlike in figure, assumed quite a womanish build and deportment, and a fresh ruddy hue. The following June she again came under my care for headache and epistaxis attendant upon the menstrual nisus. She continued under treatment (by purgatives and chalybeates) for three months more; but in the interval, and during the whole of the time she remained under my notice, there was no indication of any tendency to a return of the chorea. The catamenia had not appeared at the end of this period.

7. *Chorea, chiefly of the right side, of seven weeks' duration.*

Jane R., aged 12, a pale, delicate-looking, and slightly-made girl, of rather inactive disposition. Free from all disease until December 27, 1867. On the previous night she went to bed in her usual health, but on arising in the morning, she was, according to her mother's description, 'in one work all over, and could not keep herself still for a minute.' The speech was very defective, the mouth drawn to the right side, and she walked very lamely from weakness of the right leg. These symptoms, which were at first very severe, decreased a little in intensity; but for the last seven weeks she had experienced no improvement.

At the expiration of this period she came under my care, and the following was her condition:—The choreic movements chiefly affected the right side. She could not walk many hundred yards without assistance, progressed very slowly, and halted upon the right leg. She was unable to dress or feed herself, to sew, or to write. The right hand and wrist were in constant motion, and the upper arm frequently jerked backwards. On attempting to use this limb it was tossed upwards. The head was often jerked down to the right shoulder, the expression as frequently deformed, and the angles of the mouth in continual motion. The tongue was affected with incontrollable writhing; she only spoke when solicited to do so, and then with hesitation, indistinctly, and in a whisper.

The irregular movements often made sleep impossible for a long time, but when once asleep they ceased.

The functions were normally performed. Pulse 72. Heart's action regular, the first sound a little prolonged and sonorous.

Treatment and Results.—I gave her ℥ij. succi conii at once. Twenty minutes afterwards giddiness came on and continued for forty minutes. She continued to walk about during the whole of the time, and neither the pulse nor pupils underwent any change during the action of the medicine. But its effect upon the choreic movements was very decided. She walked so much better that the previous lameness was scarcely observable, was able to hold out the right arm quite steadily, and there was only an occasional slight flexion and twist of the wrist. The muscles of the face were quite tranquil, and the tongue much steadier.

The above-mentioned dose was repeated every morning before breakfast, for four weeks. The dose was then increased to ʒiij. every morning, and she continued this dose for four weeks more. At the end of the *second week* she was so much improved that she was able to pin her shawl, write her name legibly, carry a saucer of fluid steadily to her mouth with the right hand, and there was no halting or dragging of the right leg. At the end of the *fourth week*, the only remains of the disease were a sudden plucking-up of the right arm and slight restlessness of the wrist and tongue. She was no longer prevented from going to sleep; she could dress and feed herself, speak distinctly, and walk briskly and without fatigue a distance of four miles.

At the end of the *fifth week*, she was quite well, and could thread a needle, sew, and write as legibly as ever. She had regained strength, colour, and spirits. I thought it prudent to continue the medicine for some time longer.

It invariably produced giddiness, heaviness, as if from an inclination to sleep, and dulness. She was not, however, allowed to give way to these feelings, but kept in active motion during the whole of the time required for the operation of the medicine. ʒiij. doses caused increased heaviness of the eyelids, and giddiness, but did not disable her from walking about. I saw her from time to time up to the tenth week, and she continued quite well.

8. *Severe Chorea of the right side, of five weeks' duration. Rheumatic tendency. Speedy cure.*

Alfred V., aged 12 years, a moderately-stout and healthy-looking lad, but rather dull and inactive. At the age of six years was laid up for three months with a severe attack of rheumatic fever. Since this he has enjoyed good health, but is liable to headache. Free from nervous disorder until March 1868. About the middle of this month he became affected with choreic movements of the right side of the body, associated with lameness of the right leg, and defect of speech. He continued to get worse up to April 18 following, when he came under my care. At this time his condition was as follows:—The head drawn down towards the right shoulder, and constantly maintained in this position, and at intervals slightly jerked round in the same direction; the right corner of the mouth plucked downwards several times a minute; the protruded tongue held tolerably steady, but the speech was almost obsolete and unintelligible; the right arm was flexed at right angles, retained closely to the side, and the hand in constant motion, beating to and fro a hundred times a minute. In order to restrain the rapid motion of the right hand, the patient clutched the wrist firmly with his left hand, and so nursed the agitated member all day long. He complained of pain in the right elbow, and could not raise nor abduct the upper arm; he was unable to use the fingers, or to retain any object within the affected hand. He walked very lamely, trailing the right leg stiffly along, and this member was occasionally affected with twitchings. There was no sign of rheumatic tenderness or inflammation anywhere. The general health was good; tongue clean and moist; the cardiac sounds normal, but the systole strong from slight enlargement of the left ventricle. The boy was much depressed in spirits from want of rest; he was quite unable to assist himself, and his mother was obliged to hold the right hand for a long time every night, to enable him to go to sleep, otherwise its rapid and incessant motion kept him awake.

Treatment and Results.—I at once gave him the succus conii in doses of ʒij. twice a day. Each dose produced slight giddiness, and was followed by marked improvement. At the end of *four days* he was bright and cheerful, and after making a strong effort could raise the right arm in a straight line above the head, hold it out pretty steadily at right angles with the body, and pick up a pen and hold it in his hand. The wrist, however, continued to be jerked backwards and forwards sixty times a minute, and he still nursed it as before. The twisted head still overhung the right shoulder, and the twitching of the head and the right angle of the mouth continued. But the speech and walking were much improved.

During the next two days, the patient took ʒiij. of the succus conii twice a day, and on the *sixth day* he could walk, and even run without discernible lameness, and could now hold a pen in the hand, and make a fair attempt to write his name. It was only, however, by means of a very strong effort that he was able for a short time to restrain the play of the fingers. He still complained of headache. For the *next four days* he took ʒiij. of the succus thrice a day, and on the *tenth day* only a little restlessness of the affected hand and wrist remained; the head was now released; he walked well and briskly, and spoke freely and distinctly; the movements of the hand no longer hindered sleep; but from the weakness and unsteadiness of the affected arm, he was still unable to dress or feed himself. A severe attack of *urticaria* appeared on the eighth day, and lasted for twenty-four hours. From the *tenth* to the *seventeenth day* ʒiv. of the succus were taken thrice a day. 'Each dose made him very giddy and nearly took him off his legs.' During this period he rapidly improved, and at its termination all choreic movement had ceased, and he was able to assist himself and others, and to write as steadily as usual. In all other respects he was quite well. I saw him on the twenty-seventh, and again on the thirty-seventh day, and found that he continued well and active, but he had complained occasionally of slight rheumatic (?) pains in the knees.

9. *Chronic and obstinate Chorea, with morbid activity of the nervous system from birth. Prolonged use of Conium. Cure.*

James R., aged 6 years, a slender, but healthy-looking and very intelligent boy. Never had any illness, but was once threatened during dentition with a convulsive attack. From his birth he has been of a remarkably active, restless disposition, and during his waking hours never remained quiet. He is mischievous and destructive. If he has no other occupation, he bites his nails and tears with his teeth his pocket-handkerchiefs, and even the sheets as he lies awake in bed. He is affectionate, but although well managed, is excitable and petulant. He came under my care March 6, 1868, and for the preceding six months his natural restlessness had become extreme, and was associated with gradually increasing want of control over the movements of his limbs and tongue. Of late the whole body was in a constant writhe. In walking he goes along in a sideward direction, and the left foot is frequently jerked round behind the right ankle, so as to kick any one who walks by his right side, and often causing him to fall. This is very marked when he attempts to walk on tip-toes. He has a difficulty in getting the raised left leg down to the ground, and after some delay it is often set down in front of the right, and he then stumbles over it. As soon as the body is fairly supported on the toes, the left foot is screwed inwards and then jerked off the ground. He can only walk a very short distance. As he sits still, the left leg is occasionally jerked upwards. The left arm is similarly affected, and quite useless. If he attempts to take hold of an object with the hand, the arm is at first thrown into the air, and then as suddenly plucked downwards, and with a twist of the shoulder, brought behind the back. Without fixed attention, an object held in the left hand soon falls. The right limbs are only slightly affected, but in feeding himself he throws his food about far and wide. The head is ducked and twisted from side to side, and the levator muscles of the face are in almost constant play, tucking in the angles of the mouth and raising the eyebrows. The tongue is continually on the writhe. His speech, which is naturally ready, clear, and distinct, is now hesitating—occasionally even to stuttering—coarsely sibilant, and some days, so indistinct as to be almost unintelligible to his parents. The patient sleeps fairly, but the limbs are affected with slight jerkings. Of late he has complained of headache, and the forehead is hot. The appetite is good, and the functions regularly performed. His *weight* is just under three stone, being a little *less* than it was *this time last year.* The heart sounds are healthy; the pulse 74 to 80, often a little irregular in speed. There is no apparent cause for this disorder. He has been under medical care for some time past, with the usual treatment for chorea (including shower-baths and anthelmintic purgatives), but has derived no benefit.

Treatment and Results.—I prescribed ʒjss. of succus conii *every* day an hour before dinner, and increased the dose to ʒij. on the *third* day. On the *seventh* to ʒiij., and on the *tenth* to ʒiijss. Previous doses produced no apparent effect, but after that last mentioned he admitted that he felt a transient giddiness, and the involuntary movements seemed a little diminished. On the *fourteenth* day I increased the dose to ʒiv. He occasionally complained of slight giddiness a quarter of an hour after this dose, and he began to manifest a slight improvement during the action of the medicine. On the *twenth-first* day the dose was increased to ʒv. At first it produced within twenty minutes decided giddiness, with double vision, and a little weakness of the legs. Subsequent doses did not produce double vision, but the eyes had a dull, heavy, vacant appearance for about ten minutes. There was no tendency to sleep, and he continued of his own accord to walk about. A marked improvement was now perceptible. The headache and heat of forehead had left him, and he was now generally quieter, and had much greater control over the left leg, and consequently walked better. The left arm, however, remained almost as useless as ever. On the *twenty-fifth* day I gave the patient ʒvss. of the succus, and made him use his left hand in taking it. He held the glass pretty steadily, and keeping the arm close to the side, brought the mouth down to meet the vessel, and so managed to drink the draught. In presenting me with the glass the forearm was so suddenly thrown forwards, that the vessel was nearly jerked out of his hand. This dose produced giddiness and staggering, followed by perfect quietude for ten minutes, as if he was about falling off to sleep; but he soon got up and continued to walk about awkwardly for the next half-hour, when the effect of the medicine passed off. During its operation, and the following hour, the tongue and left arm were held steadier than I had ever seen them. From the *twenty-sixth* to the *forty-fourth* day, the little patient took ʒivss. of the succus twice a day—an hour before breakfast, and again at 4 P.M. During this time his further progress was slow; but at the end of the period he could walk and talk much better, and was beginning to use the left hand. The startings during sleep had also left him. During the *following month* he took ʒivss. of the succus thrice a day, before breakfast, at noon, and at 7 P.M. No very decided effect followed any dose, but he continued to improve, and at the end of this period an occasional hesitation of speech, and slight jerking of the left arm, were the only observable remnants of the chorea. I continued the conium *another fortnight* in doses of ʒvj. thrice a day. Every dose produced a decided effect. A quarter of an hour afterwards he was so giddy and tottering that he was obliged to lie down; the eyes were heavy, hazy, and expressionless, as if affected by sleep. But after lying quiet for twenty minutes he would get up, resume his occupations, and make no more complaint. Under the influence of these large and oft-repeated doses he rapidly and completely recovered, not only from all traces of chorea, but from the excessive restlessness which had possessed him for many months previously. During the whole of the time he was under treatment (twelve weeks) his general health improved, and he gained in strength and in weight, and there was a notable improvement in the appetite. The conium had no effect on the pulse, pupils—as far as my observations extended—nor upon any of the secretions. At first the improvement in this obstinate case was so slow that I began to despair of obtaining much relief from conium. However, after carefully watching the effect of large doses, I was encouraged to continue its use, and had the satisfaction of proving that when given in such doses as produced decided physiological effects, the morbid excitability of the motor centres was rapidly subdued. During the twelve weeks the patient was under my care he took upwards of five pints (104 fluid ounces) of the succus conii, and ʒv. or ʒvj. of the same preparation invariably produced decided hemlock symptoms, occasionally even to tottering, in several adults.

This case proves:—1st. That apart from its effect upon the motor centres, conium possesses no direct influence upon the circulatory, nutritive, or secretory functions; and 2ndly, That its use may be prolonged with safety. Its effects are transient and powerful, and it is entirely destitute of what has been called 'cumulative action.'

The two following are instances of *chorea associated with a strong tendency to convulsive action from undue sensibility of the motor centres:—*

10. Annie S., aged 3¼ years, a diminutive, pallid, nervous child. She had never walked well, and at the time she came under treatment was almost unable to do so without assistance. She was affected with great restlessness and general choreic movements—of a very spasmodic character—of the limbs. During sleep she talked, gnashed the teeth, and clenched the hands; and the arms and legs were affected with frequent jerkings. During dentition she had six or seven convulsive attacks, and it was evident that she was now strongly predisposed to a return of the fits. Under the influence of ʒss. doses of the succus conii there was a rapid improvement.

11. John H., aged 8 years, had enjoyed good health, and was free from nervous disorder, until eight months before he came under my notice, but had never been able to walk well. For the last eight months he had been affected with choreic movements of the limbs, twitchings of the facial muscles, great restlessness, and, from increasing weakness of his legs, almost an inability to walk. I prescribed ʒj. of the succus conii, and soon increased it to ʒjss. twice a day, and ultimately to ʒij. The ʒjss. doses produced slight and transient giddiness, and a perceptible diminution of the restlessness and irregular movements. ʒij. caused a little tottering, giddiness, and heaviness of the eyes, and arrested the startings of the flexor tendons, alongside of the radial pulse. The treatment was continued for *two months*, at the end of which time the child walked well, and was otherwise quite recovered.

All these cases show, and the last two most evidently, that, when muscular weakness is the result of exhaustive irritability of the motor centres, we have in conium a remedy which, in allaying and removing that irritability, is thus indirectly a restorer of muscular power.

V. **Paralysis agitans.**—The following cases will serve to illustrate the use of conium in the several varieties of this disease.

12. *General paralysis agitans from defective nutrition or atrophy of the motor centres, probably excited by ague.*

James B., a stonemason, aged 58, but looked older. He had been a healthy and powerful man up to the year before he applied to me for advice, but at this time, while residing at Sheerness, he was affected with an attack of ague, and was laid up a fortnight. The shaking fits were very severe, and seem to have made a permanent impression upon the nervous system. For four months afterwards the left arm and leg began to shake, and since this time the agitation has gradually affected the whole of the muscular system, and for the last three months the tremor has been general. Latterly his diet had been poor and insufficient; he was pallid and wrinkled, stooped very much, spoke hurriedly, walked with a short hurried and shuffling step, and complained of great bodily weakness.

Treatment and Results.—Assuming the probability of the existence of some central irritation, and doubtful how far the muscular tremor might be taken as an indication of loss of power in the motor centres, I resolved to give conium a fair trial in this case. I found that ʒiij. of the succus was sufficient to produce decided physiological effects, and prescribed this dose every other morning for two months. At first he thought that the tremor was less on the conium days, but after a persevering trial I was obliged to discontinue its use, and resort to astringent chalybeates. I describe the effects of the medicine in the patient's own words, as they were in accordance with my own observations. The conium 'is too strong, and does not agree with me; it makes me nervous, and the shaking is no better, but worse.'

13. *Local paralysis agitans, rheumatic diathesis, slight lead taint.*

John R., aged 57, a fresh-coloured, healthy-looking, and powerfully-developed smith, of temperate habits, applied to me September 1857, on account of violent tremor of the left arm of three weeks' duration. He had suffered much from muscular rheumatism, and some years ago was thus disabled for a year and a half. He uses white lead in fitting iron joints, and is occasionally exposed to its dust while cleaning it off old joints. A faint blue hue is observed along the margins of the gums. Potassii iodidi gr. viij. and potassii bicarbonatis ʒss. were prescribed to be taken thrice a day, and he continued this treatment for three months without improvement. His engagements took him away from town, and I did not see him again until the following April, at which time the disease had existed for eight months, and it was obvious now that it had steadily increased. The tremor at this time affected the entire limb from the shoulder to the wrist. He hooked the thumb within the trowsers pocket in order to steady the hand, and, as it was supported in this position, it vibrated to and fro 130 times a minute. When the arm hung down unsupported, the tremor of the whole limb was increased, and became almost unendurable. On clenching the hand and rigidly contracting the powerful muscles of the forearm he was able, by means of a strong effort, to check the motion for a few seconds, but the strain was too great to be continued, and as the effort relaxed, the rigid limb was as severely affected with tremor as when it was in a relaxed condition. There was no loss of power; the left grasp being as strong as the right, and he could raise 56 pounds with the left hand readily. A 14 lb. weight suspended from this hand failed to arrest the tremor. Just at the moment his attention was diverted to a fresh object the tremor ceased. During sleep the fingers of the affected hand became locked, and considerable force was required to open them in the morning; and the left leg was affected with painful cramps at night. Occasionally he had experienced attacks of giddiness, with staggering and momentary loss of consciousness. Passing off in the course of ten minutes, it was usually followed by sharp pain across the forehead. He would fall asleep when he attempted to read, and, when not actively employed, 'must be always sitting over the fire.' Otherwise his health was good, but he had a very decided stoop, and his gait was slowly assuming a shuffling character.

Treatment and Results.—I now treated him at intervals, and during the time that his occupations allowed him to remain in town, with large doses of hemlock; and, finding smaller doses insufficient to produce proper effects, prescribed ʒv. twice a day, and occasionally administered it in single doses of ʒvj. or ʒviij. The effects of the larger doses have been already described incidentally (p. 5 and 17). The improvement was decided, and so manifest as to be observed by his fellow-workmen. He was conscious of greater control over the agitated limb as the effects of the conium subsided. When I last saw him, at the end of January 1868, he was much improved; the general tremor was gone, the to and fro motion of the hand was very slow and feeble compared with what it had been, and the limb was occasionally free from agitation for two or three days continuously. The nightly cramp, and daily pain induced by the means employed to restrain the motion of the arm, had now left him. I have since received a report of further improvement.

Now, to whatever cause we may attribute the disease in this patient, it is pretty certain that he was lapsing into the condition described in the previous case. The result is encouraging and, taken together, these cases show that while hemlock is both theoretically and practically contraindicated in an advanced stage of the disease it may be very beneficial in the commencement. Paralysis agitans is disease which no doubt reacts readily and powerfully upon itself, a local and partial irritation, inducing exhaustion and otherwise predisposing the whole of the great motor centres to participate in the disorder. It is by checking the motion in its very earliest stages that we can alone hope to prevent the further progress of the disease.

In two other cases, one a man aged 69, the subject of chronic rheumatism; and the other a man of middle age affected by severe mercurial tremor over which the usual remedies had no control, the paralysis agitans yielded speedily under the influence of conium.

VI. **Nocturnal Cramps of the Limbs.**—Four cases, two of them very severe, were treated with conium. Speedy alleviation followed in each case.

VII. **Tetanus.**—To this disease, whether arising from inflammatory irritation of the nerve centres or from the tetanizing action of poisons such as strychnia and thebain, conium is the natural antagonist. But it must be boldly opposed. If the patient cannot swallow, from ʒvj. to ℥jss. or ℥ij. of the succus conii warmed to the temperature of the body should be injected alone into the bowel, and repeated every two, three or four hours, according to the condition of the muscles. I regret that I am unable to furnish an example of the use of conium in this disease, but I cannot too strongly insist on the dogma, that to be effectual, the physiological effects of the drug must be declared. Practically, it may require a much larger quantity than I have mentioned to produce them. Until the action of the hemlock is manifested, we may expect no amelioration.

VIII. I come now to speak of the use of hemlock in those spasmodic affections which result from derangement of the vagus nerve.

14. *Spasm of the œsophagus* occurred in a healthy young man aged 21. It originated in a scratch from a minute fishbone, and the dysphagia, and pain behind the upper rings of the trachea, were so persistent that I had reason to suppose the bone had become embedded in the mucous membrane. The passage of a full-sized œsophageal bougie always afforded temporary relief, but I failed on every occasion, and after repeated efforts, to pass the instrument until I had brought the patient under the influence of conium. Then a full-sized bougie passed at once and without resistance. I gave the patient conium in the intervals, and he slowly but perfectly recovered.

Spasmodic contractions of the stomach and œsophagus, associated with crampy pain of the stomach, eructations of large quantities of wind, and the globus hystericus. In this common nervous affection I have found conium very serviceable.

Spasmodic cough, *Laryngismus stridulus*, and *Pertussis*.—In these diseases conium will be found very beneficial. I have used it largely in the latter affection and with success. As in all other cases, we must give the remedy in full doses in order to produce a soothing effect, and in prescribing for infants and young children we must remember that ♏xv. of the succus conii by the skin, which is equal to ♏xlv. by the mouth, is only sufficient to paralyse a mouse for an hour or two (see p. 9). For a child a few weeks old I prescribe from ♏xx. to ♏xl., and for one a year old ʒj. and more, and repeat the dose thrice a day.

It is desirable in every case to watch the effect of the medicine for half an hour. A condition of quietude with a fixed sleepy appearance of the eye may be taken as a general indication that the proper dose has been reached, and we must increase the dose until this is effected.

Asthma.—I regret that I have had no experience of hemlock in this disease. It is one most suitable for topical application, and the solution prescribed at page 83 may be used with hot water as an inhalation.

IX. In organic disease or functional derangements of the spinal cord attended with excessive irritability of the reflex function, conium will be a most suitable remedy. From my experimental observations (see p. 9) I have been led to conclude that conium does not much depress the reflex function unless it be given in doses sufficient to kill the animal. In medicinal doses, and in quantities sufficient to produce complete paralysis of the voluntary muscles, the reflex function of the cord remains intact (see p. 6). Conium therefore cannot be expected to do more than quiet undue excitement of this part of the nervous system.

Paraplegia.—Conium often acts most beneficially in allaying the reflex jerkings of the limbs which are occasionally so distressing in this disease. But there are many cases in which the dose required to effect this, is too large to be safe.

15. *Concussion of the Spine.*—On stepping out of a house through a door by which he had not entered, a healthy middle-aged gentleman dropped through an open trap-door into a cellar, and falling through a considerable space, alighted upon his feet. He felt much shaken and shocked for a few minutes, but soon recovering, he went on his way and completed his usual daily work. During the next twenty-four hours he was troubled with incessant erections and profuse seminal discharges, which latterly were stained with blood. I saw him at the end of this time, and he was restless and exhausted, his legs were weak and tremulous, and the pulse was feeble,

fast, and irregular. He had never experienced symptoms such as these before. Upon the same principle that we use opium in an over-excited and exhausted state of the brain, I prescribed fl. ℥iij. of the succus conii, to be repeated at intervals of several hours, and enjoined perfect rest. This treatment was continued for six days. The emissions continued at intervals during the first day of the treatment, and then ceased. At the end of a week he was perfectly recovered, and has continued in good health up to the present time, which is five months after the accident.

The foregoing case leads me to speak of *the influence of conium upon the sexual organs.* I have given conium a full trial in every variety of cases of this kind. In those cases of exhaustion and irritability which arise from early self-abuse; in those of troublesome irritation where the patient has been suddenly deprived of the legitimate means of gratifying his desires; and in those cases of erotic tendency that arise from some obscure irritation of the lumbar portion of the spinal cord, I have never known conium fail to give relief. It is very remarkable that, while it possesses such a decided influence over the morbid conditions of the sexual function, conium should be incapable of depressing the natural function. The ancients believed that it not only repressed natural desire but actually caused atrophy of the testes and mammæ, which latter, of course, implies atrophy of the ovaries also. This does not accord with my own experience, from which I have been led to conclude that conium, in doses short of causing a poisonous action, has no more power to arrest or depress natural sexual desire than it has of arresting or depressing the respiratory function. Conium in medicinal doses can only influence, as I have before observed, morbid conditions of the spinal cord, and in doing so leaves its natural function intact.

In *Dysmenorrhœa, Ovarian irritation,* and *Ovarian tumour,* I have been unable to afford any relief by means of conium. Many such cases, however, are chiefly if not entirely due to congestion. In those which can be traced to irritation of the lumbar nerves, we may reasonably expect conium to be of service, and give it a fair trial accordingly. Upon the

Bladder conium appears to have no influence. I have never observed the slightest vesical disorder follow its use; and in a chronic case of painful spasm of the bladder I have given the drug a fair trial, but it did not afford any relief.

X. I have yet to make a few observations upon the use of hemlock in certain diseases of the eye, in cancer, and in glandular enlargements.

In inflammatory diseases of the eye, conium will hereafter become, I believe, a valued remedy. In producing complete muscular relaxation it acts beneficially in relieving pain and tension, and thus removing that irritation which aggravates the primary disease and tends to make it chronic.

I have treated six cases of strumous inflammation of the conjunctiva, more or less involving the cornea and iris, with conium, alone and unaided by external applications, successfully. The speedy relief from the photophobia, lachrymation, and spasm of the orbicularis, has often surprised me. Referring to my observations on the influence of conium upon the fifth nerve, I shall content myself with the narration of a single case as an amplification of the preceding statements.

16. *Chronic interstitial keratitis, with permanent spasm of the m. orbicularis.*

Clara J. L., aged 8 years, had been the subject of *interstitial keratitis* with a vascular zone in the sclerotic, frequent lachrymation, and intense *photophobia* for six months, during which time the eyes were closely shaded, and she was led about as one blind. The *m. orbicularis* and *corrugator supercilii* were always observed to be in a state of forcible contraction, and although she was a most patient, obedient, and intelligent child, it was quite impossible to obtain a satisfactory view of the cornea.

My friend Mr. Spencer Watson, of the Central London Ophthalmic Hospital, who referred the little patient to my care, had previously treated her with hypodermic injections of atropine to relieve the photophobia, and on one occasion administered chloroform in order to examine the cornea, and with the hope of breaking the habit of contraction which affected the orbicularis, but without success. She was also treated for a long period with syrup of iodide of iron and cods' liver oil internally, and with ointment of mercury and belladonna externally. When she came under my care, the lachrymation and photophobia were as intense as ever, and when the shade was removed from her forehead, she covered her little pallid wrinkled face with her hands.

After persevering attempts to get a satisfactory view of the cornea, we failed to do so. My treatment of this case consisted in the administration of large doses of conium, increased from ʒjss. to ʒvj., and given from three to three and a half hours after breakfast, at intervals of a few days. After the third dose, ʒij., there was a manifest improvement; the eyelids, which had been firmly closed for months, were now partially opened, and she could see to move about without guidance, and could look down under her shade at a printed page without corrugation of the orbicularis, and endeavour to make out the largest letters. On the eleventh day I could, while she was under the influence of the conium, completely explore the cornea without difficulty. Both pupils and all but a small part of the circumference of the iris were concealed by a dense bluish-white opacity of the cornea. A few of the blood-vessels of the sclerotic and conjunctiva were turgid, and there was an occasional gush of tears.

The sight was improving, and she could make out such objects as a dog or cat, and was able to go a short errand alone.

At the end of three months she had entirely discarded her shade. The eyes were free from any irritation, the opacity of the cornea had nearly disappeared, only a slight nebulosity remaining to prevent a clear view of a small part of the iridial margin. She could read the smallest type, and was now able to return to school.

This case was remarkable for the toleration of large doses of conium. Only slight giddiness, weakness under the knees, and laxity of the m. levator palpebræ, and orbicularis, followed doses of the same drug which produced much graver effects in active and powerful adults.

XI. **Cancer.**—I have given hemlock a fair and prolonged trial in a few cases.

17. A married lady, aged 46, weak, inactive, and liable for a great many years to diarrhœa and intermittent action of the heart, manifested the cancerous diathesis by the appearance of the characteristic tumour in the right breast, April 1866. The breast was wholly removed by my friend Mr. John Wood, the following July. The disease reappeared in the axillary glands nine months afterwards (April 1867), and gradually involved those of the inferior part of the neck, and of the loins. The lumbar pain obliged her to take to bed at the end of January 1868. The right lung became implicated, and she grew rapidly weaker—respiration 50, and pulse 120—and died suddenly, doubtless of syncope, shortly after eating her dinner, and just within two years from the first appearance of the disease.

Treatment and Results.—Hemlock was used internally and externally, and for a time simultaneously. ʒjss. to ʒiij. of the succus conii were taken by mouth every day, with occasional interruptions, from the time the axillary glands showed participation in the disease to the time of death—a period of eleven months. ʒij. always produced so much giddiness and languor that she was unable to walk after the dose. Lint saturated with the succus and covered with oiled silk was constantly worn in the axilla for the first two months. At the end of this time it had produced so much irritation that I was obliged to suspend its use for a time. On applying it again, diluted with an equal bulk of water, the same irritation reappeared after a time, and I then discarded the external use of the remedy altogether. It produced a dry, scaly eruption of the cuticle in crescentic patches, and a dark copper-coloured very irritable condition of the cutis, identical in appearance with the irritable variety of lepra. As often as the conium was left off, the skin returned to its healthy state.

Conium, it is clear, had no effect in arresting the progress of the disease in this case; but it appears nevertheless to have acted beneficially: 1st. In preventing and mitigating pain. Considering the extent of the disease, and the great enlargement and extreme scirrhous induration of the parts affected, the patient suffered comparatively little pain. To within six months of her death she did not require opium, and during the last three months she took only occasional doses of morphia, and never exceeded half a grain in the twenty-four hours. 2nd. The drowsiness which succeeded the morphia sleep usually passed into a calm sleep after the afternoon dose of conium, and thus the use of much opium was avoided. 3rd. The general health and appearance actually improved during the time she was taking the hemlock up to within three months of her death, a fact which attracted the notice of her friends.

18. The other case was one of spreading epithelial cancer of the rectum. The subject of it was a fleshy, healthy-looking woman, aged 40.

She suffered much pain and bloody discharge at intervals, but chiefly after defæcation. I prescribed ʒij. of the succus conii, and she took it every day for three months. Each dose produced, a quarter of an hour after taking it, 'a dreary heaviness of the eyes, dizziness, weakness of the legs, and reeling,' and she was obliged to sit down for twenty minutes, when the effects began to wear off rapidly. As she grew weaker from the extension of the disease—for the conium had no influence whatever in arresting its progress—she was obliged to diminish the dose of the medicine to ʒjss. She was so convinced of its efficacy in allaying the pain and diminishing the discharge that it was with great reluctance she would consent to intermit the use of the drug.

In *Cancer of the pylorus* I have found conium decidedly beneficial in allaying the excruciating pain referred to this part and no doubt produced by cramp of the sphincter muscle.

XII. In strumous and syphilitic enlargements of the lymphatic glands I have in several well-selected cases given conium a fair trial, but have failed to observe the slightest improvement in any case.

XIII. **In pure cerebral disease** conium in my hands has

proved useless. I have given it in a typical case of that form of *general paralysis* which accompanies dementia, from the earliest stage, until the disease was well advanced. And also in several cases of mental oppression, accompanied by cerebral headache, and tendency to give way to emotion, but without the slightest benefit. In the latter condition, hemlock is contra-indicated on account of the lassitude or want of energy which often attends it.

In *acute mania* associated with an exaggerated development of muscular power, conium is worthy of trial. And the furious maniac may doubtless be brought under the influence of conium with advantage.

In the irritable condition of the brain that often exists when an attack of *cerebral hæmorrhage* is impending I believe that conium may be advantageously employed. I do not foresee that its use is contra-indicated in any case; and when the irritation or congestion lies in the corpus striatum, conium will be of service; for by arresting the generation of nerve force in this part we shall use the best means for relieving the congestion.

Having now completed my remarks upon the use of hemlock in disease, I must acknowledge that I leave the subject in a very imperfect state. My endeavour has been to lay down a few facts which may serve to form, as far as they go, a foundation upon which we may build a more complete history. I will now devote a little space to the critical examination of the pharmaceutical preparations of hemlock. And as I cannot feel satisfied with making bare assertions as to the medicinal value of this or that preparation, I shall endeavour so to put the proofs before the reader that he will be enabled to judge of this matter for himself.

CHAPTER II.

THE PREPARATIONS OF CONIUM MACULATUM.

IN furnishing four preparations—viz., an extract, a succus, a cataplasm, and a tincture—conium maculatum occupies a prominent position in the 'British Pharmacopœia.' In a series of papers published in the 'Pharmaceutical Journal' in the year 1867, I gave a detailed account of some investigations respecting the medicinal value of the several preparations of hemlock, including an examination of the root. I proceed now to reproduce the more important of the facts contained in these papers, with such additions as I am enabled by further experience to make.

I. **Tinctura conii fructûs.**—This preparation was introduced into the first 'British Pharmacopœia' of 1864 as a substitute for the tincture of the dried leaf of the obsolete 'London Pharmacopœia.' In speaking of this preparation, Dr. Garrod (Med. Times and Gaz. vol. i. 1864, p. 168), states that 'fl. ʒv. caused the development of slight symptoms—some sensation of tightness of the forehead, and a little alteration of vision, and perhaps some increased feeling of numbness of the lower extremities.' He considers this tincture to possess at least twice the strength of the old tincture of the leaf.

As these statements were too vague to reassure me of the value of a preparation which my own experience had led me to conclude was insufficient to produce the proper physiological action of hemlock in even still larger doses than fl. ʒv., I resolved to undertake a careful enquiry into the medicinal value of this and the other preparations of hemlock.

The tincture with which the following observations were made was most carefully prepared for me by Mr. Hemingway, the distinguished pharmaceutical chemist of Portman Street, Portman Square. The following were its characters :—Reaction slightly acid, colour light greenish-brown with an internal opalescence, a strong mousy odour. Mixed with sixteen times its bulk of water, the colourless solution assumed, upon exposure to light and air for twenty-four hours, a leaf-green tinge.

In order to ascertain the physiological effects of the tincture, I selected two individuals—a weakly emaciated woman, M. A. R., aged 37, and myself.

I began, November 11th, by taking fl. ʒss., and increased the dose each day until the 20th, when I took ʒvj. After discontinuing the use of the drug for a week, I resumed it on the 28th, and, having daily increased the quantity by ʒij., took a single dose of ℥ij. on the 3rd of December.

The medicine was taken in a little water, from 1½ to 2½ hours after breakfast. Some of the mornings I took on rising ʒj. of bicarbonate of potash in a draught of water, sometimes alone, sometimes with a small proportion of tartaric acid. By this means the urine was preserved alkaline until late in the afternoon. The other mornings I purposely abstained from this or any other preparative measure.

I carefully looked for effects, but found none after any of the doses, excepting a stimulant action from the larger quantities of spirit. There was no disorder, nor diminution of muscular power. The pupils, definition in the vision of near and distant objects, the pulse and all the functions remained in their usual state, and the secretions were active and normal. During the whole of the time, I was working harder and longer than usual, and sleeping less; nevertheless there was no sense of fatigue, neither drowsiness nor tendency to inaction. Every other day I was actively engaged in body and mind, and usually walked from four to seven miles. On the alternate day I remained quiet, and was chiefly employed in study. Immediately after taking the ℥ij. of tincture on December 3, I sat down and wrote my letters, and then entered upon some microscopical investigations, and continued them, with a single break of the dinner hour, for eight hours consecutively. On this and other similar occasions I retired to bed without any of the feeling of mental fatigue which I sometimes experience after prolonged microscopical work.

The other subject of my experiments was in a very different condition. She was a pale, delicate, emaciated woman, and confined to bed by the pain and constitutional disturbance attendant upon the formation of a very large abscess in the right loin. Her pulse was 108 and feeble, and she was restless and unable to sleep. The abscess was opened on November 13, and a pint of pus discharged. The same night I ordered as an anodyne fl. ʒij. of the tincture above described, and directed the dose to be increased each night, provided, as in my own case, no effects should follow. She slept well. On the following night fl. ʒiij. were given, and there was no sleep. On the 18th she took fl. ℥ss. at night, but did not sleep well after it. On the 19th fl. ʒvij. were given, and she had a good night's rest. Having used her supply, the conium was suspended for a few days, and opiates (♏xv. to ♏xxx. tincturæ opii) administered instead. Meanwhile the abscess was closing, the appetite returning, and the health rapidly improving. On December 1st she took fl. ℥j., and on the 2nd fl. ℥jss., which exhausted my supply. On carefully examining this woman from day to day, and with special reference to the effects of conium, neither Dr. Collie, one of the resident medical officers of the hospital, nor myself, could detect any result. Great relief followed the evacuation of the matter, and her health began to improve directly afterwards, and she was soon convalescent.

Examination of the marc.—In order to make my experiments more satisfactory, I subjected the marc to the following process:—Placing it again in the percolator, I passed a solution of ʒj. of caustic potash in fl. ℥viij. of water through it, and subsequently washed it with water until it passed through colourless; fl. ℥xiv. of dark brown fluid, resembling tincture of henbane in depth of colour, were thus procured. I subjected this to distillation, drop by drop, collecting the first ounce and a half separately. I allowed fl. ℥vij. more to distil, and set this aside. I then put one-half of the marc (which had been successively exhausted by spirit and solution of potash) into the retort to the remaining fluid and distilled fl. ℥iv. more. Having satisfied myself that these three fluids differed in no respects from each other, they were mixed. The mixture gave all the reactions of a dilute aqueous solution of conia and ammonia. I carefully preserved it, and on December 4 following, I took half a drachm in the morning, and a drachm in the afternoon, and increasing the dose on the following days took a single dose of two fluid ounces on the 9th. No effect followed any of the doses.

The fruit used was a fine specimen. It was clean, and free from admixture with other umbelliferous fruits. The albumen was firm and solid, the commissure convex, the groove indicating the involution of the albumen broad and deep, and the crenations of the ridges well formed—all of which I take to be essential characters of a well-matured fruit. The powder was prepared by means of a fineish hair sieve, and *without the application of heat.* It evolved a strong heavy mousy odour.

I am informed by Mr. Hemingway that the whole of the conium fruit used in British pharmacy is obtained from Germany, and that the sample used in preparing the tincture employed in the previous investigations was grown near Prague. Doubtless the German fruit is, to say the least, equally potent with that of British growth; and, as far as our present investigations are concerned, the use of the German fruit is the more appropriate, since it was most probably that employed by Geiger in his experiments. He states (Mag. für Pharm. xxxv.) that nine pounds of the dry ripe fruits yield one ounce of conia. Accordingly one ounce of the fruit should yield three grains of conia, and the quantity contained in fl. ℥xx. of the tinctura conii fructûs—assuming the fruit to be thoroughly exhausted of the alkaloid—would be $7\frac{1}{2}$ grains = to 0·375 in fl. ℥j. Now continental physicians prescribe conia in doses of $\frac{1}{16}$ of a grain for a child, and $\frac{1}{4}$ to 1 drop for an adult.[1] Hence fl. ℥j. of the tincture would be only a medium dose for an adult—assuming, as I have said, that it contain a quantity of conia equivalent to $\frac{1}{20}$

[1] Ann. de Thérap. 1853, p. 73; Archiv. Gén. 4e sér. xxiii. 226. See also Wood and Bache's Disp. United States Pharmacop. 11th ed. p. 295.

part of the fruit employed. It appears therefore that the quantity of fruit employed in the preparation of the tincture is much too small. But even if a much larger quantity were used, it is very doubtful whether the preparation would be an efficacious one, for the active principle, although freely soluble in dilute spirit is in the fruit effectually protected from its action by the horny albumen with which it is associated—a protection which is very inadequately removed by comminution.

In order to prove the quality of the fruit used in the tincture with which my experiments were made, I subjected one ounce, finely powdered, to the following process for the extraction of conia. Having mixed it with an equal bulk of fine sand, I packed it loosely in the percolator, and passed, after previous maceration, alcoholic and aqueous solutions of caustic potash through it, and subsequently alcohol, until it dropped through colourless. By this means, fl. ℥x. of a turbid, brownish-green fluid, of the same depth of colour as the tincture of the leaf of the London Pharmacopœia, was obtained. This was exactly neutralised with sulphuric acid, and the sulphate of potash separated by filtration. The filtrate was placed in a retort, and the whole of the alcohol and the chief bulk of the water distilled off. These distillates were perfectly free, both from ammonia and conia, and also from sulphuric acid. About ʒiv. of a blackish-brown syrupy fluid remained in the retort, and to this was added fl. ʒiv. of aqueous solution of caustic potash, containing sixteen grains of the alkali. The mixture was exposed to a temperature of 248° Fahr., by means of a chloride of calcium bath, and the distillation rapidly conducted. Colourless water and minute drops of equally colourless oily fluid passed over. About fl. ʒvj. were obtained in all, and a charred black mass, which, when cold, evolved an intensely acrid and ammoniacal odour, remained in the retort. The distillate contained about two grains of conia, but I was unable to determine its exact weight, for it soon became opaquish, assumed a faint brownish tinge, and began to dissolve in the highly alkaline fluid upon which it floated. This latter assumed a brownish tinge. It possessed, but in a much greater degree, the reactions of the distillate from the marc of the tincture formerly described. It formed with iodine a colourless solution, and dissolved sulphur. When heated it became turbid and evolved the intensely acrid fumes of conia under the appearance of a white cloud. As the conia condensed again, it trickled in oily streaks down the sides of the tube. The presence of a little alcohol in the distillate doubtless rendered the conia soluble to this extent.

In operating upon so small a quantity of seed at so high a temperature, the waste of the conia is of course much greater than in experiments upon a large scale, and I believe that I am fully justified in concluding that the fruit operated upon, and used in the preparation of the tincture, possessed the full amount of conia.

The result of these experiments goes far to prove that the tinctura conii fructûs is in all proper medicinal doses, to say the least, an inert preparation. From Geiger's and Christison's experiments it appears that the fruit contains a larger quantity of conia than the other parts of the plant. But the fact that the green fruit contains a much larger quantity than the dry seems to have been overlooked. We know that the active properties of the poppy are more abundant in the circulating juices of the green fruit than in any other part of the plant; and that they diminish in proportion as the capsule becomes dry and hard. It is very probable that this is the case with the fruit of the conium also (see p. 96).

II. **Tinctura Conii P.L.** (*Tincture of the dried leaf*).—I obtained two samples of the Tinctura conii (P.L.): Messrs. J. Bell and Co. kindly furnished me with one, which I call 'Tincture No. 1;' and Mr. Hemingway prepared for my use another, which I will designate 'Tincture No. 2.' As I had in view a series of comparative experiments with the tincture of the fruit, No. 2 was prepared in December by exhausting after eight days' maceration in the percolator, ℥ijss. of fine green, strongly smelling, dried leaf (collected the same year, and carefully preserved in a tin canister in a dry place), by the passage of fl. ℥xx. of proof spirit. Thus its strength in comparison with the tincture of the P. L. was as 19 to 20; the

London process yielding only fl. ℥xix. out of the fl. ℥xx. of spirit employed. No. 1 was prepared soon after the leaves were dried, and preserved from access of light. There was no apparent difference in the two preparations. Both possessed an acid reaction; a dark greenish-brown colour, a rank odour, and its corresponding flavour with a nauseous, bitterish taste. On admixture with water both became turbid from the separation of a green resinous matter.

I began my experiments with Tincture No. 2:—

December 19, at 10.45 A.M., I took ʒij. mixed with a little water, and remained quiet all day.

Dec. 21, at 11.15 A.M., took ʒiv., and remained quiet for five or six hours afterwards.

Dec. 22, at 10.45 A.M., I took ʒvj., and was afterwards and during the rest of the day actively engaged. Walked about five miles.

Dec. 24, at 11 A.M., took ℥j., and sat still conversing with patients for the hour following, and was afterwards actively engaged until midnight, when I retired to bed free from headache or fatigue. Next day I did not take the tincture.

Dec. 26, awoke with a headache, and felt weak and poorly from broken rest, and a sharp attack of diarrhœa during the early morning. At noon I took ʒx. of the tincture, and immediately walked out a distance of three miles. No effects followed, neither was there any increase of the headache or sense of debility.

Dec. 28, at 10.45 A.M., took ʒxiij. of the tincture, and from half an hour to an hour and half afterwards experienced a slight stimulant effect.

I now began to use *Tincture No.* 1.

Dec. 29, at 10.30 A.M., I took ʒv. On the following day, at 10.30 A.M., ʒvij., and sat quiet for an hour and half afterwards.

Dec. 31, at 12.25 A.M. took ʒix. I had previously been sitting in a cold room, and felt very cold, and my pulse was only 60. I pursued my writing at the same temperature, and three-quarters of an hour after taking the conium my pulse was 72, and had increased in force; the stimulant action of the alcohol was manifest.

January 1, at 10.45 A.M., I took ʒxj. Jan. 3, at 10.30 A.M., ʒxiij.

Jan. 4, at 11 A.M., I took *a mixture* of ℥j. of Tincture No. 1, and ℥ss. of Tincture No. 2. Jan. 8, at noon, took ℥j. of No. 1, and ʒvj. of No. 2.

Jan. 10, at 11.15 A.M. took ℥j. of each of the tinctures at a draught, and remained quiet. An hour after, the stimulant and diuretic effects of alcohol were fully manifest; the pulse was 76, the pupils normal. I was actively employed during the latter half of the day; worked a considerable time with the microscope, and did not retire to bed until midnight.

Beyond the above-mentioned stimulant and diuretic action, no effects whatever followed the use of the tincture. The quantities mentioned were mixed with an equal quantity of water, and taken at a single draught from an hour and a half to three hours after breakfast, which consisted of a moderate quantity of coffee, or occasionally tea, with cold meat and bread.

On one occasion an alkali was previously taken.

I gave these tinctures, in doses varying from ʒiij. to viij., in single doses to several of my patients during convalescence from acute disease, and on the day after they were allowed to leave their beds when they were very weak and tottering, but no other effect save that of stimulation from the spirit followed in any case.

Dr. Garrod states (*loco citato*) that one of his patients took fl. ℥j. of the tincture of the carefully dried leaves, and that no symptoms followed.

The dried plant.—Feeling that it is a matter of considerable importance to determine whether the dried plant does retain any active properties, and if so in what degree, I have carefully examined the dried leaves, from a portion of which the tinctures employed in my experiments were prepared. Excepting in the *poultice*, the dried leaf is no longer used in the British Pharmacopœia; but the importance of the investigation will be recognised when it is observed that the dried plant is largely used in some other Pharmacopœias. Looking first to our nearest neighbours, I find that the French Codex contains no less than six preparations of the dried leaf, viz. :—1. An alcoholic extract; 2. A plaster made of this extract; 3. An injection, composed of an infusion of the dried leaf; 4. Powder of the dry leaves; 5. An æthereal tincture; and, lastly, 6. A tincture.

The 'Norwegian Pharmacopœia' has two preparations of conium. 1. The dried leaf, prescribed as follows:—'medium dose, 2 to 3 grains; 10 grains would be a dangerous dose.' 2. An aqueous extract of the dried leaf treated by alcohol, of which it is said:—'medium dose 1 to 2 grains; a dangerous dose, 6 grains.'

There is scarcely a Continental Pharmacopœia which does not contain these and similar preparations of conium.

The 'United States Pharmacopœia' contains four preparations of conium, three of which are derived from the dried leaf:—1, an alcoholic extract; 2, a fluid extract; and, 3, a tincture corresponding to that of the London Pharmacopœia.

It is to be observed that the dried plant is thus extensively used notwithstanding that some very competent observers have expressed doubts respecting its activity. Geiger indeed expressly states [1] that the dried leaves of hemlock do not contain any conia; and Pereira says,[2] 'no reliance can be placed on the dried leaves however carefully prepared, for they sometimes yield no conia, though they possess the proper hemlock odour and a fine green colour.' Of these two statements the latter is nearer the truth, but it implies—what I believe is untrue—that *some* dried hemlock leaves *do* possess the active properties commonly ascribed to them.

The following are my own observations upon this point:—

Examination of the dried leaves used in the preparation of the tinctures above referred to.

1. February 11, 1867. Took one ounce avoirdupois of each of the two samples of leaves, separated from leaf-stalk and in coarse powder, and packed them in thin layers alternating with layers of fine sand in a percolator. fl. ℥x. of water containing 120 grains of caustic potash was poured upon them, and maceration allowed for twenty-four hours.

[1] Magazin für Pharmacie, xxxvi.

[2] Pereira, Elem. Materia Medica, vol. ii. part ii. p. 196.

The aqueous solution was then displaced by fl. ℥viij. of dilute alcohol (equal parts of rectified spirit and water), and maceration allowed for twenty-four hours more. The spirituous fluid was next displaced by water acidulated with sulphuric acid, and percolation continued as long as the running fluid possessed colour. fl. ℥xxij. of very dark greenish-brown fluid was thus obtained. A little more acid was added to produce exact neutralisation of the alkali, and the turbid fluid filtered. Chlorophyl and sulphate of potash, destitute of conia or any of its salts, remained on the filter. The filtrate was evaporated over a water bath at a temperature under 160° F., until about ʒv. of dark brown extract, of the consistence of treacle, remained. While still warm, this was rubbed up with fl. ʒv. of solution of caustic potash (1 part in 3 of water). A very faint odour of conia was evolved. The mixture was transferred to a long tube, and shaken at intervals with an equal bulk of æther. The æther assumed a yellowish-green colour. After twenty-four hours the æthereal solution was decanted, and the extract washed with fresh portions of æther as long as it continued to dissolve anything. The mixed æthereal solutions were then distilled. Half a grain of a clear, deep sap-green, thick, oily fluid, lighter than water, remained. It possessed a mint-like odour mixed with that of conia. To the tongue it was almost as bitingly acrid as conia itself, but in minute quantity it produced, like oil of peppermint, a sharp cooling sensation. Its taste was bitter, and it possessed, in an intense degree, the nauseous flavour of the dried leaf or its tincture. It was in fact a mixture of conia and the oleo-resin of the plant, coloured by chlorophyl. It imparted to water a strong alkaline reaction. Mixed with water acidulated with sulphuric acid it refused to dissolve, but the aqueous fluid obtained a tinge of colour, and, when evaporated nearly to dryness, a dark film of syrupy fluid remained, which, when mixed with a little solution of caustic potash, evolved a distinct odour of conia.

2. An ounce of each of the two samples of dried leaves was taken and mixed with fl. ℥vss. of water acidulated with fl. ℥ss. dilute sulphuric acid, P.B. Maceration was

allowed for seven days at a temperature of 50° F. The fluid was then displaced by water. fl. ℥x. of bright sherry-coloured infusion was thus obtained. This was neutralised exactly by potash, and filtered. A modification of chlorophyl, which gave a deep yellow colour with potash, and sulphate of potash, both free from conia or any of its salts, remained on the filter. The filtrate was treated as in the first experiment, and the extract in like manner supersaturated with potash and washed with æther. A little less than half a grain of bright pale greenish-brown oily matter remained. It possessed a powerful odour, compounded of conia and the peculiar odour of the leaves with a minty addition. It smelt more of conia and less of mint than the product described in Experiment 1. Its taste was intensely biting, like that of conia itself, leaving a flavour of tobacco and peppermint, and the rank taste of the dried leaves. Treated with sulphuric acid the oily fluid partly dissolved, and the filtered solution manifested a purple tinge on evaporation, and furnished a little brown syrupy extract, which, upon the addition of potash, evolved a strong odour of conia, a distinct trace of which was obtained from the mixture by the aid of æther.

It appears from the foregoing experiments that the dried leaves do, when carefully prepared and preserved, retain a trace of conia; and it is equally conclusive that the quantity is much too small to furnish an efficient preparation.

3. In order to make my investigation complete, I subjected the leaf-stalks—primary, secondary, and tertiary—to the same process as that described in No. 1. Taking the same quantity of the leaf-stalks, viz. ℥ij., I obtained as nearly as possible the same quantity of oily matter as from the leaves. Its physical and chemical properties were identically the same as those of the oily fluid obtained from the leaves.

It will be observed that I have not followed the usual process (that of distillation) for the extraction of conia in the above experiments. I have been induced to adopt the above method in order to prevent that decomposition of the alkaloid which takes place by prolonged heating with potash.

If I had followed the prescribed processes, I should no doubt have been led to the same conclusion as Geiger—viz. that the dried leaves are altogether destitute of conia.

I am now brought to the enquiry, What is the value of the cataplasma conii, P.B.? According to the most liberal computation it contains only half a grain of conia. As far, therefore, as the active principle of hemlock is concerned, this preparation may without hesitation be set aside.

III. **Succus Conii.**—I now turn to another preparation of conium, the succus conii. This is, indeed, a most worthy representative of the famous hemlock, as I have already fully shown in the former part of this work.

The preparation which I have chiefly employed in the treatment of the cases above detailed has been prepared for me two years successively by Mr. C. F. Buckle, of 77, Gray's Inn Road, W.C. He has kindly furnished me with the following particulars respecting the herb and the preparation of the juice which I used in the early part of my investigations:—

'June 1, 1866.—Received from Mr. Gaines 56 lbs. of conium maculatum, grown in Essex. The plants were fresh and fine, and just coming into bloom. The process of pulping between finely-grooved iron rollers was commenced at once; when complete, the pulp was subjected to the pressure of a very powerful hydraulic press, and 75 per cent. of juice obtained. This was immediately mixed with the proportion of spirit prescribed by the "British Pharmacopœia" (1 volume to 3 of juice), and the mixture set aside in a cellar. The whole of the process was completed in one day. The mixture was subsequently filtered as directed, and bottled off.'

The resulting preparation was of a dark sherry-colour, possessed a delicate and agreeable herby taste and odour without acridity, and an acid reaction. The following were its characters:—Sp. gr. 1002. fl. ℥j. yielded 30 grs. of soft extract, and 0·42 grs. of pure conia. Heated with a little caustic potash, it evolved suffocating fumes of conia. Heat, alcohol, nitric acid, all precipitated albumen. The boiled and filtered juice gave reactions indicating the presence of sugar (in considerable quantity), soda, magnesia, lime, phosphoric acid (in considerable quantity), sulphuric acid (a minute proportion), chlorine. Bichloride of platinum gave a muddy molecular yellow deposit; tannic acid, a fine flocculent precipitate; perchloride of iron caused a precipitate, but neither the per- nor proto-salts produced any discoloration. fl. ℥j. of the 'succus' yielded six grains of white ash, which fused with effervescence before the blow-pipe into a porcellaneous mass, dissolved with copious effervescence in the mineral acids, and the clear acid solution gave an abundant heavy yellow crystalline precipitate with bichloride of platinum. Hence it follows that, in addition to the above constituents, the juice contains one or more vegetable acids and potash.

It is to be observed that Schrader[1] makes no mention of either soda or sugar in his analysis of the juice, and that he, De Machy, and Errhardt[2] mention nitric acid as one of its constituents. I have carefully examined the extract, and also the ash left after its combustion, and find myself in agreement with Bertrand and Baumé in being unable to discover a trace of nitrates.

A sample of the succus prepared by Messrs. Allen and Hanburys had a sp. gr. of 1015, the greater density being chiefly, if not altogether, due to the larger proportion of albumen and sugar. In all other respects this succus corresponded with that already described.

A *third sample*, prepared June 3, 1863, by Messrs. J. Bell and Co., had a density intermediate between that of the first and second samples—viz. 1005. It contained less albumen than either of the above. In all other respects it agreed with the other samples, and furnished the reactions above mentioned.

Mr. Buckle's preparation from the herb of last year, 1867, had a density of 1003·4.

The variation in colour and density is partly due to the season and locality in which the plant was grown, and partly to the degree of pressure employed. Some of the older authors direct that 'the green watery juice which flows freely on first pressing the bruised herb, should be thrown

[1] Berzelius, Traité de Chimie, vol. vi. p. 254. Berlin. Jahrbuch, 1805, s. 152.
[2] Bertrand, op. cit. p. 306.

away, and the small quantity of yellowish liquor which follows the harder pressing should be alone preserved.' I have satisfied myself that this is an erroneous idea, and that the first and last portions of the juice are equally potent.

I have used the above, and other specimens of the succus furnished by Messrs. Herring, of Aldersgate Street, and Messrs. Corbyn, of Holborn, and find, from year to year, no appreciable variation in the activity of the preparation.

I have carefully ascertained one important fact—namely, that the succus does not deteriorate on keeping. I have repeatedly used a succus, prepared by Messrs. J. Bell and Co., three, four, and five years previously, and have found it as active as specimens only a few days or weeks old.

Dose.—From two drachms to one ounce of the succus will, according to the motor activity of the individual, almost invariably produce the full physiological action of hemlock, and the beneficial effects which may be expected to follow. I usually give a child, six months old, twenty or thirty drops of the succus conii; a child over two years old, one drachm; one ten years old, from one to two drachms. For a woman, I prescribe two or three drachms; and for a man, four or five drachms. From these initial doses I ascend until the peculiar effect of hemlock is declared. Having once attained this, it is rarely necessary to increase the dose, for a dose which produces a given effect will, after six months' continuance of the medicine, usually influence the patient to the same extent at the end of that time. Care must be taken in administering conium to patients possessed of but little bodily vigour. On the other hand, there are some persons whose activity is such that ℥x. will be required to produce giddiness and muscular weakness. I have given as much as ℥xij. to one patient, and two others have occasionally taken ℥viij. without experiencing any very decided effect.

Having now clearly, and satisfactorily I hope, distinguished the active *succus* from the inert *tinctures*, and fully illustrated its use, I need only add, that it is a preparation which, in the compactness of the dose required, in the absence of any objectionable taste and odour, and in the potency and certainty of its operation, leaves nothing to be desired.

I usually prescribe it alone; if diluted, camphor water or some other antiseptic menstruum should be employed on account of the presence of albumen.

As a substitute for the cataplasma conii, P.B., a piece of lint saturated with the succus, or, if heat and moisture be required, a bran poultice containing an ounce or two of the succus, may be used.

Extractum Conii.—Having completed my examination of the tinctures and succus, I come now to the consideration of the extract. Very few medicines have attained so great a reputation and been so extensively employed as the extract of hemlock.

Introduced by Störck, in the year 1761, as a remedy of marvellous power in the removal of almost every inveterate disease to which the human frame is subject, it soon obtained admission into the Pharmacopœias; and, regarded as it is by practitioners of the present day as a powerful and useful remedy, it is still retained in almost every one of them. I myself have seen it prescribed almost daily, in doses varying from 1 to 5 grains, for the last twenty years. Nevertheless, it is to be observed that the efficacy of the extract has been questioned, and several times disproved, from the days of Störck down to our own times.

The following is the formula for the extract, to the agency of which Störck attributed his wonderful cures:—

'℞ herbæ recentis cicutæ quantum sufficiat. Exprimatur succus, isque recens lentissimo igne in vase terreo (sæpius agitando, ne amburatur) coquatur ad spissi extracti consistentiam, hoc extractum s. q. pulveris foliorum cicutæ in massam pilularem subigatur; ex qua fiant pilulæ granorum duorum.'[1]

In some cases a few grains, taken daily for two or three weeks, were sufficient to remove, as it appeared, an old-standing disease, while in others the patient swallowed ℨij. of the extract daily for four or five months without inconvenience. 'The extract of hemlock,' says Störck, 'is a remedy

[1] Essay on the Medicinal Use of Hemlock, by A. Störck, 1761, p. 14.

absolutely innocent; it does not hurt the sight, but the contrary.'

The following criticism, by an eminent contemporary of Störck, appears to me very just, and worthy of mention in this place:—

'Quin et incomprehensibile, ac plane paradoxon videtur, id statuisse. Præterquam enim quod nec in meis, nec in Breslaviensium pluribus, ea vis cicutæ confirmata fuerit, si consulam auctorem, qua namque dosi, a cicutæ extracto, hanc vim edi putet, video a granis 2 de die observasse eandem et sic porro a granis 4, ab 8, a 12, a 20, 30, 60, 120, 180, 240, idque haud rariore admodum casu sed frequenti.

'Si granum opii consuevit homini blandum conciliare, erunt alii qui indigeant dupla dosi, rariores qui triplo, quadruploque, rarissimi qui quintuplo, qui sextuplo uno die indigeant. Cicutæ autem dosis cur adeo immense augenda fit, ex comparatione cum ceteris paregoricis haud facile capitur.'[1]

Störck's observations on the use of hemlock excited so much attention that his mode of treatment was adopted in almost every country of Europe, and many of the leading practitioners of those times gave his far-famed extract ample trials. It needed but a short time to convince all observers that Störck had greatly over-estimated its virtues. Not a few, however, were satisfied that it was a remedy of considerable value. Störck, Collin,[2] Quarin,[3] F. Hoffmann,[4] Hill,[5] Rouppe,[6] Gataker,[7] Andrée,[8] W. Butter,[9] Akenside,[9] Spalowski;[10] Burrows,[11] have all advocated its use, and given us the result of their observations: but if we carefully examine their writings, we shall fail to recognise any mention of the least trace of

[1] Epistola de Cicuta, Antonius de Haen, 1766, pp. 20, 21.
[2] Observ. circa Morbos Acutos, etc. 1765.
[3] Tentamina de Cicuta, 1761.
[4] Observ. on the Internal and External Use of Hemlock, 1764.
[5] Sir J. Hill, Directions for those Afflicted with Cancers, with account of the Vienna Hemlock, 1771.
[6] De Morbis Navigantium; acced. de effectu extracti Cicutæ, etc. 1764.
[7] Essays on Medical Subjects, 1764.
[8] Obs. on Störck's Treatise, 1761.
[9] Treatise on Kinkcough, with an Appendix on Hemlock, 1773.
[10] De Cicuta, 1777.
[11] Prac. Essay on Cancers, with method of Administering Hemlock, 1767.

those effects which distinguish the action of hemlock. I believe, therefore, that we are fully justified in concluding that the extract, whether prepared in Vienna, Amsterdam, Geneva, Naples, or in London, was practically, if not absolutely, destitute of the active principle of the plant. Indeed, the impotency of the drug was occasionally recognised by some of these observers themselves, who attributed it to various causes—the wrong plant had been used; the locality in which it had been grown, or the situation in which it had been exposed, was unsuitable for the elaboration of its juices; the herb had been gathered a month too soon or too late; the whole of the watery juice of the expressed herb had been used, whereas the first portions should have been rejected and only the latter and more resinous part employed. Dr. Butter, with a more correct appreciation of the real cause, cautions against the employment of too much heat in the preparation of the extract, and gives the following directions for its preparation:—Evaporate the freshly expressed juice in a broad glazed platter over a charcoal fire, and, as soon as green clots form, stir the liquor frequently, keeping it at such a heat as will make them move about without driving them above the surface or occasioning an ebullition. Evaporate with constant stirring till the extract is of sufficient consistence to form pills. Such directions, taken in conjunction with the precaution '*ne amburatur*,' given in the previous formula, sufficiently indicate by what agency the powerful juice was reduced to an inert mass. As with the dried leaf, so with the extract, the active principle has departed and a dead inert body alone remains. The above-mentioned authors introduce us to scores of patients who are taking the extract of hemlock largely. We look from one to another to discover some evidence—no matter how slight—of its action, but we search in vain; not a trace even of its earliest and most prominent effects are anywhere visible. We can hardly admit that these effects, evanescent though they be, could have been overlooked by such a body of intelligent observers. As scholars, at least, they were acquainted with the observations of Paulus Ægineta, Dioskorides, Plato, Galen, Plinius, respecting the action of hemlock; and, as scientific facts, these observations were repeatedly advanced in the discussions which the treatise of Störck excited in those days.[1]

Passing by these earlier observers, I find the effects of hemlock practically indicated, for the first time, in the works of J. Hunter and Fothergill. Reference has been already made to an observation by the former (see p. 16). The latter, speaking of a particular patient, says, 'The dose of hemlock (extract) was gradually increased from 20 to 70 grains a day; if he took more, it either made him sickish or created a singular kind of headache and giddiness.'[2] These are, I think, real indications of the presence of hemlock. It must be observed, however, that the extract used by Dr. Fothergill was much more carefully prepared than that used by Störck and his contemporaries—precautions having been taken both to collect the plant at the proper time, when the active principle is most abundant, and to avoid prolonged exposure of the juice to a high temperature.

Whyte has very lucidly described the effects of hemlock upon his own person. After taking 15 or 20 grains of extractum cicutæ, 'I have often,' he says, 'been affected with a weakness and dazzling of my eyes, together with a giddiness and debility of my whole body, especially the muscles of my legs and arms; so that, when I attempted to walk, I was apt to stagger like a person who had drunk too much strong liquor.'—('On Nature, &c., of Nervous Disorders,' p. 22. Edinburgh: 1765.)

A medical friend of Bertrand administered ʒj. of carefully prepared extract to a patient daily for a year, without result.[3] Dr. Allbutt, of Leeds, informs me that he 'has often given the extract in doses so large as to nauseate by its mere mass, without other results.'

It thus appears conclusively that, from the time of its introduction to the present day, the extract has been regarded by many as an uncertain preparation, and it is remarkable that

[1] De Haen, op. cit. Viventius (J.) De Cicuta, Naples, 1777.

[2] Obs. on the Use of Hemlock, John Fothergill, M.D., Works, vol. ii. p. 59.

[3] Bertrand, Recueil de Mémoires de Méd., de Chir., et de Pharm. Militaires, 1ère sér. vol. ix. p. 313.

its value has not been long ago more satisfactorily determined. Christison, Geiber, Orfila, Pereira, and others, all concur in the opinion that most of the extract of conium of the shops is inert or nearly so. Pereira states that he was unable to procure any sensible quantity of conia from ℥iv. of the extract.[1] The observations on the extract are concluded in his work by the following statement, which is accepted, I believe, as a pharmaceutical axiom :—'The goodness of the extract may be determined by the disengagement of a strong odour of conia, when it is gradually triturated with liquor potassæ.' This test is so readily applied, and appears at the same time so decisive, that any more elaborate analysis seems superfluous; and yet I venture to assert that no statement can be further from the truth, no test more fallacious. Half an ounce of extract, containing but a fraction of a grain of conia, will, on trituration with caustic potash, speedily evolve a powerful and penetrating odour of conia, and the effect is usually very much heightened by the simultaneous separation of a little ammonia. A great deal too much has been inferred from this reaction, and this is the cause, I believe, why we have so long remained in a state of uncertainty respecting the virtue of the extract. A given sample has been pronounced good, because, on commixture with caustic potash, it has evolved a strong odour of conia. Attention to the following experiments will show the fallacy of such a conclusion.

I examined two extracts prepared by Mr. Buckle from a portion of the fresh juice used in making the succus, so fully described above, and largely used in the treatment of many of the cases described in the earlier pages of this work. Small quantities of the juice were operated upon, and the evaporation was therefore rapidly conducted, and the temperature never allowed to exceed 160° Fahr. In one of the extracts the chlorophyl was retained, according to the directions of the 'British Pharmacopœia;' in the other it was rejected.

Extract without the Chlorophyl.—This was of the consistence of treacle, and had a similar bright and clear, but a

[1] Elem. Mat. Med. vol. ii. pt. ii. p. 206.

richer amber-brown, colour; odour faintly approaching that of the ordinary extract, taste pleasantly sweet and acidulous, without any trace of acridity. *Triturated with caustic potash, a strong odour of conia, mixed with that of ammonia, was evolved.*

1. January 26, 1867. Took 250 grains, and having liquefied it with fl. ℥j. solution of caustic potash (gr. 32 in fl. ℥j.), transferred the mixture to a retort, and distilled from a chloride of calcium bath, at a temperature varying from 260° to 270° Fahr. 8¾ fluid drachms of colourless fluid, with a faint greasy film, passed over. fl. ℥v. water, containing 50 grains of caustic potash, were now added to the contents of the retort, and distillation continued as long as alkaline fluid passed. ℥vjss. of fluid in all was obtained. The conia was obtained from this by neutralisation with sulphuric acid, evaporation, separation of the sulphate of ammonia, decomposition of the sulphate of conia with potash, and separation of the alkaloid by æther. It weighed only 0·2 of a grain.

By the process adopted in the separation of the conia from the ordinary extract (see below), I obtained from the same quantity (250 grs.) of this extract without chlorophyl exactly one grain of bright yellowish-brown oily fluid, which almost wholly dissolved in dilute sulphuric acid. It was, therefore, nearly pure conia.

February 13, 1867, I took 5 grains of this extract. March 10: 10 grains. April 2: 15 grains. April 3: 20 grains. No effects followed either dose; nor could I obtain the slightest physiological action in the persons of two delicate women by giving the extract in the above-mentioned doses. To produce the slightest evidence of the presence of hemlock, 50 grains at least would have been required, but the doses were not further increased; for to be of any practical value, the extract should contain such a proportion of conia that its effects may be manifested after a dose of 10 or, at most, 20 grains.

Ordinary Extract of Conium of the 'British Pharmacopœia.'—The following were the characters of this extract:—smooth, dull olive-green, of a consistence sufficient for forming pills, taste acidulous, free from all bitterness and acridity, but partaking slightly of the nauseous oleo-resin of the plant. *Triturated with a little solution of caustic potash, a powerful odour, compounded of conia and ammonia, was evolved.*

2. January 22, 1867. Took 250 grains of this extract, and having liquefied it with a little water and fl. ʒiv. of solution of caustic potash (1 part to 3 of water), thoroughly washed the mixture with separate portions of æther. After distillation of the æther, there remained 1·8 grain of a dark sap-green oily matter, which partly solidified after some hours. It possessed all the physical characters of the impure conia, obtained from the dried leaf by the agency of potash and alcohol (see examination of the dried leaf). Treated with dilute sulphuric acid, a portion dissolved, leaving a remainder of oleo-resin, coloured with chlorophyl. The acid solution contained nearly 1 grain of hydrated conia.

April 7, 1867. I took 10 grains of this extract. April 10: 15 grains. April 13: 20 grains. Not the slightest effect followed any of these doses, although the conditions for their development were as favourable as could be desired.

I gave this extract in the same doses to two female patients; the one suffering from an ovarian tumour, the other from anæmic headache and dimness of sight. No effects followed its use, not even in the latter patient, who was already predisposed for its action.

3. Another sample was prepared by Messrs. J. Bell and Co., June 18, 1867, from very fine plants in full flower. The herb was subjected to a pressure of a little more than ten tons, and 25 per cent. of juice obtained: this was treated according to the process directed in the Pharmacopœia, and rapidly evaporated in divided portions at a temperature not exceeding 140° Fahr., and the process completed in the way prescribed.

I examined this extract the following day. It was smooth, firm, and of a dark olive green colour, possessed the smell of the fresh plant, had a nauseous and bitterish taste, and tinged the saliva of a deep green colour. 250 grains were treated by the process last described for the extraction of conia. The alkaline mixture was repeatedly washed with æther until that last used obtained no colour. Nine grains of rich emerald green, thick, oily extract were obtained. By means of dilute sulphuric acid 2·2 grains of tolerably pure conia were separated from this.

In a young woman very susceptible of the action of hemlock, 20 grains of this extract induced the physiological action to the same degree as fl.ʒijss. of the succus; each causing slight and transient giddiness, and weakness of the legs. 25 grains caused decided hemlock symptoms (giddiness and inability to stand) in a young man afflicted with incomplete paraplegia. In my own person, 20 grains of the extract taken a few days after it was prepared failed to produce the slightest effect under circumstances the most favourable for the action of the medicine; and in two able-bodied active youths 60 grains were required to produce slight effects.

4. In using the fresh extract, prepared as above described by Messrs. Bell, this year, 1868, I have been more careful to determine the relative power of the extract and succus, and find that 60 grains of the former is equivalent to fl.ʒiv. of the latter. One patient, who is taking 100 grains of this extract for a dose, considers its effects equal to those produced by fl.ʒvj. of the succus. Dr. Pliny Earle found that from 45 to 100 grains of extract were requisite to produce decided physiological effects in his own person.—('Amer. Jour. Med. Sc.' p. 63. July, 1845.)

From these facts it appears that the power of the extract has been, and still continues to be, greatly over-estimated. The present Pharmacopœia (1867) directs it to be given in doses of from 2 to 6 grains. Now, granting that this preparation retain the whole of the active principle, which, from my examination of the 'succus,' I place at 1·4 grain in a 100 grains; 6 grains of the extract would represent only the 0·084 of a grain of conia—a quantity insufficient to produce the effects of hemlock in a child two years old. The physiological action of hemlock is such, that doses which fall far short of producing it are of no use; and it is doubtful whether the possession of an extract containing 1 per cent. of conia would be of any advantage, since 25 grains of it would be equivalent to only fl.ʒiv. of the 'succus' of the Pharmacopœia.

But it appears from the following experiments that it is hardly possible by any process to obtain an extract from the juice of the plant, which shall contain as much as 1 per cent. of pure conia.

1. Evaporated fl.℥j. of the succus conii, P. B., over a water bath to the ordinary consistence of the extract. About an hour was required for the operation. After liberating the conia, and completely removing it, I found that it weighed 0·3 of a grain, a quantity 0·12 gr. less than I obtained by the same process from the same quantity of the succus, to which I had previously added fl.ʒss. of dilute sulphuric acid, P. B., in order to fix the conia.

2. Placed fl.℥j. of the same sample of 'succus conii' in a retort, and distilled fl.ʒiijss. by the aid of a water bath. The distillation occupied three hours. The first fl.ʒjss. passed over during the first fifteen minutes, and was collected separately. Excepting that the first fluid was chiefly spirit, the distillates did not appear to differ; both possessed a stronger odour of the plant than the succus itself; both gave out an extremely faint odour of conia on the addition of caustic potash, both were rendered faintly opalescent by the addition of nitrate of silver and of chloride of mercury. The remainder was transferred from the retort to an evaporating dish, and exposed to the heat of a water bath for another hour. The syrupy residue was then mixed with potash, and thoroughly washed with æther. 0·19 gr. only of conia was obtained, being 0·11 gr. less than was obtained by the first experiment, and less than half the quantity contained in the ounce of 'succus.'

3. Exposed fl.℥j. of the 'succus' upon a plate in a glass-house with a south aspect, and where the natural temperature ranged at the time from 70° to 90° Fahr. After thirty-four hours the small syrupy residue was treated with potash and washed with æther; 0·25 of a grain of conia was obtained.

Mr. Judd, in his experiments upon cats, found that it required 80 grains of Squire's extract, carefully prepared by spontaneous evaporation, to kill one of these animals.—('Trans. Med. Bot. Soc.' p. 127. Lond.)

Two facts appear from these experiments—first, that the active principle of the plant is to a certain extent vaporisable even at a natural temperature of 70° to 90° Fahr.; and, secondly, that prolonged exposure to a high temperature is accompanied by a progressive diminution of the conia, the alkaloid being converted, as Dr. Christison has pointed out, into ammonia and some other secondary product.

It has been doubted whether the Athenian state-poison was wholly derived from the hemlock; I see no reason—on account of the expression '*μικρὸν πάνυ καταπότιον*, a very little dose'[1]—for doing so. The inspissation of the juice was effected, according to Dioskorides, by exposing it to the sun; and by this means a syrup may be prepared, of which, assuming the Greek plant to be equally powerful with that grown in these temperate regions, a tablespoonful or two would doubtless prove a fatal dose.

An efficient extract may be prepared from the green and nearly ripe fruits by means of alcohol, and the avoidance of a temperature above 160° Fahr. (see p. 92). Dr. Fountain experienced well-marked hemlock symptoms after taking 12 grains of an alcoholic extract of the fruit prepared at a low temperature.—('Amer. Jour. Med. Soc.' Jan. 1846.)

Conia.—Finding that the ordinary processes for the separation of conia fail where only very small quantities of the alkaloid are present, I have been led, after many trials, to adopt the following process, by which I believe we may succeed in isolating a mere trace of conia. The substance known or suspected to contain conia is comminuted and exhausted, after a few days' maceration in a percolator, by water acidulated with $\frac{1}{50}$ of its bulk of sulphuric acid; the filtrate is spread out in a thin layer upon flat dishes, and allowed to evaporate to the consistence of a thin syrup in a warm, dry room, or at the distance of three or four feet before a fire; the residue is mixed with an equal bulk of strong solution of caustic potash (1 part to 3 of water), transferred to a long tube, and agitated with its bulk of æther several times during twenty-four hours. The æther is then decanted, and

[1] Theophrastos, Hist. Plant. iv. 8, p. 298, ed. Schneider.

the alkaline mixture washed again and again with fresh portions of æther, until the conia is completely removed. Two, or at most three, washings are sufficient for this purpose. On distillation of the æthereal solution, the conia, more or less pure, remains. The impure conia is next shaken with a small quantity of dilute sulphuric acid, which separates the alkaloid from oily or resinous impurities. From this solution of sulphate of conia the base is separated in the usual way—viz. evaporation to a syrupy consistence, mixture with caustic potash, washing the mixture with æther, evaporation of the æther, and finally distillation of the conia in a current of hydrogen, which may of course be omitted when we only want to determine the presence of conia.

If spirit be used in the exhausting process, the æthereal extract will be contaminated by much fatty and resinous matters ; hence the advantage of using a watery solution.

I satisfied myself that æther will entirely remove the conia from an alkalised extract, by the following experiment :—I took fl. ℥iij. of a dark brown mixture of extract and caustic potash, from which the conia had been completely removed by æther, and added to it a drop of conia dissolved in an excess of a dilute solution of sulphuric acid. After thorough admixture, it was set by for a few hours, and then, finding that it still contained an excess of caustic potash, it was mixed with one-third its bulk of æther, and agitated for two minutes. The æther was removed as soon as it had separated, and the mixture washed again with the same volume of æther, and decanted without delay. On evaporating the æthereal solutions, nearly the whole of the original drop of conia was recovered. In this case the process was hastily performed, and no pains taken to thoroughly wash the exhausted extract.

In searching for conia in organic mixtures, the same process may be adopted. The contents of the stomach may be digested for a few days with a sufficiently large quantity of sulphuric or oxalic acid to prevent decomposition, then strained, evaporated spontaneously, and treated as above. Before we conclude that conia is present, we must isolate an oily matter which possesses, in addition to a conia odour, an intensely sharp biting taste, and which dissolves readily, with loss of odour, in a drop of dilute sulphuric acid. We must pour off this drop into a clean tube, and re-develope from it a strong conia odour by the addition of a little strong solution of caustic potash. If we depend on the odour alone we may fall into error (see p. 18).

Conia is not suitable for internal use either by the stomach or the skin. In some experiments made by Mr. Frederick and Mr. Alexander Mavor and myself we could not produce hemlock symptoms in the horse by the hypodermic injection of conia, either alone or neutralised by acetic acid. In one experiment we injected ♏xiv. of conia dissolved in spirit and water, and introduced by two punctures into opposite sides of the body. In another, ♏xviij. of conia, neutralised with acetic acid, and diluted with spirit and water to make ʒjss. of solution, was injected. But no results followed these or smaller doses. The local inflammation produced by the conia no doubt prevented its absorption. Abscesses resulted in both of the animals subjected to these operations. Now ♏xviij. of conia are equivalent to about forty ounces of the succus conii; and it has been shown that sixteen ounces of the succus are sufficient to produce full hemlock symptoms in a young horse.

For experiments with conia upon animals, the reader is referred to Geiger,[1] Christison,[2] Orfila,[3] Pohlmann,[4] Leonidas van Praag[5] (frogs), Kölliker,[6] Guttmann.[7]

Schroff[8] gave conia to three healthy human adults, in doses varying from $\frac{1}{20}$th of a grain to $1\frac{1}{3}$rd of a grain. The same symptoms as those produced by the fresh herb followed *plus* some gastro-intestinal irritation.

He supposed that this must be caused by an irritant

[1] Magazin für Pharmacie, vol. xxxv. p. 259.
[2] Philosoph. Trans. Edinb., vol. xiii.
[3] Traité de Toxicologie, 4me édit., vol. ii. p. 423.
[4] Journal für Pharmacie, vol. i. p. 44.
[5] Ibid.
[6] Verh. d. phys.-med. Ges. zu Würzburg, 1859, vol. ix. part 2, p. 55, and Virchow's Archiv, vol. x. p. 235.
[7] Berliner klin. Wochenschr., Nos. 5 and 6. 1866.
[8] Reil, Mat. Med. der chem. Pflanzenstoffe, p. 135.

principle which has not yet been isolated; but there is no need of this supposition, since all the effects he mentions are readily produced by conia alone, which is a violent local irritant.

If we cannot conveniently administer conium by the mouth or bowel—in cases of poisoning by strychnia, for example—we may then resort to the subcutaneous use of conia, and ♏xx. of the solution prescribed below, for the vapor coniæ may be injected at intervals into various parts of the skin until the required effects are produced. In tetanus the subcutaneous use of conia is more objectionable than in any other condition, inasmuch as it is adding another source of irritation to the nervous system. But we may sometimes be obliged to use it in spite of these objections.

Vapor Coniæ.—The use of the extract in the formation of the vapor is objectionable, for two reasons: first, the quantity of conia contained in the portion of mixture prescribed, is too small to relieve spasm; and, secondly, any influence which a minute portion of the alkaloid might possess, would probably be more than neutralised by the simultaneous evolution of ammonia from the alkalised extract.

In the following form these objections do not exist, and the dose of conia can be readily graduated:—

Conia, 1 grain.

Alcohol, 1½ fluid drachm. Dissolve the conia in ʒss. of the alcohol, and add the remainder mixed with—

Water, 2½ fluid drachms.

20 minims contain $\frac{1}{12}$th of a grain of conia.

Protected from strong light, it appears that this solution may be kept unimpaired for a very long time. I have already preserved it unchanged for eighteen months.

Hemlock Root.—All that is known of the root of the hemlock is contained in the following:—

(*a*) Theofrastos[1] says that in the case of other roots the juice is weaker than the fruit, but that of *κώνειον* is stronger, and rids a man of life easily and quickly when given as a potion in a very small quantity.

[1] Hist. Plant. IV. viii. p. 298, ed. Schneider.

(*b*) 'Two priests ate hemlock roots by mistake; they became raving mad, and mistaking themselves for geese plunged into the water. For three years they suffered with partial palsy and violent pain.'[1]

(*c*) A vinedresser and his wife ate hemlock roots by mistake for parsneps, and went to bed. They awoke in the middle of the night quite mad, and took to running about the house in the dark, bruising themselves severely against the walls. They recovered under suitable treatment.[2]

(*d*) Störck makes the following extraordinary statement:—'The fresh root, when it is cut in pieces, emits a milk which is acrid and bitter to the taste. On rubbing a drop or two of it on the end of the tongue, it presently became stiff, swollen, and very painful, and soon afterwards I lost the power of speaking.' Again, 'If the powder of the root of hemlock be made into pills with a sufficient quantity of mucilage of gum tragacanth, a medicine is produced of great efficacy, but which requires great circumspection in the use of it.'[3]

(*e*) Gmelin quotes an instance in which 4 ounces of the juice of the root were taken without injury. But the plant which furnished it does not appear to have been properly identified.[4]

(*f*) On April 22, Orfila gave 1½ ounce of the fresh root to a small dog. No effect followed. The next day he introduced into the stomach of another dog 1 ounce of the same sample of the root, bruised, together with about 8 ounces of the fresh juice of the root. No effect followed.[5]

(*g*) Dr. Christison 'found that 4½ ounces of the juice, the produce of 12 ounces of the roots collected in November, had no effect on a dog; and that 4 ounces obtained from 10 ounces of the roots in the middle of June, when the plant was coming into flower, merely caused diarrhœa and languor. The alcoholic extract of the juice obtained from 6 ounces of

[1] Kircher, Wimber, Wirkung der Arzneimittel und Gifte, lor. ii. 172. Pereira, Mat. Med. vol. ii. part ii. p. 201.

[2] Petri A. Matthioli, Commentarii in sex libros Dioscoridis, p. 736, ed. Venetiis, 1582. Gmelin's Pflanzengifte, p. 604. Orfila, Toxicol. Gén. ii. p. 426, 4me édit. Christison, Trans. Roy. Soc. Edin., vol. xiii. p. 396.

[3] A. Störck, Essay on the Medicinal Use of Hemlock, pp. 8, 12.

[4] Op. cit. sec. 605.

[5] Orfila, Traité de Toxicologie, 4me édit. ii. p. 423.

the roots, on the last day of May, killed a rabbit in 37 minutes, when introduced in a state of emulsion between the skin and muscles of the back, and the effects were analogous to those obtained with the extract of the leaves.'[1]

With statements so conflicting as these, it seemed desirable that a careful examination of the root should be made at the season when it is in full vigour, and about to put forth its leaves.

In procuring for me a quantity of fine roots, and in providing me with an 'extract' derived therefrom, Mr. Hemingway gave me every inducement to make such an investigation, and my obligations are due to him for affording me these and other facilities in the execution of my task.

The roots were removed from the ground on January 9 of the present year, during a short intermission of severe frost. They were large and well developed, many being more than two feet long, and, near the crown, an inch in diameter. They were all carefully examined and identified, and I have at this present time some fine plants of hemlock growing from a few of the roots, of the identity of which I was doubtful (see p. 93). The young yellowish-green leaves were beginning to shoot from the crown, and here and there one could be found an inch long. These hemlock roots had the same sweet taste and pleasant flavour as the roots of the carrot, and, side by side, there was in this respect little to distinguish them. The hemlock roots were equally sweet, but the carrot roots had a stronger flavour. After chewing the hemlock root for a few seconds, a numbing sensation like that produced by pyrethrum, but milder, declared the difference. When bruised and in bulk, the hemlock root, moreover, had a rankish odour, approaching to that of the recent leaves.

The roots were well washed and set aside to drain, at a temperature of about 38° Fahr., and thirty-six hours after they were removed from the ground, and reduced to a coarse pulp by twice passing them between finely grooved iron rollers. The pulp was then placed in a number of horsehair bags, and subjected to a pressure of 110 tons, by means of a

[1] Christison On Poisons, 4th edit. p. 856.

powerful hydraulic press. 9¾ pounds of the crushed root yielded 5½ pounds of juice, or about 56 per cent. The process was conducted at Mr. Buckle's establishment, 77 Gray's Inn Road, and I have to express my thanks to that gentleman for his kind and ready help on this occasion. One portion of the juice was immediately converted, by the addition of one part of rectified spirit to every three parts of the juice, into a preparation corresponding to the succus conii of the 'British Pharmacopœia'—a 'succus conii radicis.' A small portion of the crude juice was preserved for separate examination; the rest was at once carefully evaporated down to the consistence of an extract, at a temperature below 100° Fahr.

The Crude Juice.—The following were the characters of the crude juice:—A turbid brownish-white fluid, of sp. gr. 1022·8, having a decided acid reaction, a carrot odour, and sweet carrot taste, leaving a slight numbing sensation on the tongue. Heated with caustic potash it evolved an odour of conia, but not so strong as that from the juice of the leaves. After standing at a temperature of 32° Fahr. for 36 hours the juice remained opalescent. On boiling, a cloud of albumen, equal, after standing 24 hours, to ⅛th of the bulk of the fluid, separated. The supernatant fluid was bright and of a faint greenish-brown tinge. It gave reactions indicating the presence of a large quantity of *sugar*; of *chlorine*; *phosphoric acid* in abundance; *sulphuric acid*, a mere trace; *soda*; *lime*; and *magnesia*. Fl. ℥j. of the crude juice yielded 20 grains of extract; and this quantity of extract 1·3 grain of ash. The ash was with difficulty fusible, and refused to run into a compact porcellaneous mass, like that derived from the juice of the leaf. It dissolved with brisk effervescence in dilute hydrochloric acid. The solution contained abundance of *potash*; and a trace of iron, derived no doubt, from the rollers used in crushing the root. No trace of nitric acid could be detected. Fl. ℥viij. of the crude juice, heated to 150° Fahr. to precipitate the albumen, and filtered, were mixed with ♏xxx. strong sulphuric acid, and set by. Five months afterwards the mixture was unchanged, and fl. ℥j. was taken, mixed with a considerable excess of caustic potash, and twice washed with æther. After separation of the æther and its distillation, there remained less than half a grain of soft solid oily matter. It had a sharp minty tobacco taste, and a strong alkaline reaction. Stirred with a little dilute sulphuric acid it refused to dissolve, but on pouring off the acid and adding an excess of caustic potash to it, an odour of conia, as strong as that from a solution of a small drop of pure conia in four drachms of water, was evolved.

The extract was prepared from the crude juice, as I have stated. Fifty-five fluid ounces yielded 2½ ounces avoir. (1093 grains). A very powerful hemlock odour was evolved during the whole of the process. The extract was chiefly composed of sugar; it was of a drab colour, possessed a faint odour, and very sweet saltish taste, otherwise resembling the extract of the leaf. 30 grains of it taken internally produced no effect, but the sensation of numbness before mentioned remained upon the tongue after swallowing it. It was excessively tenacious, and could be drawn out into long threads.

I. *January* 26, 1867.—Liquified 250 grains of the extract with a little water and fl. ℥j. of solution of caustic potash (gr. xxxvj. in fl. ℥j.), a strong odour of conia and ammonia was immediately evolved. The mixture was transferred to a retort, and distilled from a chloride of calcium bath at a temperature varying from 220° to 250° Fahr. A fluid ounce of highly alkaline colourless fluid was obtained. By Geiger's process 0·3 of a grain of oily matter, smelling strongly of conia, and 3½ grains of nearly colourless sulphate of ammonia, were obtained. Only a very slight darkening occurred during the process of evaporation of the neutralised distillate, but a very powerful acrid odour of conia was evolved on mixing the residue with caustic potash. After lying by in a corked tube for a few days, the conia product consolidated into stellate groups of almost colourless minute crystals; and when, after the lapse of three months, it was dissolved in æther the solution had a neutral reaction and a taste free from the biting acridity of conia. In fact it contained but a faint trace of conia, and appeared to be principally composed of one of the three substances about to be described.

A dark brown dry mass, evolving an intensely acrid odour somewhat resembling the empyreumatic oil of tobacco, remained in the retort. It was liquified with 1½ ounce of water, and washed with a mixture of 1 part of chloroform and 5 parts of æther. After separation and distillation of the æthereal solution, a little clear brown fluid remained, and on allowing it to evaporate spontaneously, a partly waxy and partly crystalline substance, of a rich brown colour, was obtained. I will call this the 'æthereal extract;' it is composed of three distinct bodies—two crystalline neutral principles, and a resinous substance. As I find no mention of them in chemical works, I will briefly describe them under the names of rhizoconine, rhizoconylene, and conamarine, names which I use merely for the sake of distinction, and without any reference to the relationship of these bodies.

Rhizoconine is readily separated from the 'æthereal extract' by means of alcohol. The residue obtained by evaporation of the alcoholic solution is treated with æther, which dissolves out the rhizoconine. It is easily soluble in alcohol, æther, and chloroform; the solutions are neutral to test paper. From alcohol and chloroform rhizoconine separates partly in the form of indistinct squarish crystalline masses, and partly as a soft waxy matter. From an æthereal solution it is deposited in the form of rich yellowish-brown radiating transparent prisms of considerable length, but of soft consistence. However obtained, it possesses a very diffusive, persistent, and slightly pungent odour, strongly resembling that of a dirty tobacco-pipe. Compared with nicotylia, the odour is heavier and somewhat peculiar. Its taste is at first slightly bitter and minty, but it soon becomes subacrid and tobacco-like, and leaves a slight but persistent numbing sensation upon the tongue. Without undergoing solution to any appreciable extent, it imparts to water and to the dilute mineral acids its characteristic taste. The aqueous solution is neutral, and gives no precipitate with either nitrate of silver, chloride of mercury, acetate of lead, or sulphate of copper. The strong mineral acids have no particular effect upon it: rubbed with strong sulphuric acid a dark brown muddy mixture results. It fuses at 160° Fahr.

and above 500° gives off abundant white fumes of a disagreeable odour; it then chars and burns, leaving no ash. Boiled with solution of caustic potash, no alkaline vapours arise.

Rhizoconylene.—This is a colourless crystalline body, obtained from the 'æthereal extract' by means of alcohol, which, dissolving out the rhizoconine and conamarine, leaves the rhizoconylene. It is insoluble in water and nearly so in cold alcohol, but at 175° Fahr. the latter takes up about $\frac{1}{30}$. It dissolves readily in both æther and chloroform, and separates from the former in long, dry, brilliant prisms, apparently possessing a rectangular base. From hot alcohol it is deposited in hard brilliant stellæ, composed of short but very sharply acuminated prisms. Its solutions are neutral, and it is destitute of taste and odour. In the cold, the strongest sulphuric, nitric, and hydrochloric acids have no action upon it. When boiled with solution of caustic potash, no alkaline fumes are evolved. The crystals melt at about 212° Fahr., and assume a brown colour. Above 600° Fahr. the fused mass is wholly dissipated into white fumes possessing a faint, somewhat fatty odour.

Conamarine.—When the 'æthereal extract' is treated with alcohol, rhizoconine and conamarine are dissolved away, leaving the rhizoconylene. The alcoholic solution of the former is evaporated to dryness, and the two bodies separated by means of æther, which dissolves away the rhizoconine. Conamarine is an intensely bitter brownish-green resin, freely soluble in alcohol and chloroform, but wholly insoluble in æther. When heated with solution of caustic potash, it evolves an offensive odour like that of the urine of a carnivorous animal.

With this introduction, I must leave these substances to some one better qualified than myself to determine their composition and relationship. I thought at first that they were the products of the decomposition of conia at a high temperature in the presence of caustic potash and organic matters; but the following observations upon the extract of the root induce me to regard them now as natural constituents of the plant, related more closely perhaps to the oleo-resin of the plant than to its active principle.

II. Having liquified a portion of the extract of the root (described above), I washed it thoroughly with a mixture of æther and chloroform, and after separation distilled off the latter by means of a hot water bath. A small quantity of light yellowish-brown oily matter of neutral reaction remained. It had a faint, fatty, non-characteristic odour, and a warm, slightly bitter, rancid taste. Heated with caustic potash it evolved no odour of conia. After this mixture had been retained at a temperature of 212° Fahr. for some minutes, it was shaken with æther, which dissolved out a little brown subcrystalline matter having a faint odour and taste of rhizoconine.

III. April 10, 1867.—250 grains of the same extract of the root were liquified with a little water, and fl. ℥iv. of solution of caustic potash (1 part to 3 of water). The mixture was thoroughly washed with separate portions of æther, and the latter decanted. After distillation of the æther there remained 1·8 grain of light brown oily matter, which after standing-by for a few hours became in great part solid, from the formation of beautiful dendritic masses of crystals. It had a mixed odour of conia and tobacco, and an acrid cooling minty taste, becoming very bitter and tobacco-like. It contained but the faintest trace of conia, and was composed, like the 'æthereal extract,' from the retort remainder, and in about the same proportions, of rhizoconine, rhizoconylene, and conamarine—the first of these bodies being, in both cases, the most abundant, and the last the least so. After separation of the conia and conamarine, the remainder (about 1 grain) was dissolved in a mixture of 15 minims of alcohol and 5 of æther, and injected beneath the skin of a cat. It produced no effect whatever.

IV. As it occurred to me that the conia might be stored up in the root in some insoluble combination, I took the whole of the roots from which the juice had been expressed, weighing now 4¼ pounds, and pulped them the same day with 30 ounces of hot water holding 4 ounces of caustic potash in solution. The whole of the house was filled with a very powerful and disagreeable mousy odour. The mixture was set aside for 24 hours. At the end of this time 17 fluid ounces of dark blackish-brown grumous fluid was obtained by pressure. It was distilled from a chloride of calcium bath, at a temperature between 218° and 220° Fahr., and 14½ ounces of clear colourless fluid were obtained. It presented a slight greasy film, was strongly alkaline, and had all the other physical and chemical characters of a mixed solution of ammonia and conia. By Geiger's process 15 grains of sulphate of ammonia and a small drop of nearly pure conia were separated from this distillate after neutralisation with sulphuric acid.

From the foregoing observations it appears:—1. That as compared with the other parts of the plant, the root contains only a very small proportion of conia. 2. That in the careful preparation of an extract from the juice, this small quantity of conia is almost wholly lost. 3. That the root contains in addition to a bitter resin, two neutral bodies which, at a temperature between 220° and 250° Fahr., are capable of volatilisation with water (see Experiment I.). 4. That these latter bodies do not appear to possess any active poisonous properties.

They exist in all other parts of the plant, although apparently in much smaller quantities than in the root, for I have obtained them from the retort remainder after the distillation of conia both from the leaves and the fruit. In the latter case they were associated with conhydrin.

Rhizoconine is an interesting body, inasmuch as it resembles nicotylia in some of its physical characters. The slight acridity and carrot flavour of the root are undoubtedly due to this substance.

I come now to the prime object of my inquiries—the medicinal value of the root. Within an hour after the expression of the juice I took fl. ʒss. of it without result. But a full answer to this inquiry will be found in the following experiments with the—

Succus conii radicis prepared as above described. The mixture of crude juice and spirit deposited some dirty white albumen. The filtered product was quite clear, and of a delicate yellowish-brown tinge, and it completely retains the original odour and taste of the root at the present time

April 21st, I took fl. ʒij. of the 'succus;' on the 26th, fl. ʒiv.; May 2nd, fl. ʒvj., and set out walking; on the 10th of the same month I swallowed fl. ℥j. of the 'succus' and remained perfectly quiet, watching for some effect. None, however, followed this or any of the previous doses.

I subsequently gave a patient, a young man aged seventeen, whose general health was good, ℥j. of the succus; and to another, a man of middle age, suffering from debility of the sexual organs, fl. ℥jss. No effects followed in either case.

With another patient I continued its use every alternate day, in doses increased from fl. ʒj. to fl. ʒviij. for some weeks, without any result.

It is conclusive, therefore, that in medicinal doses the root is quite inert. And it appears equally certain from the experiments of Christison and Orfila (*f* and *g*) that it is not poisonous, even when taken in such quantities as would supply the place of vegetables in an ordinary meal. No properly authenticated case of poisoning by hemlock root has been recorded; and the time has arrived, I think, when, in reference to a history of hemlock, such accounts as those given by Matthioli and Kircher (*b* and *c*), and so often adduced by medical authors, should be treated as mere fables. Such cases given, even with an understood 'valeant quantum valent,' do much more harm than good. In the present instance they have, probably more than any other cause, served to obscure and thus retard that clear knowledge of the physiological action of hemlock which I believe we now possess. As to Störck's statement (*d*), if it be not the effect of a greatly excited imagination, it must certainly be referred to some other plant, perhaps aconite. And yet such a statement, applied to the root of that plant, would still be an exaggeration.

The green Fruit.—By far the most efficient preparations of hemlock may be obtained from the green and nearly ripe fruits, as the following observations will show:—

Three-fourths of an ounce of green but nearly ripe fruit was collected, July 22, 1868, during the continuance of a long season of heat and drought, from a plant sprung up in a little soil on the south side of my house here in London, from the self-sown seed of one of the plants referred to (p. 85). The fruit was immediately crushed to a pulp, and mixed with ℥iijss. of a mixture of equal parts of alcohol and water. Maceration was allowed for a month at 70° Fahr. The product was then filtered, and ℥iij. ʒij. of bright yellowish-brown tincture, possessing a strong odour of conia, obtained. A fluid ounce, evaporated over a steam bath for twenty minutes at a temperature below 160° Fahr., yielded ten grains of a bright yellowish-brown, brittle, but very deliquescent extract.

The Tincture.—1. Mary L., aged 36, in whom ʒiijss. of the 'succus conii' produced no appreciable effects, took 50 minims of this tincture, and it caused, within half an hour, hemlock giddiness; fl. ʒj. produced more decided symptoms; and fl. ʒjss. was followed by double vision for several minutes, great giddiness, muscular weakness causing slight tottering, and dilatation of the pupils from $\frac{1}{7}$ to $\frac{1}{6}$. The giddiness and muscular weakness lasted for 1¼ hour.

2. In a strong man fl. ʒij. produced effects equivalent to ʒvj. of the succus.

The Extract.—1. Three grains, taken an hour before breakfast two days after it was prepared, was followed in my own person, after half an hour, by full hemlock symptoms, and for 1½ hour the muscular weakness continued to such a degree that, after walking upstairs, the knees ached from sheer weariness, and I could hardly prevent myself from tottering, and was glad to come to a state of rest. The effect was equivalent to about ʒivss. of the succus.

2. The same dose produced equal effects in another active adult.

3. Mrs. M., aged 46, able-bodied, but weakly, also took three grains. After half an hour, severe giddiness, muscular relaxation of the orbicularis, drooping of the eyelids almost amounting to ptosis, and nearly inability to walk, came on, and continued for 1½ hour.

The use of the green fruit will no doubt be always attended

with more or less uncertainty, unless, indeed, we can depend upon its being collected at the proper season, and used for the preparation of tincture and extract the same day. An extract as potent as that described above would undoubtedly be a valuable addition to our Pharmacopœia, but rapidity of evaporation at a low temperature will be required to obtain it.

Just as this sheet was going to press (October 1868), I received, through Messrs. Allen & Hanburys, a parcel of the unripe fruit of conium maculatum, from Dr. E. R. Squibb, of Brooklyn, and Dr. William Manlius Smith, of Manlius, New York, together with a paper, entitled, 'An attempt to answer the question, Which part of conium is the best for medicinal use.'[1] My grateful acknowledgments are due to these gentlemen for two most acceptable contributions. 'The only experiment made by Dr. Smith to test the therapeutic effects of the conium, consisted in swallowing 16 minims of Squibb's fluid extract,' ♏i. of which represents a grain of the undried fruits. The effects were those which have been so often described as following a moderate dose of the succus. Dr. Smith concludes that the immature fruit of conium is the most appropriate part of the plant for use.

The fruits had a rank odour of conia. They were green, unripe, elongated, and wanted the rotundity of the ripe fruit already described. By weight, 1140 of the former corresponded to 950 of the latter. Having prepared a tincture from these fruits, of the strength and according to the directions of the British Pharmacopœia, I carefully examined its properties. They were briefly as follows:—fl. ℥j. contained 0·5 grain of conia, and yielded 10 grains of a resinous, deliquescent extract, precisely agreeing in its characters with that described in the previous page.

Both preparations were very powerful; 3 grains of the extract produced effects in my own person exactly equal to those which followed the same dose of the extract of my own fruit. The tincture was not so strong (less fruit having been used in its preparation), but ʒiv. were fully equal to ʒvj. of the succus conii P.B. Hence it is conclusive that the green unripe fruit may be dried and preserved for a time without deterioration. The examination of hemlock fruit, grown on the other side of the Atlantic, has afforded me much interest and pleasure.

[1] Reprinted from the Transactions of the N.Y. State Medical Soc. for 1867.

CHAPTER III.

ON THE ACTION AND USES OF CONIUM IN COMBINATION WITH OPIUM, HYOSCIAMUS, AND BELLADONNA.

I. **CONIUM and OPIUM.**—Conium alone, as I have shown (pp. 13, 26), strongly predisposes the brain for sleep by allaying excitement of the motor centres, and thus removing restlessness or agitation of the muscular system. Opium, alone, occasionally fails to procure sleep on account of its tendency to induce, or to increase, existing excitement of the motor centres (see 'ACTION OF OPIUM').

This nascent tendency of opium to excite convulsive action is antagonised by conium. What is wanting in the one drug is complemented by the other; and together they form a most perfect combination for bringing repose to every part of the body.

Action.—As might be expected, conium and opium intensify each other's action. The following will serve as an illustration:—John W. W., aged 22, took ℥v. of *succus conii before breakfast* every second or third day for several weeks; and the medicine invariably produced the following effects:—*After an hour*, giddiness, considerable muscular weakness and a feeling of looseness of the joints, and a heavy sleepy sensation over the eyes. These symptoms continued for twenty minutes, and then passed off. By dint of a little exertion, he was always able to continue his work during the action of the hemlock.

♏xv. *tinctura opii* taken alone, at the same hour, and under the same circumstances, produced so slight a feeling of heaviness and sleepiness, that the patient did not attribute the symptoms to the medicine. But on giving a *combination of* ℥v. *of the succus conii and* ♏xv. *tinctura opii*, at the same hour of the day and under the same circumstances, the effects were so powerful that the patient was obliged to relinquish his occupation, and sit down for an hour and a half. Giddiness, heaviness, and weakness of the legs came on *half an hour* after the dose, and continued to increase in intensity for the next half hour, during which time he 'was excessively heavy and drowsy, and could not fully open the eyelids; it was only with very great exertion that he could rise from the chair; the knee joints seemed fixed and yet loose, and he made false steps.' Heaviness and unsteadiness of gait continued for another hour, at the end of which time he was able to resume his work. These effects followed the above-mentioned dose on three different occasions.

Four drachms of the succus combined with five minims of tinctura opii usually produced decided symptoms in other patients, although no appreciable effects followed the administration of either separately.

Uses.—Two advantages result from the use of these drugs when judiciously combined: first, the use of large quantities of opium will be avoided; and, secondly, one of the disagreeable and disappointing effects of opium will be neutralised, and this drug thus rendered subservient to those whose idiosyncrasies are aroused by it to such a degree as to render its exhibition, alone, not only useless, but injurious.

In *delirium tremens* and *acute mania* the hypnotic influence of opium is often foiled by its excitant action. In a state of *insomnia* and *threatening convulsions* in children we are compelled to use opium; but on account of its tendency to excite convulsive movement its use is often contra-indicated. It is in these and such-like cases that the combination of opium and conium will be most serviceable. Thus, in a case of delirium tremens, instead of giving ʒj. of tincture of opium, we may prescribe from ʒiv. to ʒvj. or ʒviij. of hemlock juice, with ♏xx. or ♏xxx. of laudanum. To a child in the above-mentioned condition, and from one to two years old, we may give ʒj. succi conii and ♏iij. tincturæ opii.

Conium and Hyosciamus.—There is much similarity in the combined action of these two drugs to that of hemlock and opium. Next to opium, henbane is the most powerful hypnotic we possess; but we often meet with patients who are more readily impressed by its latent power to provoke insomnia and restlessness than by its hypnotic action. In such cases, and in many others in which it is desirable to induce the action of the two drugs, a combination of them will be found most effective.

Action.—Hemlock and henbane prolong and to some extent intensify each other's action, the akinesiant power of conium predisposing for the hypnotic action of the henbane, while the general sedative influence of the latter assists the action of the former. Given together, the effects of the conium are first declared, and then quickly follow those of the hyosciamus; and while the hemlock appears to accelerate the action of the henbane, the latter prolongs the influence of the conium for an hour or more beyond the usual period of its action.

I have frequently prescribed a mixture of equal parts of *tinctura or succus hyosciami and succus conii*, in doses ranging from ʒij. to ʒx.

The following were the effects of *four drachms* of the mixture upon John M., aged 17:—*After an hour*, a little feeling of dryness of the mouth and throat, trifling dilatation of the pupil, a full, soft, and regular pulse of 56, and slight drowsiness lasting for the following hour. *An ounce* of the mixture caused, in addition to the above symptoms, a heaviness and drowsiness, as if he had been drinking too much spirituous liquor, a little actual dryness of the mouth, a dilatation of the pupils from $\frac{1}{7}$ to $\frac{1}{5}$, and a general feeling of languor, with decided weakness of the legs.

In another individual, Geo. W., aged 18, an epileptic patient, *an ounce* of the mixture failed during 2¼ hours to produce more effect than a little giddiness and dimness of sight, a dry clammy state of the mouth, a dilatation of the pupils from $\frac{1}{8}$ to $\frac{1}{4}$, and a *rise* of the pulse 18 beats. *Ten drachms* caused, after fifteen minutes, giddiness, a heavy weight upon the eyelids, and weakness of the legs—effects resulting from the action of the conium. Towards the end of an hour the henbane symptoms—general dryness of the mouth and fauces, with clamminess and henbane fœtor, dilatation of the pupils, and an acceleration of the pulse 16 beats—were developed in addition. At the end of 2½ hours, the effects of both medicines persisted, and the pulse had fallen 48 beats, now numbering only 40 (see 'ACTION OF HENBANE'), but it was regular and of full volume and power.

The general effects were lethargy and drowsiness, and during the half-hour when the action of the medicine was at its height, the patient had to use the greatest exertion to keep the eye open, and continue his occupation. At the end of 2½ hours, on the particular occasion referred to, he walked a distance of two miles; and by the time that he got to the end of his journey the giddiness and muscular weakness had passed off, but the henbane symptoms continued an hour longer, at least.

Conium and Belladonna.—The remarks which have now been made with regard to the combination of conium and henbane, strictly apply to a combination of conium and belladonna in equivalent doses.

PAPAVER SOMNIFERUM—OPIUM.

CHAPTER IV.

A CORRECT knowledge of the physiological action of opium and its numerous principles has yet to be attained. The effects, indeed, of this drug are so opposite, and, with regard to constitutional peculiarities, so uncertain, that we are constantly reminded of the necessity, not simply of knowing the possible action of the drug, but of being able to predict which of its actions will prevail in the particular individual for whom we are called upon to prescribe. In a word, we want to be able to recognise the compatibility or incompatibility which subsists between opium on the one hand, and constitutional peculiarity on the other.

Such knowledge is doubtless attainable, although it may be at present concealed in a maze of obscure nervous actions. And, believing that the study of the comparative effects of the drug upon different animals will lead most directly to this fuller and more comprehensive knowledge that we require, I have taken this direction; and hope in the following pages to set before the reader such evidence of the action of each of the active principles of opium as shall give him a tolerably correct view of their medicinal value, and of the indications or contra-indications for their use.

I propose to examine in detail, and in the following order —Morphia, Narceine, Meconine, Cryptopia, Codeia, and Thebaia.

Physiological Action.—In determining the toxical effect of these substances, I have found the mouse very useful, inasmuch as very small quantities are required to bring the animal fully under their influence, while at the same time the action of each is very uniform.

I. **MORPHIA.**—The action of this alkaloid has been studied in the horse, the dog, the mouse, and in man. For full opportunities of making observations upon the horse, I am indebted to my friends the Messrs. Mavor, of Park Street. The experiments on this animal recorded in the following pages were undertaken, in the year 1867, by these gentlemen and myself, in order to determine the combined action of opium and belladonna.

On the Horse.—In this animal the hypnotic effect of morphia is altogether superseded by its excitant action upon the cerebrum and motor centres. Restlessness and delirium are the prominent features.

A *single grain* of acetate of morphia, and less, used subcutaneously, has no appreciable effect upon the horse.

A dose of *two grains* produces, at the end of a quarter of an hour, in an animal predisposed for its action, a little drowsiness, manifested by an occasional falling of the eyelids. The pupil remains unchanged, and all effect passes away in the course of another quarter of an hour. In an animal unsusceptible of its hypnotic action, *three grains* have no apparent effect.

Obs. 1.—The following symptoms followed the subcutaneous injection of *four grains* in two animals of opposite disposition—the one a grey horse, about six years old, of a quiet and stubborn temperament; the other, a weakly and excitable thoroughbred colt, two years old: pulses, 32 and 36 respectively. An acceleration of the pulse occurred in fifteen minutes. *An hour after* the injection, the acceleration had attained its maximum of 20 and 28 beats, respectively, in the two animals. There was considerable restlessness, indicated in the one by a constant stepping from side to side, chiefly with the forelegs; and in the other, by as constant a pawing of the ground; and they were impatient of restraint. There was increased moisture of the mouth, a rise in the temperature of the skin, a tendency to sweat, and slight dilatation of the pupils. All of these symptoms continued, and at the end of *seven hours* the pulse of the stronger animal was 12 beats more than before the injection, while that of the weaker and more excitable one still retained its maximum acceleration of 20 beats. Shortly after this time the restlessness subsided. The general effect upon both animals was a powerful stimulant action; the force and volume as well as the number of pulsations were increased. The mucous membrane of the mouth and tongue remained clean and healthy throughout, and moistened with a superabundant clear fluid. The restlessness never amounted to delirium, nor was there any increased injection of the membranes of the eye. There was a complete retention of secretions as long as the animals remained under observation. There were no after-effects.

Obs. 2.—After an interval of six days, *six grains of acetate of morphia*, dissolved in ʒjss. of water, was injected by two punctures into each of these horses, their pulses being 36 and 44 respectively. Precisely the same effects followed as when four grains were given, but in a more intense degree. *At the end of an hour*, the pulse of the grey horse was 72, being exactly doubled; that of the colt was only five beats less. Both were in a high state of excitement—the mouth very clean and pink, but hot, and wet almost to slobbering; the head and skin hot, the conjunctival membrane injected and wet, and the pupils slightly dilated. *At the end of* 4½ *hours*, the pulses had attained their maximum acceleration, and numbered 85 and 90 respectively; but the volume and power were a little diminished. *At the end of the seventh hour*, the pulse in both was 60, and the restlessness was now abating. Both animals had staled freely. Pupils unchanged.

Throughout the whole of this time, and for an hour subsequently, the horses were in a state of great restlessness, and trod their clean straw beds to a complete litter. Their mouths were hot and wet, and the skin warm and almost sensibly perspiring. Next day, and nineteen hours after the doses were given, the animals appeared in their usual health; they had dunged rather copiously, and the dung was a little greener and moister than usual. The pulse of the grey horse was 42, that of the colt 60. Pupils natural, mouth clean and wet.

Obs. 3.—*Twelve grains of acetate of morphia,* dissolved in ℥iij. of water, was administered by three punctures to each of the two horses. The effects were as follows:—In the grey horse, *after five minutes,* strong erection of the penis, and copious emission of semen; then gaping. *After twenty minutes* the lids began to droop and the head to nod, and he continued very drowsy for the *next* 2½ *hours.* He remained standing still and very quiet the whole of this time, breathing slowly and heavily, and every now and then falling off to sleep, and snoring slightly for a few minutes. The pulse meantime rose from 44 to 72, and became full; the pupils were unchanged, the mouth clean and wet, and the forehead warmer than usual. *After the third hour* the animal became very restless, constantly pawing the ground with the hoof of the left foreleg; but a considerable amount of drowsiness and stupor still continued. The pulse rose quickly to 96, and was regular and of good power; the pupils were slightly dilated, and appeared to be fixed; the membranes of the eye injected, the mouth hot, but clean and pink, and wet to slobbering; the temperature of the skin was raised, and its secretion increased, but not to such a degree as to wet the hair. He continued in this state of restlessness and delirium till the *sixth hour,* when the symptoms began to decline. Half an hour afterwards he took food and water.

In the brown colt, not the slightest sedative effect was produced. *Twenty-five minutes after the dose,* the muscles of the trunk were affected with a fine tremor and some rigidity, lasting only a few seconds; and five minutes afterwards he began stepping from side to side, and continued thus in restless motion for the next *nine hours.* The pulse at the same time rose rapidly, and at the end of an hour attained its maximum acceleration, and numbered 96. The restlessness greatly increased, and *at the end of the fourth hour* he was walking rapidly to and fro, bathed in sweat and slobbering from the mouth; the heart was much excited, the facial pulse 96, and weak. The pupils were unchanged, being rather dilated; the membranes of the eye injected. There was no appearance of stupor or delirium, but an intense excitement, with apparent distress, giving way to exhaustion. Towards the ninth hour the restlessness and other symptoms gradually subsided, and at this time he took food and water.

Ten hours afterwards, both animals were in their usual condition. There was neither constipation nor any other after-effect.

In the one horse a strong hypnotic effect was soon induced, but it was cut short by the development of a more powerful excitant action; and, thus disturbed and overruled, it was converted into a stupefying and deliriant influence.

In the other animal, the rapid development of the excitant action appeared to prevent altogether any manifestation of a hypnotic action.

Obs. 4.—*Thirty-six grains of acetate of morphia* dissolved in ʒvij. of water, and introduced by three punctures into the subcutaneous tissue of a powerful full-conditioned hunter, about seven years old, caused the following symptoms:—*Fifteen minutes after* the morphia was given, and continuing up to the end of the *third hour*, great somnolency, and a gradual rise of the pulse from 36 to 96, and it became full and thrilling; pupils moderately dilated and fixed. The drowsiness and stupor were very great, but they did not pass into sleep. At first he continued to pick up a little hay, and masticate it at intervals in a slow mechanical way, and with the eyelids closed. *After twenty-five minutes* he ceased eating, but continued to stand still; pointing his ears in the direction of any sound, and occasionally turning the head round to the left. On attempting to move he staggered, and once nearly fell down. The soporific effects attained their maximum at about *forty-five minutes* after the dose, and he occasionally leaned against the wall with the whole of his weight, or forwards over the rail—moving at short intervals slowly and awkwardly. *An hour* after the dose a fine rigid tremor affected the head, neck, shoulders, and tail, but passed off in a few minutes. The eyelids were also very tremulous, and he blinked on the approach of a light; the pupils were dilated and fixed. The respiration was slow, and occasionally interrupted by a sigh. Such was his condition up to the end of the *third hour*, when great restlessness and severe delirium set in, and, soon attaining its maximum, continued unabated for *seven hours*. It then began to decline, but the animal continued under the influence of the morphia for at least twenty-four hours longer, and at the end of this time he was greatly exhausted.

Between the *third and tenth hours* he continued to walk rapidly, and sometimes even run, round his stall, and always in the same direction—viz. from right to left. The shoulders, sides, and haunches were damped with perspiration; the pulse was 96, quite regular, full and thrilling; the respiration accelerated. The animal was perfectly blind, and allowed a candle to approach near enough to singe the eyelashes; the pupils were dilated and fixed; the membranes of the eye and nose so intensely injected as to present the appearance of crimson plush; the mucous membrane of the mouth pink and wet with a superabundance of thin glairy mucus.

Eighteen hours after the dose the pulse was 75, regular, very soft, and of moderate volume; the pupils moderately dilated, but contracting on exposure to a strong light; the membranes of the eye and nose of a dark crimson colour, and the vessels turgid, as if the blood were stagnant in them; the tongue was red, clean, and moist—the skin warm. Some restlessness and impatience of restraint still remained, and alternated throughout the day with intervals of drowsiness.

Given by the stomach, very large quantities of opium or its active principle are required to produce any decided effect upon the horse, as the following observations will show:—

(*a*.) The grey horse swallowed twelve grains of acetate of morphia in a pint of water. Excepting an acceleration of the pulse eight beats, from the second to the fifth hour, absolutely no effects followed. There was neither restlessness nor any diminution of the natural liveliness. The secretions were normal and abundant, the appetite good.

(*b*.) The same animal took ʒiv. of powdered opium. Up to the seventh hour there was no change, but thenceforward, up to the eighteenth hour, there was an acceleration of the pulse sixteen beats. No sedative or excitant effect followed, neither was there any interference with the secretions.

(*c.*) The brown colt took ℥iv. of laudanum in a pint of water, and

(*d.*) A bay mare took ℥iv. of another sample of laudanum. There were absolutely no effects in either case.

From the foregoing facts the following conclusions are deducible:—

1. That morphia has two distinct and antagonistic actions upon the horse—viz., an excitant and a soporific.

2. That of these two actions the former is generally, if not always, so much the more powerful as to prevent the effectual working of the latter.

3. That it happens with the horse, as with man and other animals, that some individuals are so susceptible of the excitant action, and are so powerfully affected by it, as to be totally uninfluenced by the hypnotic action.

4. That morphia acts upon both the cerebro-spinal and the sympathetic nervous systems—the soporific effects resulting from its action upon the cerebral hemispheres; the excitant from excessive stimulation of the corpora striata and spinal cord; and the acceleration of the heart's action partly to direct stimulation of the sympathetic nerves, and partly to an indirect stimulation of the same centres resulting from the excitement of muscular movement.

How far the muscular movements are due to excitation of the motor centres in the brain is not very evident. That the spinal cord is implicated appears to be indicated by the rhythmical character of the movements; the animal scrapes the ground with the same hoof for hours together, begins and ends with a regular tread from side to side, or goes round and round continuously in the same direction. There is an evident tendency to forward movement together with inaction of the hind-legs.

5. That the vascular excitement, if intense and prolonged, ends in dilatation of the capillaries, general congestion, imperfect oxidation of the blood, and weakness of the heart.

6. That opium has no contracting influence upon the

pupil, any tendency to such that may exist being overcome by a stimulant action upon the sympathetic, commonly resulting in moderate dilatation.

7. That in moderate doses morphia and opium are tonic and stimulant, and do not diminish the activity of the excretory processes. In large doses they increase the secretions of the mouth and skin.

On the Dog.—The action of morphia and opium upon the dog differs in no respect from their action on man.

In the dog, as in man, there are two classes of individuals, each of which is affected differently by opium. In the one class, the hypnotic effects of the drug are readily and constantly induced, and, if the dose be sufficient, profound narcotism results. If, however, the dose be larger than is necessary for this purpose, the narcotism sometimes gives place to delirium with stupor. In the other class it is difficult, and in many cases altogether impossible, to induce sleep or narcotism by any dose excepting such as would destroy life.

In this class, opium or morphia produces the most distressing effects—viz. faintness, prolonged nausea, and retching, with intervals of dreamy delirious somnolency, rarely or never passing into sleep, which only comes to the exhausted body when the protracted effects of the drug have passed off.

The following were the effects of morphia upon an animal —a *brown bitch* weighing about twenty-five pounds, that well represents the first of the two classes above distinguished:—

Obs. 5.—*Half a grain* of acetate of morphia was injected beneath the skin three-quarters of an hour after food. Between the *third and fourth minutes*, vomited a large quantity of food, then passed a little urine, and quietly lay down. At the end of *half an hour* was lying on the belly, with the legs extended before and behind, and the nose on the rug, in a state of complete narcotism, in which she continued for the *next* 3¼ *hours*. During the whole of this time, she lay perfectly motionless in the position in which I had placed her—viz., upon the side, with the forelegs extended forwards, and the hind-legs moderately flexed. *The pulse* fell during the first half-hour from 120 to 52, and beat regularly. It soon increased again, and for the next five hours ranged between 65 and 75. During the whole of the time it manifested a remarkable irregularity, gradually becoming slower towards the end of every expiration, and then, with the descent of the diaphragm, receiving a sudden acceleration for two or even three beats. The rate of the expiratory pulse was 60, gradually diminishing, and the rate of the inspiratory pulse was about 100. The movements of the heart were in fact governed by the breathing; and by counting the number of accelerations that the pulse received, I counted the number of inspirations. *The respirations*, which were 20 at the end of the first half-hour, soon decreased to 14 or 15, and continued at this rate during the whole of the time. The breathing was shallow, and some inspirations fuller than the others. *The pupils*, at the distance of a yard from a south window, and with moderate daylight, measured the $\frac{1}{8}$th of an inch in diameter, and contracted a little on the close approach of a taper. The nose was dry, the tip of the tongue a little protruded, pink, and cool, and dripping wet with clear alkaline mucous fluid. *The muscular system* was quite flaccid, and at each inspiration the disengaged hind-leg was slightly jerked towards the abdomen. Suspended by the ears, or dragged along by a leg, the animal made no sound or motion, took no notice of the approach of anyone or of any disturbances around her, and made no other response to loud and continued calls than a slight movement of the ear and forepaws. On blowing upon the face, she put up the forepaws over the nose; and winked a little when the finger was placed upon the cornea, unless care was taken to avoid the cilia.

Two hours and a quarter after the injection, and whilst lying as above indicated, she passed, involuntarily, a large, lax, greenish motion of an alkaline reaction. *Three hours and three-quarters* after the injection she awoke, got up and walked a few paces; the pupils dilated in the dark, and contracted to $\frac{1}{6}$ on approach of a taper.

For the *next* $3\frac{1}{2}$ *hours* she continued to sleep tranquilly, without once changing her posture, but looking up occasionally when called or disturbed by noise. At the end of this time, i.e. 7¼ *hours after the injection*, she awoke with a start and became a little restless, and whined occasionally during the next 1½ hour. The discomfort appeared to be due to distension of the bladder; for having voided urine for the first time during nine hours, she appeared to be comfortable, but still refused food.

Obs. 6.—After an interval of some weeks the injection of *half a grain* of acetate of morphia was repeated. Immediate vomiting; rapid diminution of the pulse; irregular and shallow breathing, interrupted by deep sighing inspirations; contraction of the pupils; profuse salivation; flaccidity of the muscles, and falling of the lower jaw; and the same respiratory pulse—resulted as in the first experiment. *At the end of three hours*, the following was the condition:—Lying upon the side in the position described in Obs. 5; narcotism passing off, but sleeping soundly; pupils contracted to $\frac{1}{7}$ in moderate light, but dilating to $\frac{1}{5}$ when aroused. Pulse 66, irregular, weak. When roughly handled she raised the head and looked round, but, immediately laying it down again, relapsed into sound sleep. When thoroughly aroused she walked round the room, and then, dropping down, fell off again to sleep. At this time atropia was injected (see 'ACTION OF OPIUM AND BELLAD.').

Obs. 7.—*One grain* of the acetate was injected beneath the skin of the same animal after the lapse of a fortnight. Between two and three minutes afterwards she vomited a quantity of partially-digested food and much glairy mucus, and then lay down and seemed sleepy. Between the fifteenth and twenty-fifth minutes she quietly passed from sleep into a state of narcotism, the pupils at the same time slowly contracting from $\frac{1}{4}$ to $\frac{1}{8}$; the pulse was reduced to 48; the respiration 15 to 18, interrupted once or twice a minute by deep sighs; the lower jaw had fallen, and a little clear alkaline fluid was trickling from the protruded tip of the tongue. The eyelids were closed. On raising an upper lid, so as to expose the pupil to a bright light, it contracted to $\frac{1}{10}$. On removing the light it dilated to $\frac{1}{7}$, and finally adjusted itself to $\frac{1}{8}$, as before. *At the end of* 1¼ *hour*, the pulse was 42 one minute, 44 the next, and very irregular from the cause explained in Obs. 5: but, inasmuch as the breathing was more irregular in this experiment, the irregularity of the pulse was proportionately increased. Respiration 8 to 14, occasionally interrupted by a long-drawn sighing inspiration. Pupil as at twenty-fifth minute; but after endeavouring to arouse her by strong shaking, they dilated to $\frac{1}{5}$; as soon as she was allowed to remain undisturbed they returned to $\frac{1}{8}$, in a light which allowed them to dilate to $\frac{1}{4}$ before the injection.

At this time atropia was injected (see 'ACTION OF OPIUM AND BELLAD.').

The effects of *opium* upon this animal will be seen from the following experiments:—

Obs. 8.—Injected ♏xx. *of tincture of opium* beneath the skin, the pulse being 120, and the pupils contracting to $\frac{1}{8}$ at the light, and dilating to $\frac{1}{3}$. Between the fifth and fifteenth minutes she was very qualmish, and at last vomited a heap of food. The diaphragm continued to act spasmodically for three minutes more. She then lay down with the head on the ground. *Seven minutes* afterwards, the bowels acted, clear mucus ran from the mouth, and the pulse was 112. *Three-quarters of an hour* after the injection, the pulse was 78 and irregular. Respiration irregular and sighing; glairy alkaline mucus continued to drop from the mouth. The pupils were broadly dilated in moderate light. Slight somnolency was now manifested for the first time, with a little twitching of the hind-legs. As the somnolency did not increase, I injected, *at the end of the hour,* ♏xx. *more of tincture of opium.* A quarter of an hour afterwards the pulse was 72, and very regular in its respiratory irregularity—i. e. the expiratory rate was 40, and the inspiratory rate about 120. Respiration 16 and regular. She continued in the same position, and there was no further contraction of the pupils. Every now and then she raised the head and looked about her, and then slowly laid it down again upon the carpet and closed the eyes. *Fifteen minutes* afterwards, i. e. *half an hour after the second injection,* she became decidedly drowsy, and the pupils, from being broadly dilated, contracted to $\frac{1}{4}$.

The drowsiness continued for an hour, but did not amount to sound and continuous sleep at any time; and at the end of this period, upon being a little disturbed, she got up and walked away, and began to retch. From this time the drowsiness began to pass off, and she continued to walk about, listen to footsteps on the other side of the door, and pry about. The pulse was now 78, and less irregular; the respiration 18. At this time atropia was injected (see 'ACTION OF OPIUM AND BELLAD.').

Obs. 9.—Injected ♏xl. tincturæ opii by two punctures, one on either side of the body. From *the seventh to the tenth minutes* vomiting, until the stomach was emptied. *At the fifteenth minute* lay down and remained quiet, with the eyes open. Pulse 130, regular. Respiration shallow, irregular. During the *next half-hour*, the respiration was much disturbed, and occasionally suspended for fifteen seconds; after a long-drawn inspiration, it gradually became slower, and ultimately sank to 15. The pulse meanwhile was depressed to 72, and influenced by the breathings. Glairy mucus continued to trickle from the mouth.

From the time of the injection up to the *end of the third hour*, the bitch did not have any sound or continuous sleep; she remained dull and quiet, and when called, or disturbed by the approach of anyone, would get up and walk slowly into the next room. When left undisturbed she fell off into a nap, from which a slight noise awoke her. At the *end of* 3¼ *hours* the somnolency had almost entirely passed off, and she was walking about most of the time. She made an attempt to push her way through the door as often as it was opened. She could run, and jumped well off a chair. Pulse 80, respiratory. Respiration 15, irregular. Pupils contracted to ⅛ at the diffuse light of day, and dilated widely towards the dusky side of the room.

At this time hyosciamia was injected (see 'ACTION OF OPIUM AND HENBANE').

The effects of the drug upon the other class of individuals, in whom the hypnotic action of opium is counteracted by its depressent and excitant influence, are well seen in the following experiments with two dogs of excitable temperament—

the one a Scotch terrier weighing about twenty-five pounds, and the other a stout beagle weighing about forty pounds:

Obs. 10.—*Half a grain of acetate of morphia* was injected beneath the skin of the terrier-dog. *Within three minutes* he vomited a quantity of undigested food, and soon became drowsy, and sank on the belly, with the legs sprawling. In this state he remained perfectly quiet and drowsy for half an hour, and then became restless, starting off with a whine, and dragging the seemingly paralysed hind-legs after him, and then, after going a few paces, falling heavily down. When urged to do so, he used the hind-legs, and trotted away without difficulty. The restlessness increased, continued for nine or ten hours, and then gradually gave way to a condition of ordinary repose. During the whole of this time the animal was evidently suffering from nausea and headache. There was continual rumbling of fluid in the intestines, and he frequently dropped a little loose motion. The cardiac systoles ranged from 67 to 72, and presented the characters which in former observations I have called '*respiratory.*' The heart's action was feeble throughout, and the quick inspiratory beats could not be felt in the brachial artery, where the pulse numbered only 56 to 64. The respirations varied from 17 to 24. Pupils unchanged.

Seven hours after the injection the cardiac systoles were 72, the brachial and femoral pulse 54. The dog was still in a state of great distress, whining, gulping up a little mucus, and voiding a little frothy fæcal matter. Throughout, the wretched animal failed to get a quarter of an hour's unbroken sleep, and during the latter part of the time there was no apparent tendency to it. Enfeebled and exhausted, he lay on the belly, with the muzzle upon the ground, the legs lax and sprawling, and the eyes open. Now he would throw the head from one side to the other in an impatient manner; and now, impelled by his distressed feelings, or by some direct excitement to motion, he would get up, drag his thighs along the ground for a few paces, and then throw himself helplessly down. This experiment was repeated on another occasion with the same results.

Obs. 11.—*Three-fourths of a grain* was followed by a repetition of the same symptoms in a more intense degree. Nausea came on in two minutes, and, after two minutes more, copious vomiting, and the pulse went down within ten minutes from 128 to 96. He then became very restless, curling himself up first in one direction and then in the other, putting the paws impatiently over the ears, and rubbing his head in the straw; now sitting upon the haunches, and, after a few seconds, throwing himself impatiently along upon the ground; now getting up and running a few paces, and so changing his posture constantly, and at the same time keeping up a continual subdued whine. Between the *twentieth and fiftieth minutes* he lay still, but quite awake, and was startled by the least noise, but at the end of this time the restlessness and distress increased. He was scared by the approach of anyone, and ultimately became very noisy and delirious. The cardiac pulsations ranged from 54 to 70, but many were not distinguishable in the brachial or femoral arteries, and the heart's action manifested the irregularity already described, in a very marked degree. *Eight hours and three-quarters* after the injection he was still wakeful, restless, frightened, and whining. Ten hours later on he was quite recovered.

Obs. 12.—*Eighty minims of tincture of opium administered by mouth* produced the same effects, but in a much milder degree, and without giving sound sleep at any time.

Obs. 13.—In the beagle-dog the subcutaneous injection of *half-a-grain* of acetate of morphia produced exactly the same effects as in the terrier (*Obs.* 10), but to a worse degree. *Within three minutes* the animal vomited the few scraps of food that remained in the stomach, and during the next twenty-five minutes he vomited again and again, until nothing but frothy mucus tinged with bile came away. The intestines were emptied at the same time. The usual hard mortar-like fæces were first voided; then from time to time a quantity of soft blackish-green matter; afterwards a large, loose, and frothy light ochre-coloured stool; and last of all, and during the latter part of the action of the medicine, a little white jelly-like mucus.

For the *next six hours* he continued in a wretched state,

evidently suffering from nausea, faintness, and restlessness. There was a strong tendency to sleep, but only short dozes were obtained, from which even the movement of my chair disturbed him. He kept up a subdued whine almost continuously, and changed position every few minutes, dragging the body and hind-legs along a few paces by means of the fore-legs, and then throwing himself heavily and helplessly along. He was able, however, to use the hind-legs when aroused. Sometimes, as I approached him, he rolled over on the back, and, as the legs were laxly flexed in the air, they were frequently twitched. At first the breathing was short and irregular; inspirations 50; but during the last three hours of the action of the drug they numbered only 12 to 14, and the respiratory movements were almost wholly abdominal, and very shallow.

The pulse soon fell to 64, and manifested the respiratory character very distinctly; *from the third to the fifth hour* it was excessively feeble.

When dozing, the pupils measured $\frac{1}{6}''$ in a light which allowed of their dilatation to $\frac{1}{3}$ before the injection; but the moment he was aroused they assumed the latter dimension.

At the seventh hour he had so far recovered as to appear pretty much in his usual condition when noticed or fondled; but, left to himself, he gave way to a little restlessness and whining, and during the whole of the succeeding night he moaned and whined most piteously. The next day he was quite well.

Such are the effects of opium and its chief alkaloid upon the dog, and they are particularly instructive. It appears—

1. That opium acts both upon the brain and the spinal cord.

2. That upon the cerebrum, its action results, first, and in many individuals almost exclusively, in hypnosis; and subsequently, in others, in active delirium.

3. That upon the corpora striata it exerts a similar influence, but resulting in animals fully under the influence of its hypnotic action in temporary paralysis.

4. That its action upon the cranio-spinal axis is partly excitant, indicated by contraction of the pupils, derangement

of the vagus, and restlessness; and partly depressent as indicated by the feebleness of the movements.

5. That in consequence of the derangement of the functions of the vagus nerve the action of the heart is depressed and enfeebled.

6. That individuals of the canine species are differently affected according as they are susceptible of the hypnotic or the excitant action of the drug.

On the Mouse.—The action of morphia on this animal is uniform and characteristic. It consists essentially in forced exercise associated and apparently dependent upon a cramped condition of the spine. Hypnosis is altogether an after effect, and narcotism only occurs after a dangerous dose.

The mouse runs about nimbly for a minute or two after the subcutaneous injection of from $\frac{1}{20}$ to $\frac{1}{12}$ of a grain of the acetate. Then sudden and frequent pauses are made, and the animal remains for a few seconds at a time in a fixed attitude, as if of attention to some internal sensation. During the first, second, or third pause the tail begins to bend slowly and stiffly upwards from its root until it is elevated in a single curve or double undulation over the back and at some distance from it. The spine is at the same time shortened by rigid flexures, whereby the head is a little raised and the face bent downwards to a right angle with the body, the back elevated into a high hump, and the lumbar and sacral regions so depressed as to bring the perineum in contact with the surface upon which the animal rests, the sharp upward curvature of the caudal extremity of the spine giving at the same time a backward and upward inclination to the hinder part of the pelvis and the attached limbs. The cranio-spinal axis is in short affected by three curves, an anterior and a posterior, corresponding to the junction of the head and tail respectively, and the summits of which are in opposite directions; and an intermediate and more considerable one affecting the thoracic vertebræ, cramping the chest and abdomen together, and giving to the back a contour exactly resembling that of the hog.

The movements now become slow, stiff, and creeping, and the animal is arrested every four or five paces by inaction of the hind-legs, the thighs being closely appressed to the sides of the pelvis by the downward inclination of the sacral part of the spine, and apparently useless, while the legs and feet are extended downwards in a straight line with each other at right angles to the thighs, and the hinder parts of the body are thus supported upon the extended and divergent toes.

After a few seconds the body is very slowly advanced, as if by an under-curving action of the anterior part of the spine and a simultaneous pressure upon the hinder part of the dorsal curve. The feet remaining stationary meanwhile, are thus left in the rear until, the forward balance of the body preponderating, the little animal is impelled a few paces forwards and the cramped, lagging action of the hind-legs is for a moment interrupted. As soon as the body comes fairly to rest, the forward impetus again commences and soon prevails; and thus the mouse is kept in a state of cramped rest and forced motion as if by two antagonistic powers exerting their influence simultaneously upon the anterior and posterior parts of the body respectively. Sometimes the one power prevails; and the mouse is urged round and round the limits of his cell at a slow, measured, laborious pace, as if upon a treadmill, and showing the soles of the feet at every step; and sometimes the other power is in the ascendant, and the animal is arrested in his progress for several minutes at a time, while the hinder legs first one and then the other, are stiffly extended backwards in a straight line with the body, or raised upwards in the air. But sooner or later the anterior power so far prevails that a slow, laborious, forced creep is maintained for hours together.

It is probable that some amount of delirium accompanies this condition, but there is not the slightest tendency to somnolency. The hearing continues excessively acute; at first the animal is startled by the least noise, and the abovementioned phenomena are effaced for the moment, and he runs a short distance as usual, but soon coming to a pause relapses into his former cramped condition. After a time, however, stupor comes on to a greater or less degree, dependent upon the dose, and then the animal, unless strongly aroused, takes no notice of interruptions. As often as the depressed face comes in contact with the finger or any other object, the little animal leans against it with all its weight, and rests for a while.

Such are the general phenomena presented by a mouse under the influence of morphia. I now proceed to indicate its other effects, by briefly giving the results of one or two illustrative experiments.

Obs. 14.—Injected $\frac{1}{16}$ of a grain of acetate of morphia beneath the skin of a full-grown male mouse. Respiration, 160. After *four minutes*, the spine was drawn up into the 'hog's back,' the tail curled upwards over the back, the forced movements began, and the respiration was panting. At the *end of an hour* the respiration was shallow and irregular, varying from 140 to 180, and apparently arrested when he was brought to a standstill by the posterior extension of the hind-legs. After 1 *hr.* 20 *m.*, the pupils dilated, and the refracting media presented a milky opalescence; the movements became slower, and one or other hind-leg was stiffly extended backwards and upwards in a line with the body, at every few steps. The animal was dull and stupid, but the hearing remained acute. Respiration 100 to 140, shallow and irregular. He continued in this state for *another hour*, going constantly round and round the interior of the glass shade. 2½ *hours after* the injection, the lagging extension of the hind-legs began to cease, the tail dropped, the hump-back slowly disappeared, and the mouse began to run nimbly about. At the 5*th hour* he was quite recovered, and took some food, the pupils contracted, and the animal continued active and lively, without manifesting the slightest tendency to sleep, up to the end of the 8*th hour*. As yet no excreta had been passed. The next day he was lively and well, and there was abundance both of fæces and urine; the latter was freely acid.

Obs. 15.—Injected at intervals during three hours ½ of a grain of acetate of morphia beneath the skin of a very vigorous adult mouse, beginning with $\frac{1}{16}$ of a grain. Respiration, 160. After *four minutes*, the spine was cramped, the tail curled, &c., and the respiration panting. At the *seventh minute*, the respiration had fallen to 100, and the forward impulse was sometimes a little jerky. 1¾ *hour* after the first injection, and shortly after the second, the pupils were dilated and the animal appeared to be quite blind, but the hearing remained very acute, and he continued to go round and round with lagging, laboured steps, stopping as often as the head came in contact with any firm object, which seemed to give him the support that he needed, and he pressed firmly against it. As often as he started afresh, the vibrissæ were strongly worked. As there was no increase of the effects, the rest of the morphia was now injected. The movements afterwards became slower, and the pauses more frequent, and shortly *after the third hour* a slight touch caused him to fall over on the side, and for the *next four hours* he remained completely narcotised, motionless, and insensible to all external impressions; the limbs were quite flaccid, but the tail and spine were as stiffly curved as ever, and could only be partially straightened by the exertion of considerable force, and when the extending force was removed they returned to their previous condition; no reflex movements could be excited, the respirations had fallen to 90, and were shallow, occasionally irregular, and wholly diaphragmatic. The action of the heart was regular, and the pulsations 300 at least.

Towards the end of the *sixth hour* the respirations rose to 110 or 120. At the *seventh hour* the legs began to be affected three or four times a minute, with involuntary jerkings, and, as he lay upon his side, they were sufficiently strong to carry the body backwards a little distance. He continued in the same state with occasional jerkings of the limbs until the *ninth hour*, when the narcotism gave way, and he recovered the couching posture, and steadied himself by pushing the head firmly against a support. But he was unable to retain this position, for the spine was suddenly and powerfully extended five or six times in the minute, and the rump and hind-legs were thus thrown into the air, and the animal rolled over. He recovered his position, however, almost as soon as the spine resumed its curvature.

The respirations were now only 80. The spasmodic movements continued severe for about half an hour, and then diminished, and the mouse began to walk on all fours, and became almost as restless as he was before he was narcotised. As often, however, as he came to a state of rest, the twitchings, which were now almost wholly confined to the tail, would set him in motion again. Respirations 120, regular.

From this time he continued to improve; the spine recovered from its cramped condition, and the twitchings and restlessness gradually ceased.

At the *twelfth hour* the symptoms of spinal irritation gave way to somnolency, and he sat dozing with the eyes closed, and apparently pretty comfortable, occasionally turning round and sniffing for food, and putting the ears back when called. Throughout the next day up to 6 P.M. he was very sleepy, but otherwise in his usual condition, and took a little food. Respirations at 8 A.M. 140. During the rest of the day from 180 to 200.

He passed, apparently with some difficulty, a little high coloured urine at the nineteenth and twenty-fourth hours.

At the *thirtieth hour* the effects of the morphia had passed off, and he was quite lively and active. The day following he took food eagerly, and as a consequence the excretions became very free. It is now six weeks since the experiment, and the little animal remains a pet in the household.

Obs. 16.—Took another adult male mouse, and injected $\frac{1}{10}$ *of a grain* of acetate of morphia beneath the skin. The same symptoms followed as in the preceding experiments. The respiration manifested the usual irregularity, the rate of breathing being sometimes 180, sometimes 100.

At the *end of half an hour* I injected $\frac{1}{8}$ *of a grain more. Ten minutes afterwards* the animal was completely narcotized and insensible, and no reflex movements could be excited except a sluggish contraction of the eyelids when the cornea was touched. The breathing very irregular, sometimes decreasing to 50, and then accelerated to a faint diaphragmatic panting, numbering 220 times a minute. Pulse 200, quite regular.

One and a half hour after the second injection the little animal manifested indications of recovery: I therefore injected more morphia, so as to make in all $\frac{1}{3}$ *of a grain*. The animal again passed into a state of narcotism.

The respiration, which had been wholly diaphragmatic for the previous hour, became rapidly shallower, and the temperature of the body fell.

Thirty-five minutes after the third injection the respirations were only 70, shallow and intermittent. A few minutes after, the small pulsatile diaphragmatic breathing intermitted, and the chest walls were then fully expanded 60 times a minute by the extraordinary muscles of inspiration.

At the end of a minute or so this breathing ceased, and was replaced by the pulsations of the diaphragm, and these were thrice relieved by the deeper inspirations, which were accompanied by a faint sobbing noise, and a depression of the head.

This alternation of diaphragmatic and extraordinary respiratory movements occurred 4 times within as many minutes. The latter then became slower and slower, and when they intermitted, were no longer replaced by contractions of the diaphragm.

During the last 14 minutes of life the respiration was wholly sustained by the extraordinary muscles of inspiration, and these ultimately decreased to 5 in the minute, and then finally ceased.

At the moment of death the contractions of the heart as indicated by the motion of the hairs over the cardiac region were 60. Three minutes after the breathing had ceased I counted 44, and after three minutes more 10 systoles were appreciable.

Twelve minutes after the breathing had finally ceased, the chest was opened. The lungs were of a salmon colour, and collapsed above and behind the heart. The right heart was distended with dark fluid blood; the left contracted and empty. Both ventricles contracted regularly 20 times a minute; the right auricle 55; the left a faint occasional movement. On relieving the right auricle of blood the pulsations were 100, regular and strong.

Twenty minutes after the breathing ceased, the ventricles ceased to contract, and 3 minutes later on the right auricle was motionless.

The pupils had dilated during the last quarter of an hour of life, and now remained so. The refracting media of the eye had a milky opaque appearance; the bladder was distended. The stomach was moderately full and firmly contracted.

Briefly summed up, the following are the effects of morphia upon the mouse in the order in which they occur:—1. Cramp of the spine. 2. Restlessness with forced and apparently involuntary motion of the weakened limbs. 3. Irregularity of the breathing. 4. Acceleration of the pulse. 5. Stupor. 6. Coma, with diaphragmatic breathing. 7. Death by apnœa from inaction of the lungs, or 8. Recovery, with convulsive movements of the hinder part of the body, and return of restlessness. It appears to me that these phenomena may be thus explained:—The cerebro-spinal nervous system is, so to speak, steeped in the influence of morphia, and its functions are at once benumbed and excited; the action on the motor centres and their conductors being the exact counterpart of that on the brain so often and so clearly manifested in the delirious stupor which results from the struggle between sleep and mental excitement.

The motor centres are clearly under the influence of excitement, and yet it is equally apparent that the motor power is restrained and depressed; the muscles nearest to the cranio-spinal axis, and those therefore which receive the nervous impressions soonest, are affected with persistent cramp; while those most distant from it are flaccid and almost paralysed. It appears that the conductivity of the nerves is impaired, and that the impressions being retarded and perhaps accumulated, radiate upon the nearest objects, the spinal muscles, only a faint irritation being conveyed to the more remote muscles of the limbs. It is reasonable to suppose that it is a law of nervous action, if it be not simply a result of anatomical structure, that the central portions of the muscular system are more readily and directly affected by central impressions than those which belong to the mere appendages of the body. And it follows that when the whole of the conducting fibres are affected by a general impairment of their conducting power, the muscles of the limbs are the first to show it.

Cramp is the maximum effect of motor excitement; and yet, in the case under consideration, it is associated with a condition of other muscles bordering on paralysis. Something then, we must conclude, intervenes between the roots of the longer motor nerves and their ultimate ramifications, to prevent the uniform diffusion of the cramping impressions which, originating in the excited centres of motion, are expended upon the spinal and thoracic muscles. That the obstruction is due to diminished conductivity of the nerve cords still further appears from the fact, that as the torpifying effects of the morphia pass off, the hinder parts of the body are for the first time affected with decided convulsive movements. The nerves having recovered their conducting function, allow the currents to pass without hindrance, and those tetanising impressions which throughout had been antagonising and counteracting the hypnotic action, are now, when on their wane, free to pass and manifest themselves in slight convulsive movements.

The nervous systems of the horse and mouse show that, physiologically, morphia is a compound substance, and that its constituents are tetanus and hypnosis—effects which in these animals are nearly counterbalanced, the preponderance being in favour of the former. But so slight is the advantage, that little more than active delirium and restlessness usually results. If the hypnotic action were much weaker, or altogether eliminated, as may perhaps occur in some animals, then, indeed, morphia would not differ in its action from thebaia or strychnia.

This general view of the action of morphia is strictly in accordance with the respiratory phenomena. A cramped, contracted, and at intervals fixed condition of the chest, is one of the earliest effects of the drug. At a later stage the breathing is wholly maintained by the diaphragm; and last of all, when this ceases to contract, the external muscles of the chest-wall—the extraordinary muscles of inspiration—carry on the process. The organs supplied by the vagus appear to be equally cramped and depressed in function; the lungs after death are bloodless and collapsed, the stomach firmly contracted. The heart, however, still pulsates, its left cavities are contracted and empty, and the right distended with black blood.

On Man.—The effects of morphia and opium upon man include all those which we have seen follow its use in the horse, the dog, and the mouse. Like the brown colt or the mouse, there are some persons who experience only the excitant and deliriant effects. Others, like the subject of *Obs.* 4, p. 103, are distracted between somnolency and mental and bodily excitement. Many, as were the two dogs (see p. 110), are distressed beyond measure, and alarmingly depressed by a remedy from which they had expected nothing but comfort and repose. But a larger number than perhaps all the other classes together are, like the bitch in *Obs.* 5, p. 106, happily influenced by its hypnotic action, to the almost complete neutralisation of its excitant, deliriant, or depressent effects.

An explanation of this difference in the effects of morphia upon particular animals and individuals will be considered hereafter.

Man, like the canine species, may, with respect to the action of opium, be subdivided generally into two classes :—1. Those who are readily influenced by its hypnotic action, and who suffer little or no inconvenience from its excitant effects upon the brain or its depressent action upon the vagus; and, 2. Those who are distressed by its deliriant or depressent effects, or both, to such a degree that its hypnotic action is altogether counteracted until it has passed away.

Every practitioner is well aware of these general facts, but only those who happen to have been taught by experience can, I believe, realise the importance they assume when, as is so frequently done in the present day, morphia is introduced beneath the skin.

The Report of the Committee on the Hypodermic method of Injection ('Medico-Chirurgical Trans.,' vol. L., p. 586) mentions several cases in which alarming symptoms were induced by the subcutaneous use of morphia, and two in which the injection of half a grain proved fatal. At page 638 the case of a lady aged 24 is given, in which the injection of $\frac{1}{4}$ of a grain was followed for 4 hours by repeated attacks of syncope, during which the countenance was livid and anxious, and the patient was pulseless and at the point of death. Slighter attacks recurred during the next two days, and she remained so exhausted for a fortnight that she was unable to leave her couch.

Mr. F. Woodhouse Braine records the following case in the 'Medical Times and Gazette,' January 1868 :—Mrs. H. C., aged 35, in good health, was injected for insomnia, from neuralgia, of 72 hours' duration, with $\frac{1}{3}$ grain of morphia. In 15 seconds tightness of the chest and difficulty of breathing were complained of, and she asked to be raised up, and said she felt as if she were dying. The face and lips now became pale, the speech indistinct, pulse irregular; there was some spasm of the facial muscles, and she fell back to all appearance dead. The face was blanched, the pulse and respiration imperceptible. Insensibility continued for three minutes, and then, cold affusion, ammonia, and artificial respiration having been used, one or two feeble beats, and a shallow inspiration or two, showed returning animation, and she soon became conscious. Pulse regular, but feeble. Respiration slow, fingers numbed, and the thumbs drawn into the palms. The symptoms passed off in about six minutes, leaving her very ill. There was no nausea nor attempt to vomit throughout.

At a subsequent page of the same volume Mr. Arthur Roberts records a similar observation :—The patient was a gentleman who had been several times before injected with morphia, but, on the particular occasion referred to, its use was followed within a few minutes by intense flushing of the face, vomiting, and then a dead faint and struggling for breath. The pulse was scarcely perceptible.

I will now record as briefly as possible some results of the hypodermic use of morphia and opium, partly to illustrate what has just been said concerning its action, and partly in reference to the combined action of opium and belladonna, which will be considered in a subsequent chapter.

Obs. 17.—Michael E., aged 48, strong, but disabled by sciatica. Pulse 78. Pupils $\frac{1}{10}''$ diameter at light, towards the dark side of the room dilating to $\frac{1}{8}''$. Tongue a little furred, and not very moist.

INJECTED $\frac{1}{4}$ GRAIN MORPHIÆ ACET. :—

After 30 *minutes*, pulse unchanged. Pupils $\frac{1}{12}$, dilate to $\frac{1}{8}$. At first experienced a little nausea, after 15 minutes drowsiness, and since then sleep.

After 1 *hour* 20 *minutes*, pulse unchanged, nausea gone, felt a little giddy, and continued heavy and sleepy.

After 2 *hours* 18 *minutes*, pulse fallen 6 beats, full, and regular. Pupils as at 30 minutes. Mouth unchanged throughout. Had continued drowsy, and still remained so. Now walked home, a distance of three miles, felt very weak going along, and on getting home 'broke out into a cold perspiration, and was low-spirited, faint, and almost ready to drop.' Recovered in about half an hour after taking a cup of tea; but still feeling poorly and sleepy, went to bed and slept about two hours.

The injection was repeated on another occasion with exactly the same results.

Obs. 18.—John L., aged 54, an able-bodied man, suffering from severe facial neuralgia. Pulse 74, regular, good volume and power. Pupils $\frac{1}{8}''$. Tongue clean and moist.

INJECTED $\frac{1}{6}$ GRAIN MORPHIÆ ACET. by the skin. Produced great somnolency after 15 minutes, and

After $2\frac{1}{2}$ *hours*, pulse reduced 19 beats, being now 55, unchanged in volume and power. Pupils $\frac{1}{12}$. Had continued very drowsy, and slept at intervals. Walked cautiously, and felt giddy. No nauseating or other effects. Walked home, and continued sleepy for two or three hours.

Obs. 19.—Samuel M., aged 49, able-bodied and well nourished, but for 15 years a great sufferer from facial neuralgia. Pulse 74, of fair volume and power. Respiration

19-20. Pupils at light, $\frac{1}{9}''$; sideways, $\frac{1}{6}$; tongue moist, and when the pain came on there was much slobbering.

(*a*) INJECTED $\frac{1}{6}$ GRAIN ACETATE OF MORPHIA. Somnolency followed, and after half an hour he fell asleep and had a short nap.

After 2 *hours.* Pulse 70, a little fuller and stronger. Pupil $\frac{1}{12}''$. Somnolency had continued; and he still felt a little sleepy.

(*b*) INJECTED $\frac{1}{4}$ GRAIN, three hours after tea; urine acid, specific gravity 1022, about ℥ij. in the hour.

After 15 *minutes* great somnolency.

After 30 *minutes.* Pulse accelerated 6 beats, unchanged in volume and power; intermitted thrice in two minutes. Pupils $\frac{1}{10}$. Somnolency continues.

After $\frac{3}{4}$ *hour.* Dozing. Respiration 15, regular. Pulse accelerated 8 beats; no intermittence. Pupil $\frac{1}{12}''$; sideways $\frac{1}{10}$.

After 1 *hour.* Pulse accelerated only 1 beat of diminished volume.

After $2\frac{1}{4}$ *hours.* Pulse decreased 8 beats, and of less volume and power than before the injection. Pupils as at 30 minutes. Somnolency continued, and he was very comfortable. Head, face, and hands cool.

The somnolency was at no time so great that he could not prevent sleep, and when he gave way to it he passed into a gentle slumber from which a slight noise awoke him.

During the $2\frac{1}{4}$ hours ℥iv. of bright acid urine, specific gravity 1020, was secreted.

(*c*) INJECTED $\frac{1}{2}$ GRAIN two hours after tea; secreted during the previous hour ℥ij. pale acid urine, specific gravity 1009·2.

After 10 *minutes* began to feel very sleepy. Pulse accelerated 16 beats. Respiration 16, regular. Eyes suffused. Pupil $\frac{1}{12}$; sideways $\frac{1}{8}$.

After $\frac{1}{2}$ *hour.* Pulse accelerated 20 beats, increased in power. Respiration 14, regular. Tongue unchanged. Membranes of eye much injected; cheeks hot and flushed. Scalp a little less so.

After 1 *hour.* Pulse attained its maximum acceleration

of 26 beats, strong and regular, and of good volume; a pulse indicative of strong stimulation. Respiration 14, regular. Pupil $\frac{1}{12}$; sideways $\frac{1}{10}$; a general diffusion of warmth throughout the body. Eyes injected; face and head hot and flushed.

After 2 *hours* the maximum acceleration of the pulse, and the other symptoms continued unchanged.

After 3 *hours*. The symptoms began to decline; the pulse had fallen from 100 to 96, and had nearly returned to its original volume and power. Respiration still 14. Pupils $\frac{1}{10}$; sideways $\frac{1}{8}$. Less heat and flushing of head and face. Somnolency less, but continued very sleepy.

After 3½ *hours*. Pulse 94, still a little fuller and stronger. Respiration 14, regular. Anterior part of the tongue dryish.

Had slept throughout comfortably and a little more soundly than in ordinary sleep. Was not conscious of dreaming, but muttered a good deal. During the 3 hours following the injection, ℥vss. of bright acid urine, specific gravity 1022·0, was secreted. The only after-effect was a little sleepiness during the next 48 hours.

Obs. 20.—Mr. John W., aged 33, the subject of severe lumbar neuralgia. Pulse 78. Pupil $\frac{1}{6}''$. Tongue clean and wet. Loss of vision of one eye from smallpox.

(*a*) INJECTED ¼ GRAIN MORPHIÆ ACET.:—

After 50 *minutes*. Pulse decreased 4 beats. Some drowsiness and giddiness.

After 1¼ *hour*. Pulse decreased 10 beats, unchanged in volume and power. Pupil $\frac{1}{7}''$; began to feel light-headed and sleepy a ¼ of an hour previously. Face slightly flushed.

After 3 *hours*. Pulse 60; decreased 16 beats, of normal volume and power. Pupil $\frac{1}{7}$. Dozed, and had some sound sleep since the last date.

No change throughout in the mouth or vision; continued to read the newspaper for 1¼ hour succeeding the injection. Now walked away a distance of four miles.

(*b*) Some weeks after, INJECTED ½ GRAIN. Somnolency followed within 5 minutes.

After 1¼ *hour*. Pulse decreased 4 beats; unchanged in volume and power. Had slept soundly for ½ an hour, and was now very sleepy and giddy. Pupil $\frac{1}{8}$; vision for near and distant objects unimpaired; in reading the newspaper he could detect no change in his vision, and he clearly discerned an inconspicuous lightning conductor, a rod of the thickness of the little finger, by the side of a chimney pot at a distance of about 70 yards. Sulphate of atropia $\frac{1}{20}$ grain was now injected, and at the end of another hour the pupil was dilated to $\frac{1}{4}''$; and he was unable to read at the ordinary distance, but when the newspaper was held at full arm's length he could read the smallest type without difficulty. He could still see the lightning conductor, but less distinctly.

Obs. 21.—Charles V., aged 32, a strong man, but disabled by chronic lumbago. Pulse 76. Pupils at light $\frac{1}{10}''$; sideways $\frac{1}{7}$; at dark side of room $\frac{1}{6}$.

INJECTED $\frac{1}{4}$ GRAIN ACETATE MORPHIA 2 hours after breakfast. Passed ℥ix. urine, acid, specific gravity 1010·4.

After 30 *minutes.* Pulse unchanged. Pupil $\frac{1}{12}$; sideways $\frac{1}{10}$; and at dark side of room $\frac{1}{7}$. Tongue clean and moist; no drowsiness.

After 1½ *hour.* Pulse decreased 16 beats, regular; of unchanged volume and power.

After 2 *hours.* Pulse decreased 20 beats. Passed ℥xviijss. urine, alkaline, specific gravity 1008·8.

After 4 *hours.* Pulse decreased 22 beats, being now 54, regular, and of good volume and power. Pupils continued as at 30 minutes. Tongue and mouth remained unchanged, moist, and clean. No nausea or somnolency throughout. Passed ℥vij. more of urine, freely acid, specific gravity 1009·2. Now walked home, and between the fifth and sixth hour felt a little nausea, but no somnolency. The injection was repeated on two occasions with the same results; viz., decrease in the frequency of the pulse and contraction of the pupils; but in addition, a little heat and flushing of the face and slight giddiness. There was no apparent somnolency on either occasion.

Obs. 22.—Dr. ——, aged 65. A loud systolic bruit over the sternum from disease of the aortic valves, great œdema of the legs, and neuralgic pains of the limbs. For the relief of the latter symptom he began the subcutaneous use of morphia in the usual doses, and increased them until he took a grain every eight, or even every five hours, i. e. from 3 to 5 grains of the acetate by the skin every 24 hours. The drug never produced other than calming effects. After he had taken the quantity just stated daily for six months, the following was his condition. From one to three hours after the injection of a grain of the drug, pulse full, strong, regular, from 90 to 100; countenance ruddy, bright, and cheerful; and the patient abounded with his natural vivacity and mental vigour. From 4 to 8 hours after the injection, pulse from 80 to 90, weak and flaccid, but still regular; and he was haggard, depressed, and irritable. The morphia soothed the pains, which seemed now to be more imaginary than real; but it never induced a trace of hypnosis, muscular spasm, or contraction of the pupil. The tongue always preternaturally clean and smooth, resembling recent muscle in colour, and in the appearance of superficial translucency, moist, or even wet, but sometimes inclining to dry and glaze. Urine free, bright, acid, specific gravity ranging from 1020 to 1024, often containing an excess of uric acid. Bowels acting regularly with the aid of warm water enemata. Appetite fair; daily allowance of wine from one to two bottles. The morphia, in this case, simply acted as a stimulant and tonic.

Obs. 23.—Mary B., aged 46, well nourished, but very lame from sciatica. Pulse 80. Pupils at light $\frac{1}{7}''$; aside, barely $\frac{1}{6}$. Tongue clean and moist.

INJECTED $\frac{1}{6}$ GRAIN MORPHIÆ ACET. behind the trochanter.

After 10 *minutes*, giddiness and somnolency; then fell asleep for $\frac{3}{4}$ hour.

After 1 *hour*, awoke, and complained of being very giddy, and said that everything was running round. Cheeks and forehead hot and flushed, hands hot, 'felt hot all over.' Pulse increased in volume and power, evidently under the influence of strong stimulation. Pupils at light $\frac{1}{8}''$; aside $\frac{1}{6}$. Tongue clean and wet.

After 2 *hours*. Had felt very drowsy and sickish. Pulse 74, regular; of the same volume and power as before the injection. Pupils at light $\frac{1}{8}''$; aside $\frac{1}{7}''$. Tongue moist, a slight whitish fur, indented. During the *next* 5 *hours* was distressed with nausea, faintness, and constant retchings, with alternate flushes of heat and cold. Atropia, alone and in combination with morphia (see Chap. vi.), was subsequently injected at intervals of a week; and at the end of a month $\frac{1}{6}$ of a grain of morphia was again injected alone, and with exactly the same results.

Obs. 24.—Mrs. N., aged 40, a spare, very excitable, loquacious little woman, afflicted with severe neuralgic pains in the right shoulder. Pulse 100. Pupils $\frac{1}{10}$. Tongue clean and moist. General health good.

INJECTED $\frac{1}{7}$ GR. MORPHIÆ ACET. into the affected part.

After 40 *minutes*, looked very heavy, and felt 'dreadfully tipsy.' Pupils $\frac{1}{12}$.

After 1 *hour*. Pulse decreased 4 beats, regular. Drowsiness gone off; now felt very ill with 'an indescribable sensation worse to bear than any pain' behind the lower half of the sternum. Was sick and faint.

After 2 *hours*. Pulse decreased 30 beats, feeble, and slightly irregular. Had suffered constant nausea, and vomited twice, and the bowels had acted loosely twice; was pale and cold, and almost unable to walk. During the next four hours she continued in this state, and did not recover from weakness and tendency to nausea and faintings for a week. She subsequently told me that morphia always had the above-described effect upon her, that six drops of laudanum by stomach would make her sick and faint, and that she could readily distinguish between the action of laudanum and morphia; the latter always producing a more intense feeling of distress in the præcordia.

Obs. 25.—Mrs. E. W., aged 34, stout and healthy, pregnant 4 months, of hysterical temperament, and liable to neuralgia; suffered greatly for two years with inveterate burning pain of the index, middle, and ring fingers, originating in the bed of the nail of the middle finger. The ends of these fingers were turgid and hot, and pressure upon the root of the nail of the middle finger gave great pain. There was no known or apparent cause for the pain. It was so severe that she kept the fingers immersed in a jug of water which she carried about with her. Pulse 80. Pupils $\frac{1}{10}''$ at the light, $\frac{1}{6}''$ towards the dark side of the room. Tongue clean and moist.

(*a*) INJECTED $\frac{1}{15}$ GRAIN MORPHIÆ ACET. into the arm. Drowsiness in 15 minutes.

After $\frac{3}{4}$ *hour.* Pulse accelerated 6 beats. Pupils at light $\frac{1}{12}$; from the light $\frac{1}{8}$.

After 1 *hour.* Felt drowsy and giddy. Pulse 86 still.

After 2 *hours.* Pulse 80, regular; of good volume and power. Pupils as at $\frac{3}{4}$ hour. The giddiness and drowsiness now passed off. Tongue and mouth unchanged; and there was no nausea or faintness throughout.

Some days afterwards, and when in her usual health:—

(*b*) INJECTED ♏ vj. TINCTURÆ OPII into the arm.

After 1 *hour.* Pulse accelerated 10 beats; no somnolency; had felt uncomfortable, and was now faint, and gave way to hysterical crying and sobbing.

After 2 *hours.* Pulse weak, accelerated 16 beats. Had vomited twice, and now retched violently, and vomited mucus mixed with bile. Face pale and anxious.

After 5 *hours.* Pulse small, feeble, and intermitting, accelerated 20 beats. Had vomited bilious matter twice, and experienced constant nausea, with considerable drowsiness, but was restless, distressed, and unable to sleep on account of illusory objects which continued to pass before the vision as soon as she closed the eyes; surface cold. Nausea, repeated action of the bowels, somnolency, giddiness and faintness, with violent retching and vomiting of yellow phlegm, continued up to the *eleventh hour*, when she went to sleep for the first time, and slept throughout the night with much dreaming and wandering. The pupils and tongue remained unchanged throughout.

(*c*) A week afterwards, and having twice injected atropine in the interim:—

INJECTED $\frac{1}{12}$ GR. MORPHIÆ ACET. into the affected finger.

After $\frac{3}{4}$ *hour.* Pulse weaker. Face slightly flushed. Had felt faint, tremulous, and sick for 15 minutes.

After $1\frac{3}{4}$ *hour.* Had napped for 20 minutes. Pulse feeble, decreased 4 beats; intermitted occasionally. Felt sick and faint. Face pale and anxious, and was generally cold and tremulous. Respiration irregular, with frequent sighs.

After 3½ *hours.* Had slept half an hour more, but still continued very sick and faint. Pulse very small, weak, and intermittent; accelerated 6 beats. The pupils and mouth remained unchanged throughout. She now walked home. Slight drowsiness continued for a short time, but there was no more nausea. A month later on, and after repeated injections of atropia alone and in combination with morphia, $\frac{1}{12}$ of a grain of morph. acet. was again used alone, and with the same results, but the vomiting and retching continued for 6 hours. (See ACTION OF OPIUM AND BELLADONNA.)

Such are the results of the action of morphia when administered by the areolar tissue; and unfortunately, they are commonly thus varied and uncertain.

As far as my own experience goes the action is constant for the same individual, and the effects of larger doses only differ in degree from those of smaller ones. If, for example, a very moderate dose induce nausea and syncope, a larger one will produce these effects in a more intense degree.

It is but reasonable to expect that the particular action, whatever it may be, would be more rapid, if not more intense in proportion as the drug is more directly and rapidly admitted into the blood. But if venous bleeding of the puncture is to be taken as an indication of the direct admission of the agent into the blood, my own observations have not fulfilled these expectations. In no case where bleeding has followed the punctures in using morphia or any other drug, have I ever noticed a more rapid or intense development of the symptoms.

The effects of morphia and opium are precisely the same when *given by the stomach.* The following examples will be sufficient to complete the view of the action of opium.

Obs. 26.—Gave *one grain of powdered opium* to a gentleman aged 32, entirely free from nervous symptoms. The dose was followed by insomnia and restlessness during the whole of the night, and he knew from previous experience that this was the effect of the opium. Thinking that a larger dose would be more effectual, I prescribed on another occasion *two grains*. This produced a mixture of somnolence and delirium with so much restlessness that it was necessary to keep him constantly aroused to prevent him from getting out of bed.

Nausea and headache are very common after-effects, and in some cases, are very distressing; thus—

Obs. 27.—♏xv. of tincture of opium prescribed for insomnia, in a lady aged 68, of highly excitable temperament, commonly produced disturbed sleep; and, after eight or nine hours, intense headache, nausea, and vomiting, which continued throughout the chief part of the following day.

Obs. 28.—An old gentleman of very abstemious habits, never took opium until after an attack of paralysis in his 75th year, when it was required to procure sleep. For the year following, he took from ½ to 1 grain of powdered opium every night. During the next two years, he required only occasional doses of a grain; but at the end of this time it was necessary to take a grain every night. This failed after a time to produce the required effect, and morphia was substituted, and from time to time increased, until he took 1 grain every night. After continuing the dose for six months, its effect was similar to that recorded in *Obs.* 22. It produced a quieting effect, but the patient was never conscious of sound sleep as a result of its use. Its action upon the heart was stimulant and tonic, and the tongue from being constantly furred, became preternaturally clean and smooth, like raw flesh, and had a great tendency to become dry and glazed. The appetite remained good, and the secretions were as active as when no opiate was taken.

Having now, I think, advanced sufficient evidence of the effects of morphia upon individuals of different constitution, I proceed, with the help of those already deduced from observations upon the lower animals, to draw precise conclusions as to its action upon man. And—

1. With regard to the *Brain*. The *general* phenomena are the same in all animals and every variety of constitution, and they are plainly the result of two opposite effects—a hypnotic, which includes anæsthesia; and an excitant, which includes cramp and convulsion.

What has been said at p. 105 of the action of the drug on the horse is strictly applicable to man. The *specific* and *individual* effects are determined by peculiarities of nervous constitution. In some, these two effects on the nervous system are so equally balanced, that in moderate doses the drug has no very decided action (*Obs.* 21, p. 127), or at most, only a tonic and stimulant one (*Obs.* 22), effects which in increased doses may rise to active delirium (*Obs.* 2, p. 101).

In others, the hypnotic effect prevails, and the stimulant action is apparently confined to the heart (*Obs.* 19 *c.* p. 125). In a third class, the excitant action counteracts to a greater or less degree the hypnotic, and insomnia with restlessness or delirium results (*Obs.* 26, p. 131). Amongst animals none perhaps are more emotional and excitable than the horse and the dog, and as a rule, both suffer severely from the effects of opium. Of human kind, women are more liable to its excitant action than men; and amongst women individuals of a highly emotional, excitable, and energetic temperament are those to whom opium in any form is a very distressing remedy, and when hypodermically used, a most dangerous one.

If we go further and enquire into the ultimate cause of this variation in the action of morphia, we shall doubtless find it in that universal law of nature whereby like responds to like.

The unequal vibrations of two powerful sounds destroy each other's effect in some degree, but in proportion as one or the other prevails, so is its power to excite a sympathetic note in an accordant object within its influence. So it is with morphia; and let the nervous system of a given individual be simultaneously exposed to its two antagonistic actions, and it will respond to that one most in harmony with itself.

If the individual be calm, unemotional, unexcitable, or dull, heavy, and stolid, his mind will naturally sympathise with the hypnotic impressions.

If he be of the reverse disposition, the actions of his brain will vibrate in unison with the excitant impulses. On the whole, these two influences appear to be pretty equally poised, and it is probable that a very little is sufficient to disturb the balance. The mere habit, or accidental tendency to dreaming, for example, by initiating a preponderance on behalf of an action which rapidly passes from sleeping to waking delirium, is perhaps the simple clue wanting, to enable us in many cases to predict or explain the variation of the action of morphia.

2. *The Spinal Cord.*—Little need be added to what has already been said (pp. 105, 114) of the action of morphia upon this part of the nervous system. Those fearful symptoms which are sometimes the immediate result of its introduction beneath the skin; the syncope, nausea, and general distress which ensue a little later on; and the fatal issue of the prolonged narcotising effect of poisonous doses, are all traceable to its influence upon the spinal cord.

At first, life is in danger from *spasmodic cramp of the respiratory muscles*, associated probably with a similar condition of the muscular constituents of the lungs themselves. The chest is fixed, the patient gasps for breath, the right heart becomes distended, and the pulse rapidly sinks until an inspiration relieves the cardiac distension and snatches the patient from the jaws of death.

These effects may be readily observed in the mouse: on man they are fully illustrated in the case cited p. 123.

Derangement of the Vagus Nerve is the main source of the distress which follows; its sentient branches are blunted, and the lungs no longer invite a flow of blood; its motor branches convey only cramp to the muscular parts, the mechanical suction power of the chest is depressed, and the lungs themselves tend to collapse. The right heart soon becomes distended, cardiac distress is complained of, the pulse loses force and volume, intermits, and syncope results.

The attendant symptoms, violent retching and vomiting, are the direct consequences of disorderly contractions of the stomach; and the spasmodic impressions passing beyond this viscus, are often conveyed to the lower portions of the digestive canal, and cause repeated evacuations (*Obs.* 13, p. 112; *Obs.* 24, p. 129).

In the dog, vomiting is the immediate and invariable consequence of the action of opium and morphia.

Under the influence of morphia *the whole respiratory function is depressed,* and the breathing becomes more or less irregular and interrupted by sighs. During the action of a medicinal dose the respiratory movements decrease $\frac{1}{4}$ or $\frac{1}{3}$ (*Obs.* 19 *c.* p. 125), and after a poisonous dose they are progressively reduced in number and extent, until they become imperceptible and ultimately cease altogether.

3. *The Sympathetic.*—The heart retains its vigour unimpaired throughout, and but for the physiological and mechanical impediments which exist to the flow of blood through the lungs, the circulation would be maintained with full vigour. Even when death has followed the protracted action of the drug, the movements of the heart continue, and may be revived or accelerated by relieving the distended right cavities, and maintained for half an hour after the respirations have ceased. I have verified this observation many times, and have never found an exception in any case. The fact is equally true of opium as a whole, and of all its constituents. The effect of the breathing upon the pulse, which I have fully described in *Obs.* 5, p. 106, proves still more directly that the heart is in a constant state of repression whenever the excitant action of the drug upon the spinal cord is in the ascendant. (See ACTION OF BELLADONNA ON THE VAGUS, in a subsequent chapter.)

When these effects upon the respiratory functions generally, and the vagus in particular, are neutralised by other actions, then the stimulant effect of opium upon the sympathetic is manifest, and the pulse is increased in rapidity, volume and force. (See *Obs.* 1–4, p. 100 ; and *Obs.* 19 *c.* p. 125.)

Under these circumstances, the effect in man seems to be identical with that of henbane ; the morphia causing in small doses a diminution in the rapidity, with increase in the volume and power, of the pulse ; and in larger doses, considerable acceleration, sometimes rivalling the effect of belladonna itself. (See ACTION OF HENBANE AND BELLADONNA respectively, upon the Sympathetic.)

After the subcutaneous injection of morphia in the frog,

the circulation of the web goes on for hours with increased rapidity and without any appreciable change in the calibre of the bloodvessels. When the web is soaked in a strong solution (gr. j. to ʒj. of water) for several hours, no change is observable until about the end of the third hour, when slight dilatation of the bloodvessels with retardation of the flow and adhesion of the spherical corpuscles to the walls of the capillaries, is observable. In the subject of the observation here referred to, the spinal cord had been divided in the occipital region eleven days previously.

The result of the intense and prolonged action of morphia is congestion and stasis from paralysis or loss of conductivity of the vaso-motor nerves. (See *Obs.* 4, p. 103.)

4. *On Nutrition.*—During the action of opium, digestion and absorption are retarded, and, if a large dose be taken by the mouth, for a time delayed. When the vagus is much deranged there is necessarily a considerable depression of all the functions. And so long as any hypnotic and anæsthetic influence remain, the *excretions are retained.* But when the effects are simply hypnotic and stimulant as in *Obs.* 19, p. 124, or when the patient becomes habituated to its use, the *secretions and excretions are abundant and free.* (See *Obs.* 22; also the effects upon the horse.) The influence of opium on the renal secretion will be further illustrated in a subsequent chapter.

5. *Action on the Pupils.*—Mr. Richard Hughes in his remarks 'On the Signification of the Contraction and Dilatation of the Pupil by Opium and Belladonna respectively,' thus lucidly explains the action of opium:—Since excitation of the third nerve, and depression of the sympathetic, are alike followed by contraction of the pupil, 'it is very unlikely that opium, whose full effect is depressent to the whole of the cerebro-spinal system, should act as an excitant to one nerve proceeding from that system. It is most probable, therefore, that opium contracts the pupil by such a depressent influence on the sympathetic nerve, as that the third still excited by the stimulus of light through the optic, or without this even, is more than a balance for it.'[1] Such a

[1] Lond. Med. Review, vol. i. 1860–1, p. 92.

theory proceeds from wrong premises. I have, I believe, satisfactorily shown that opium is a powerful excitant to both the cerebro-spinal and the sympathetic nervous systems; but that, owing to the diminished conductivity of the nerve fibres, due to the hypnotic action of the drug, the restrained impulses are conveyed only a short distance, and radiate upon the nearest muscles, which are thus kept in a state of partial cramp. It thus happens that while the limbs are weak and flaccid the pupils are strongly contracted. In the horse, the stimulant effect upon the sympathetic is much more decided than in man, and instead of contraction there is dilatation of the pupil. And yet I think it will not be denied that even in this animal the upper part of the motor tract is, to say the least, as strongly stimulated as the sympathetic. How then can this difference in the action on the pupil be explained? Consistently with what I have before advanced respecting the action of morphia, I venture to offer the following solution of the problem. In both horse and man the distance between the centre of the third nerve and the circular fibres of the iris is greater than that between the particular ganglion of the sympathetic, from which the iris derives its nerves and the radiating fibres. In man, however, the difference is less, and so short is the distance the excitant impressions from the third centre have to travel, that they meet with little or no hindrance, and radiate upon the iris in sufficient force to counteract the impressions conveyed to its dilator muscle by the sympathetic. In the horse, on the other hand, the stimulant effect upon the sympathetic is proportionately stronger than in man, and as the greater length of the third nerve is sufficient to offer a considerable resistance to the passage of motor impulses, the influence of the sympathetic prevails, and dilatation of the pupil results.

In man, contraction of the pupil is the most constant of all the effects of opium, and comes on in ten or fifteen minutes after the subcutaneous use of the drug. It occurs independently of hypnosis. (*Obs.* 21, p. 127.)

I have only twice failed to observe any change, and on both occasions the sympathetic *appeared* to be greatly depressed on account of the derangement of the vagus. (See *Obs.* 25 *b. c.* p. 130)

In the dog, the sympathetic and spinal influences are more equally balanced than in the horse. The pupil only contracts when the animal is under the influence of sleep, or is narcotised; and even in the latter condition, it will be found to respond to any considerable variation in the light; and when disturbed or aroused, dilatation immediately ensues.

Another point worthy of notice in reference to the action of the pupil, is the dilatation which occurs at the moment of death from opium, or, in the comatose condition which precedes it. The full active dilatation which is uniformly observable when death is imminent, taken in conjunction with the continued action of the heart after death, must be accepted as complete proof of the enduring activity of the sympathetic, after the irritability of the cerebro-spinal system is exhausted.[1] Thus it appears conclusively that the sympathetic is powerfully stimulated by the action of morphia, and its function in no degree depressed thereby.

In the dog, indeed, the contraction appears to be more a consequence of sleep than of the direct action of opium.

In some observations on the alteration of the accommodation of the eye by belladonna and opium, Professor R. von Graefe[2] states, that morphia produces a spasm of accommodation, in consequence of which the space allowed for accommodation becomes greatly limited, and myopia results, distant objects being indistinctly seen; but if concave glasses be used this is obviated.

My own observations upon this point (see *Obs.* 20 *b.* p. 127) do not lead me to agreement with the latter part of this statement. The patient upon whom the observations were made was very intelligent, and defect of the one eye made him critical with respect to any alteration in the capacity of his vision.

Excepting in three instances, the *same* sample of pure acetate of morphia was used in each of the preceding obser-

[1] See Note 1. Appendix.
[2] Congrès périod. internat d'Ophthal. 1862; Med. Times and Gaz. 1861, p. 533.

vations. An equally pure preparation was employed in the exceptions alluded to.

II. **NARCEINE or NARCEIA.**—Much confusion exists with regard both to the physical properties and the physiological action of this substance, as will appear from the following statements:—

(*a*) M. Claude Bernard[1] regards narceine as 'the most somniferous substance of opium;' and as a soporific he therefore places it first, and morphia second. 'In equal doses,' he states, 'it causes a much more profound sleep than codeia; but animals are not, however, stupefied by a leaden sleep, as with morphia, and they remain quite alive to painful sensations. That which characterises the narceic sleep is, that it is calm and profound. Dogs reduced to a profound sleep of several hours' duration make no resistance when hurt; and if they complain, they do not seek to escape or to bite. I injected,' he says, 'from 7 *to* 8 *centigrammes* (=1·2 *grain*) *of narceine, dissolved in* 2 *centimetre cubes* (=38 *grain measures*) *of water*, beneath the skin of a dog. At the end of about a quarter of an hour the animal was taken with a sleep so profound that I was obliged to produce the animal at the next meeting, in order to convince the President and some members of the Society that the dog was not dead.' He concludes that narceine is free from any excitant or convulsivant action, and remarks that after a poisonous dose animals die in a state of muscular relaxation.

(*b*) Dr. Béhier[2] of the Hôpital de la Pitié, in applying the results of Bernard's experiments to therapeutical uses, gave narceine to a number of his patients. Except in two cases, the narceine was given by mouth in the form of pills, and in the two cases ♏xx. to ♏xxx. (= gr. 1/5 to about 1/3) of a solution of 1 part of narceine in 100 of distilled water was used subcutaneously. The results were as follows: The injections calmed pain, but invariably caused dysuria. Given

[1] Bulletin gén. de Thérapeutique, tome lxvii. 1864, p. 193, and Compt. rend. Acad. de Sciences, 1864, vol. 59.

[2] Bul. gén. de Thérap. t. lxvii. p. 151, and Annuaire de Thérap. Bouchardat, 1865.

by the mouth, it allayed cough and diminished expectoration. With females it sometimes determined vomiting at the moment sleep was interrupted.

(*c*) Dr. Debout[1] used a solution of narceine in syrup, in the proportion of 0·15 *of a grain, to half an ounce.* He took it in increasing doses, and found the hypnotic effect commenced when the dose reached half a grain. 'The sleep,' he says, 'is calm and interrupted by the least noise; but one goes to sleep again immediately. On awaking there is freedom from the heaviness of the head experienced after the use of morphia,' and there is less liability to vomiting and constipation. The whole of the effect is limited to a somniferous influence, and its *calming and hypnotic* effects are *superior* to those of *codeia,* and *almost equal* to those of *morphia.* It has no influence upon the respiratory movements. When the dose exceeds ¾ of a grain, it causes some dysuria.

(*d*) Dr. A. Eulenberg[2] administered hydrochlorate of narceine in doses of ⅙ to ⅓ grain by the stomach, and ⅛ to ¼ of a grain by the subcutaneous tissue, and found it serviceable as a sedative and hypnotic. He considers it a valuable remedy in cases where morphia is not tolerated, or in which it has lost its power. He could discover no effects upon the urinary organs.

(*e*) Dr. Liné[3] concludes that narceine possesses the greatest hypnotic power of all the alkaloids of opium. In equal doses, neither morphia nor codeia produce a sleep so prolonged and complete as narceine. *Anuria* is so constant a result of its use that he recommends it in *enuresis.*

(*f*) Dr. J. M. da Costa[4] gave a solution of narceine in water, acidulated with hydrochloric acid, in doses varying from ½ a grain to 2½ grains, by the stomach, and found that it did not produce sleep. That in the same doses as morphia it is totally destitute of soporific and anodyne properties. That in larger doses it is uncertain, and often probably

[1] Bul. gén. de Thérap. t. lxvii. p. 145.
[2] Schmidt's Jahrbücher der gesammten Medicin, August and October, 1866.
[3] Journal de Pharmacie et de Chimie, 1866, sér. iv. tome iii. p. 386.
[4] Pennsylvania Hosp. Rep. 1868, p. 177.

inert. That it has no influence upon the pupils, and is destitute of any excitant action.

(*g*) Dr. Oettinger[1] regards narceine as a pure hypnotic. He states that it causes muscular weakness, sleepiness, and, in a slight degree, obtuseness of sensation, and also in small doses retardation of the pulse. As a substitute for morphia, it may be given by the mouth in doses of half or one grain, in powder or in solution. It is unsuitable for hypodermic use, because there is no combination in which it is readily soluble in warm water, and thus rendered fit for introduction by one injection; and the solutions at present used cause great pain and irritation.

Apart from the discrepancies which appear on reading over the above statements, a primary question arises as to the identity if not the purity of the drug employed. One need not hesitate to conclude that the substance used by M. Bernard was not Pelletier's narceine. His own words are conclusive on this point. 'La narceine,' he says, 'étant plus soluble (than morphia and codeia), je l'ai souvent employée directement dans des solutions (1 part in 20 parts of water) à la même dose.' (Op. cit. p. 194.) Again, he used a solution of 1 part of narceine in 38 parts of water in the experiment upon the dog.

Dr. Béhier, we have seen, recommends a solution of 1 part of narceine to 100 parts of water. And yet both of these investigators appear to have been in possession of narceine of acknowledged purity, and prepared by the same manufacturer as that employed by Dr. Debout.

Dr. da Costa is, I find, equally perplexed upon this point; and he thinks that the activity of the narceine employed by Béhier is due to admixture with codeia, which greatly increases the solubility of narceine.

Mr. T. Smith, of Edinburgh, informs me that meconine makes narceine more soluble; that these two principles separate together from the solution containing them; and that if the mixture of the two be not thoroughly extracted by æther, the narceine obtained is impure, and much more

[1] Schmidt's Jahrbücher der gesammten Medicin, August and October, 1866.

soluble than it should be. Mr. Smith has kindly furnished me with a sample of pure narceine prepared by himself. My investigations have been made with this and another specimen of pure narceine prepared by Messrs. Morson & Son, of London.

As I am wishful to avoid confusion, I will briefly describe the characters of these preparations.

Characters of Narceine.—A compressed, asbestos-like mass of soft acicular crystals, very light and colourless. Soluble in 100 parts of boiling water; begins to separate at 170° Fahr.; and at 150° the fluid is thick with silky crystals; at 60° the water retains 0·25 part, 1 part therefore requires 400 parts of water at 60° for solution. It is still less soluble in alcohol. Insoluble in chloroform and æther. Soluble in 33 parts of Price's glycerine at 212°, and in 66 parts at 60°. Above 212° Fahr., narceine is soluble in 12 parts of glycerine, and if one or two drops of HCl be added to the solution while hot, no separation of narceine takes place on cooling, and the fluid remains as bright and colourless as glycerine itself. HCl sp. gr. 1·17 diluted with from 5 to 10 parts of water, colours it azure blue. In stronger acid, it silently disappears without colour. HNO_6 developes a greenish-orange colour in aqueous solutions. $H_2S_2O_8$ dissolves it with the production of a rich amber colour, which, rapidly passing through greenish-orange, changes to that of port wine; on heating, this changes to a chocolate brown, which, after prolonged heating, assumes a dingy purple hue. Diluted with 100 parts of water, the liquid has a brownish tinge, and contains dust-like brown flocculi. After the addition of an excess of ammonia or potash, a quantity of dark brown flocculi settle from a colourless fluid. In these reactions narceine closely resembles meconine and cryptopia.

Such are the characters of what is here recognised as pure narceine. Its solutions are as bitter as morphia. It is incapable of neutralising HCl to any appreciable extent, and even the solutions given below are as acid as they would be if the narceine were altogether absent. Anderson's narceine appears to be a different substance. By means of HCl, short thick irregular prisms, easily soluble in water and alcohol, are obtained from it. And with sulphuric acid it forms a dark red solution, which turns green when warmed.[1]

Physiological Action. — The solutions with which I have made the following observations were thus formed :—

Solution 1.—Narceine, 5 grains.
Hydrochloric acid, 10 minims.
Water, 10 fluid drachms. Dissolve by the aid of a water-bath.

120 parts contain 1 part of narceine. If more narceine be used, it separates on cooling.

Solution 2.—Narceine, 5 grains.
Price's glycerine, 70 minims.
Hydrochloric acid (sp. gr. 1·17), 1½ minims.
Water sufficient to make the whole measure 100 minims.

Heat the narceine and glycerine together with agitation for a minute over a spirit-lamp, and to the clear and bright solution, while still warm, add the hydrochloric acid diluted with the water (about 25 minims). 20 parts of the solution contain 1 part of narceine.

Both of these solutions are suitable for internal use. No. 1 is too weak for hypodermic use, while No. 2 is perhaps the strongest and least irritating that can be formed for this purpose. Both solutions are freely acid, and, when introduced beneath the skin, cause smarting and sufficient irritation, subsequently, to produce a little inflammatory hardening of the integument. Abscess is liable to follow their use with those in whom suppurative action is easily induced. If it be thought necessary, Solution 2 may be diluted with a little warm water before it is injected.

The following are my own observations on the action of pure narceine in the forms above given :—

On the Dog.—*Obs.* 29. I first of all repeated Bernard's experiment, and injected beneath the skin of the beagle dog (p. 111) ♏xxv. of Sol. 2 (=1·25 grain narceine). I watched the animal carefully for two hours, but no effect

[1] Gmelin's Chemistry, Cavendish Soc. Trans. vol. xvii. pp. 598 and 600.

whatever followed. At the end of this time I injected beneath another part of the skin a quantity of the solution equal to 3·75 grains more of narceine, making in all 5 grains, and kept the animal under observation for the following twelve hours. Throughout the whole of this time a stranger would have failed to recognise any effect, but the drug certainly exercised a tranquilising influence. When all was quiet, the dog lay down and became quiet too, and occasionally appeared to be sleeping lightly. But if I moved my chair, he would get up and frolic around me, and the slightest invitation would bring him up to me with a full bound. The pulse, pupils, and tongue remained unchanged. Fæces and urine were passed before the first injection, but there were no evacuations between this time and the fifteenth hour. At the sixth hour he ate a hearty dinner. Three days afterwards a little abscess burst in the site of the second puncture. Besides this there were no after effects.

On the Mouse.—Obs. 30. Injected ♏x. Solution 1 = $\frac{1}{12}$ *of a grain* beneath the skin of a strong adult mouse. Respiration 140. *After* 15 *minutes* he became dull and drowsy, respiration 120; and remained in a comfortable sleepy state for the *next three hours*, the eyes closed, the snout resting on the table, occasionally turning the head and pointing the ears when called or disturbed by noise; reluctant to stir, and allowing me to take him up; but when aroused, struggling strongly and remaining for a minute quite active and alert, but soon relapsing again into a comfortable sleep. Between the third and fourth hours, the little animal picked up a bit of food and nibbled it, but soon fell asleep over it. Shortly after he began to move about oftener, and the effect of the medicine slowly wore off, but the mouse continued a little dull and sleepy for two hours more.

The effect was simply hypnotic and apparently grateful. The sleep was scarcely deeper than natural, and when aroused, he seemed as active and intelligent as when disturbed from ordinary sleep. At the end of 3½ hours, the respiration had fallen from 140 to 100, and remained regular. There were no after effects, and the next day the animal was lively and well.

Obs. 31.—Injected ♏xv. Solution 1 = ⅛ *grain,* beneath the skin of another active adult mouse. Respiration 160. Somnolency followed within seven minutes. *After* 15 *minutes,* was quite asleep. Respiration fallen to 140, regular. During *the next hour and quarter,* continued sleeping, lying heavily on the belly, with the legs outspread and the eyes closed. Respiration 120; was easily awakened, and turned the ears in the direction of the disturbing sound, and made feeble efforts to crawl, dragging the hind legs along. When taken up, he became completely awake and struggled vigorously, and when put down again he used the hind legs well. Then, after a minute or two, he fell asleep again. At the end of this time, i.e. *after an hour and a half,* the breathing had become shallower, and the respiration only 80. During the next 15 minutes, the respiratory movements rapidly decreased and became almost imperceptible. The breathing was at last relieved by extraordinary inspiratory efforts, which heaved the chest fifteen times a minute. Shortly afterwards, 2 *hours after the injection,* the animal died without a struggle.

The chest was opened two minutes after the breathing had ceased. The lungs were collapsed above and behind the heart, and of a salmon colour. The right heart was distended with black fluid blood, the left contracted and empty. The right ventricle contracted 30 times, and the right auricle 60 times, a minute; on relieving the latter of a little blood, its contractions were doubled.

Obs. 32.—Injected ½ *grain,* dissolved in ♏lx. of hot water, beneath the skin of a vigorous adult mouse, by four punctures, in divided doses of ⅛ grain, and at intervals of half an hour. Respiration 160. *The first dose* had merely a tranquilising effect, and at the end of the half-hour he was very comfortable and disinclined to move, but showed no tendency to sleep. A minute after *the second dose,* he couched down, closed the eyes, and continued drowsy, bestirring himself occasionally, and then prying about his new place of confinement and jumping lazily into the air to escape from it; but, in the intervals, he slept tranquilly. *After the third dose,* the hypnotic effect was still more marked.

On putting him down, he crawled lazily a few paces, then couched, closed the eyes, and remained motionless until disturbed by some noise. Then he resumed his prying for a few seconds, and even made one or two feeble and clumsy efforts to jump upwards. *At the end of an hour and a half,* the respiration was still 160, and he was so dull and sleepy that he allowed me to take him up, and made but slight resistance when handled. Ten minutes *after the fourth dose,* he was sleeping soundly in the couching position, with the snout resting on the table, and took no notice when called. Respiration irregular, ranging from 130 to 120, and sometimes falling as low as 100. Heart's action regular. Pupils unchanged. The little animal appeared very comfortable, and remained in the state described for 1½ hour. He made a slight movement occasionally, apparently to ease his posture, and then the whole body was affected with general tremor like that of paralysis agitans; and as it somewhat disturbed his balance, he was often obliged to creep a pace or two to preserve it. The tremor, however, only ceased when he came to a state of rest. It was induced by the slightest disturbance.

At the end of three hours I took him up, and, in doing so, aroused him, and he struggled with good power, and, on regaining his liberty, walked without tremor. Respiration at this time 110. Heart's action 240, regular. Pupils unchanged. Left alone, he soon fell asleep again, and continued in this state with tremor for three hours more, the respiration falling meanwhile to 80, and the cardiac contractions to 160. At the end of this time, i.e.

Six and a half hours after the first injection, the somnolency appeared to have passed off, and the mouse seemed uncomfortable, now and then changing his position hurriedly. On calling to him he rolled over and over in a struggling convulsion, preserving consciousness, and trying in vain to regain his feet. During the next five minutes he had eight convulsions, each lasting a few seconds, and assuming a more decided convulsive character than the preceding one; the body being at last extended backwards, and the limbs and tail affected with strong vibratile motion. The breathing appeared to remain unaffected. After the eighth attack he regained the couching posture, but lay along upon the belly with the legs outspread. Respiration 100 to 70, shallow. Took no notice when called, but, when disturbed, made an ineffectual effort to stir. He remained in this condition until

The seventh hour, when he rolled over again in convulsions, after which the respiration was 16, and extraordinary. The convulsions were renewed after a few seconds, and the respiratory movements finally ceased during the attack.

The chest was opened two minutes after death. The lungs were partially collapsed and salmon-coloured; the right heart distended with black fluid blood. Contractions had ceased, and were not revived on relieving the auricle of blood. The abdominal viscera presented a most interesting physiological study. The greatly distended stomach first invited attention. It contained a drachm of clear watery fluid of acid reaction, and free from both narceine and urea. The urinary bladder was moderately distended, and appeared white and opaque; the contents were semi-solid, of chalky whiteness, and were composed of long hair-like crystals insoluble in cold water and the dilute acids. The pelves of the kidneys were filled with the same cretaceous-looking matter, and, on examining sections of the kidneys, the straight tubuli, from the cortex to the pelvis, were, in both glands, found stuffed with the deposit and quite opaque. The convoluted tubuli were entirely free, and lined with normal epithelium. On now examining the sites of the four punctures in the skin of the neck and back, I found a similar chalky-looking matter diffused over considerable areas of the subcutaneous areolar tissue, which was itself healthy and free from the least trace of ecchymosis or inflammatory action. This deposit was the unabsorbed portion of the injected narceine; and on comparing it with that from the bladder and pelves of the kidneys, I found that they were identical both as to crystalline form and chemical reaction, and both moreover furnished the reactions of pure narceine as described above.

The urinary and other organs were perfectly healthy, and

presented no appearance of irritation. Blood from the right side of the heart was free from crystalline deposit.

It thus appears that just as the absorbents of a cereal take up flint by its roots, and carry it in solution through the living portion of its structure to lay it upon the stem, so, in the case before us, the absorbents of the mammalian took up solid narceine from the skin, and carried it in solution to the kidneys, where, as soon as it passed the boundary line of vital action—the terminal extremities of the secreting tubes of the kidney—it returned to its original insoluble state.

In no previous experiment have I observed the slightest tendency to convulsive action follow the use of narceine; and the fatal result in the present case may fairly, I think,

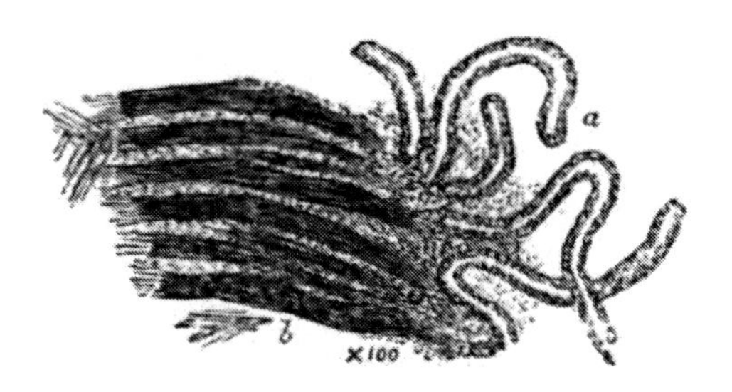

be attributed to the mechanical suppression of urine. The figure represents a section of one of the kidneys at the junction of the convoluted (*a*) and straight tubuli (*b*). The latter are impacted with crystals of narceine.

The little animal had not taken water for at least eleven hours before its death, and during the whole of the time he was under the influence of narceine (seven hours) neither urine nor fæcal matter was excreted. There was, in fact, complete retention as well as suppression; the urethra itself being blocked up with the narceine, which appeared at the external office. The drachm of water which had been introduced beneath the skin with the narceine was poured out into the stomach.

On Man.—I have frequently administered narceine, *by mouth*, in the solid form and in solution, and in doses varying from ½ grain to 5 grains. I have given it to people of different constitutions, and in the largest dose mentioned, an hour before breakfast, and when the body has been at rest; but it has failed to produce the slightest hypnotic or any other effect. The following will serve as an example:—Miss L., aged 37, took ʒx. of Solution 1=5 grs. narceine every other morning, on rising, and an hour before breakfast, for a week. Neither somnolency, heaviness, dysuria, nor indeed any appreciable effect, followed either dose. I then prescribed the ⅛ of a grain of acetate of morphia in lieu of the 5 grains of narceine, and gave this on alternate mornings at the same time, and under the same circumstances during the following week. She stated that on each occasion it caused heaviness and sleepiness throughout the whole of the day.

I have always found narceine, when given *by the skin*, produce a quieting, and usually a slight hypnotic effect. In doses of ¾ of a grain to 1 grain, the hypnotic effect in some individuals equals that of the ⅛ of a grain of acetate of morphia. Dysuria has never been a consequence of its use in my practice. I give the following details of two observations, in order that the effects of narceine may be compared with those of other principles of opium in the same individuals:—

Obs. 33.—Injected ¾ of a grain of narceine, ♏ xv. Sol. 2, diluted with ♏ xv. warm water, beneath the skin of Samuel M., aged 49, an able-bodied man (see *Obs.* 19, p. 124). Pulse 72. Respiration 20. Pupils 1/9. Tongue moist. After 15 *minutes*, felt comfortable and sleepy. After 45 *minutes*, was very sleepy, but had not slept. Pulse 72, unchanged. Respiration 19. Pupils unchanged. Experienced the same kind of sleepiness as after morphia. After 2¼ *hours*, had slept ¼ of an hour, and was still sleepy. Pulse 70. Respiration 17. Pupils, tongue, and force of pulse unchanged. He now passed, without the least difficulty, ℥ivss. of normal urine, sp. gr. 1020. This had been secreted during the 2¼ hours that followed the injection. He walked home, a distance of two miles; and, as it was 11 P.M., went to bed and slept soundly. Excepting the formation of a tender swelling at the seat of puncture, there were no after-effects. ♏ xxx. Sol. 1 = ¼ of a grain, produced on a previous occasion a proportionately slight hypnotic effect.

Obs. 34.—Injected ♏ xx. Sol. 2 (= 1 grain narceine), diluted with an equal quantity of hot water, beneath the skin of John L., aged 54, an able-bodied but weakly man (see *Obs.* 18, p. 124). The hypodermic tumour soon disappeared. A calm, comfortable feeling, with a slight tendency to sleep, followed in about ¼ of an hour, and continued for 4 or 5 hours; but the hypnotic effect was less than in the preceding case. The pulse, pupils, and tongue were unchanged. The injected part remained tender and a little puffy for two days.

On previous occasions ♏ xxiv. Sol. 1 = 0·2 grain produced a pleasant quieting effect, hardly amounting to sleepiness; and ♏ xlviij. = 0·4 grain, introduced by two punctures, one in each arm, was followed by the same comfortable tranquilising effect, and at the end of an hour a little sleepiness, which soon passed off.

Although this patient was liable to retention of urine, and required to be relieved by catheter on two occasions after the injection of cryptopia, no retention or dysuria followed either dose of narceine.

I conclude—1. That narceine is a *pure hypnotic*, but its action is so feeble that, when taken by the stomach, more than 5 grains are required to induce a slight tendency to sleep; and when introduced by the skin 1 grain is equivalent to only the ⅛ of a grain at most of a salt of morphia.

2. That it is impossible to reduce this quantity to the state of a non-irritating solution of such bulk that it may be introduced by one or two punctures without risk of inducing subcutaneous inflammation.

3. Granting that an efficient dose may be introduced beneath the skin without inconvenience, evidence is wanting that it possesses any advantage over morphia.

4. That narceine is, therefore, practically useless as a medicine.

5. Narceine is eliminated by the kidneys, and if sufficient

be administered by the skin, mechanical suppression of urine may result from the insolubility of the substance.[1]

6. That narceine, like morphia, kills by depressing and ultimately paralysing the respiratory movements.

III. **MECONINE or OPIANYLE.** — For the specimen of this substance with which the following observations were made, I am indebted to Mr. Morson, of Southampton Row, who has accumulated a large quantity of it in the manufacture of the other principles of opium. Mr. Morson informs me that opium yields from 0·1 to 0·2 per centum of this substance. The following are its *characters*:—Fine white silky prisms, in bulk resembling commercial sulphate of quinia, but being a little more coarsely crystalline. When carefully heated, it melts and sublimes without discoloration. It is freely soluble in hot water and in hot glycerine, from both of which, on cooling, nearly the whole is deposited in the original form. Readily and freely soluble in chloroform, from which on evaporation it separates in beautiful snowy-white bushy expansions. Less soluble in æther and in alcohol, and separating from them in its original form of long slender prisms.

All its solutions have a mild bitter taste, followed by a peculiar dryish sensation.

Meconine dissolves readily in sulphuric acid, forming a bright amber-tinted solution. When heated, the mixture slowly assumes a green colour, which deepens to the richest emerald. If the heat be continued, the green colour persists for some minutes, and then, losing its transparency and brightness, changes first to indigo-blue, then to neutral tint, which, on further application of the heat, passes slowly into purple, and this at last gives way to the deepest and richest claret, and the thick fluid ceases to exhibit any further variation of colour. Diluted with 100 times its bulk of

[1] Anuria, it appears, frequently follows its use, and if the quantity of urine secreted at the time of its admission beneath the skin happen to be small, it is quite possible that this result may be due to the deposition of narceine in the straight tubuli. Half a grain of this very bulky substance, when mixed with a large quantity of water, completely obstructs its passage through a tube one-fiftieth of an inch in diameter.

water, an almost colourless solution, containing a scanty brown dust-like precipitate, results. The addition of an excess of ammonia developes a rich purple colour in both fluid and precipitate, and a quantity of smooth dark violet-coloured matter is ultimately deposited, which when agitated with strong solution of ammonia, forms beautiful satiny undulations.

PHYSIOLOGICAL ACTION.—I am not aware that any observations have been made on the action of meconine. Orfila states (op. cit. p. 210) that it does not appear to be poisonous, for he injected 20 centigrammes (3 grains), dissolved in water, into the jugular vein of a dog with impunity.

The following are the results of my own observations:—

On the Horse.—Obs. 35. *Injected* 14 *grains*, partly dissolved and partly suspended in water, beneath the skin of the brown colt (see p. 100); and on another occasion gave him 20 *grains by mouth*, dissolved in a pint of water. The animal was carefully watched for several hours after both doses, but no effects were observed.

On the Dog.—Obs. 36. *Injected two grains* at 4 P.M. beneath the skin of the beagle (p. 143). *After half an hour*, a decided tranquilising effect, equal to that produced by double the quantity of narceine, was observable (see *Obs.* 29, p. 143). The dog was notably quieter than usual the whole of the next day. The pulse and pupils were unaffected, and the secretions free.

On the Mouse.—Obs. 37. *Injected* $\frac{1}{10}$ *grain*, dissolved in ♏ x. of water, beneath the skin of the mouse used in *Obs.* 15, p. 116. *After a minute*, he became very quiet and dull. Respiration 120, slightly irregular and a little laboured, and every 30 or 40 seconds interrupted by a slight catch, as if from spasm of the diaphragm. *After* 6 *minutes*, the respiration fell to 110, and *after* 20 *minutes* the catch had ceased, and the respiration had risen to 140°, and was more regular. He remained in the same position, and gradually fell off into a comfortable doze. *After three-quarters of an hour* he awoke, cleaned his fur, and pried about, occasionally raising himself upon the hind legs. At this time I caught him after a little trouble, and *injected* $\frac{1}{8}$ *grain more* of meconine, dissolved in ♏ xij. of hot water. The drowsiness returned immediately, and after turning round once or twice, and making lazy attempts to rub the forepaws together and sit upon the haunches, he couched upon the belly, partially closed the eyes, and went off to sleep, and continued to sleep soundly and without motion for an hour and a-half. I then took him up and placed him in his old cage. At first he seemed thoroughly awakened, picked up a bit of food, cleaned his fur, and walked about; but he soon became drowsy again, and continued to sleep soundly for the *next eight hours*, occasionally waking up upon some disturbance, and walking half round the cage and taking a little food and water, and being to all appearance thoroughly awakened. But he fell asleep again in a few seconds, and slept as heavily and composedly as before. At the seventh hour he passed a little urine. He did not completely recover from the hypnotic effects of the drug until *the twelfth hour* after the first injection.

During the whole of the time the little animal was under the influence of the meconine, he was evidently very comfortable, and the effect throughout was most charming. When the sleep was deepest he still moved the ears and vibrissæ in response to my calls, but only stirred when actually pushed along, and when taken up he made no resistance nor effort to escape until he was thoroughly aroused, when it was evident that there was no loss of intelligence or muscular movement. The brain, in fact, was tranquil, nor was there the slightest tendency to excitement or convulsive movement throughout, and the pulse and pupils continued unchanged. After the second injection, the breathing was affected, as at first, but there was no distress, and the catching passed off in a minute or two. (See Note 1, Appendix.)

On Man.—I have taken meconine *by mouth*, in powder and solution, in doses varying from the $\frac{1}{16}$ of a grain to 5 grains, and have given it in the same and larger doses to several patients, but neither in myself nor others have I obtained the slightest hypnotic or other effect.

By the skin, I have given it in doses ranging from $\frac{1}{2}$ grain to 2 grains, which is the largest quantity that can be conveniently used hypodermically. The following solution is the one which I have generally used :—Meconine, 3 grains; alcohol and glycerine, of each 45 minims; dissolve by the aid of a gentle heat. 30 minims contain 1 grain. A little meconine separates at a low temperature, but the warmth of the hand is sufficient to effect resolution.

Obs. 38.—Subject, Samuel M. (see *Obs.* 33, p. 149). Pulse 75, respiration 20, pupils, $\frac{1}{8}''$.

(*a*) *Injected* $\frac{1}{2}$ *grain* at 8 P.M. Slight somnolency followed in about half an hour. *After* 1 *hour*, pulse 66, respiration 16; somnolency continued, and he felt very comfortable. *After* 2$\frac{1}{2}$ *hours*, pulse 66, respiration 16 to 17, regular; had continued decidedly sleepy, but the effect was passing off. Tongue and pupils unchanged throughout. Went home, but did not sleep well during the night. Slept most of the next day. No diminution or retention of the secretions.

(*b*) *Injected* 1 *grain* by two punctures, six days after the last dose. Somnolency came on after twenty minutes, and continued. *After* 2 *hours*, was very comfortable, and had had a few light dozes. Pulse 65, regular, not quite so full as before the injection. Respiration 17. Tongue and pupils unchanged throughout. Somnolency continued. Went home and slept well all night, and was sleepy, as usual after his opiate injections, all the next day. Secretions quite free.

(*c*) *Injected* 2 *grains*, dissolved in 50 minims of warm alcohol, and introduced by two punctures, one in each arm. *After* 1 *hour*, pulse 66, respiration 18. Felt very comfortable and sleepy, but not more so than on the last occasion. *After* 2$\frac{1}{4}$ *hours*, pulse 62, respiration 17 to 18. Somnolency increased during the last hour, but he had not actually slept since the injection. Tongue, temperature, membranes of eye, pupils, and countenance, all unchanged. Went home and slept well, and felt sleepy throughout the next day. Secretions free.

Obs. 39.—Mr. James C., aged 42. Severe rheumatic fever, pain, and insomnia.

(*a*) *Injected* $\frac{1}{2}$ *grain* = ♏ xv. of the above solution at 7.30 P.M. It produced a very grateful tranquilising effect for 14 hours, and he slept lightly all night. The secretions were unaltered. After three days, (*b*) *Injected* 1½ *grain* by two punctures, at 2.30 P.M. Pulse 120, full and throbbing. *After* 8 *hours*, pulse 118, a little softer. Secreted ℥x. dark brown urine since 2.30 P.M. Had been very quiet and comfortable, and slept at intervals. Passed a comfortable night, but did not sleep soundly. Two nights afterwards, a sedative was again required, and I gave gr. xij. pulvis ipecacuanhæ co. = 1⅕ grain powdered opium. This procured such a sound refreshing night's sleep as he had not derived before either from narceine (½ to 1½ grain used subcutaneously) or the meconine.

It appears from the foregoing that the action of meconine, like that of narceine, is simply tranquilising and hypnotic, and is not followed by any unpleasant results. If this can be said of narceine, then the actions of the two principles are identical. Of the two, meconine appears to be slightly more active, an advantage probably due to its greater solubility. To the use of narceine there is a great, and I believe an insuperable, objection. The substance itself is free from irritant action (see p. 147), but as it separates from water at a temperature considerably above that of the body, it is necessary to form more permanent solutions, and this cannot be done without the aid of acids; and as these are not neutralised by narceine, but, on the contrary, seem to have their irritant action increased when mixed with it, such solutions are very liable to cause cellular inflammation. This objection fortunately does not apply to meconine, which is both free from irritant action itself, and does not require an acid to facilitate its solution. I have never observed the least irritation follow the use of the solution above prescribed, nor when alcohol and water, or alcohol alone, have been employed as the vehicles.

Much, therefore, of the laudation that has been bestowed upon narceine, may with more justice be transferred to meconine, which I believe will prove a really useful medicine. By the stomach, meconine, like narceine, has no effect, or so slight a one as to be inappreciable.[1] By the

[1] It appears doubtful whether either of these substances is absorbed by the stomach.

skin, the maximum effect is reached by a dose of one or at most two grains.

Compared with opium, the hypnotic effect of meconine is so feeble that it cannot be expected to take the place either of the crude drug or of morphia in cases which require but moderate doses of these to produce sleep. But in children, and in those who yield readily to a soporific influence, half a grain of meconine will generally, I believe, be found effectual. In the former class of persons, where convulsions impend, and in the latter, when morphia disagrees, meconine promises to be a valuable remedy, and the necessity of introducing it by the skin appears to be the only objection to its use.

IV. CRYPTOPIA. — This interesting alkaloid was lately discovered by Messrs. T. and H. Smith, of Edinburgh, who have published a description of it, together with an analysis by Dr. Cook, in the 'Pharmaceutical Journal,' April 1867. It appears to be the least abundant of all the constituents of opium. At present, four or five tons of opium have yielded only the same number of ounces of the alkaloid.

Mr. Smith has kindly furnished me with a quantity of the pure substance, and I have thus been enabled to examine its properties pretty fully. As observations upon the action of any substance are worthless unless its characters are clearly ascertained, I will preface my remarks on the physiological action of the substance by a description of the

Characters of Cryptopia.—Colourless crystals composed of broken and entire well-defined six-sided prisms.

100 parts of water boiled with a large excess and filtered while hot, yield, on spontaneous evaporation, a stain weighing 0·3 of a part, and presenting, under a low magnifying power, minute radiated granules. Soluble in chloroform, very slightly soluble in alcohol, still less soluble in æther, which like water takes up a mere trace. Separates from each of these fluids in a crystalline form—from chloroform in long radiating fibres, and from the other two in short acuminated six-sided prisms or plates. Readily soluble in water slightly acidulated with hydrochloric or acetic acids,

and separating therefrom in satiny tufts of hairlike crystals. The hydrochlorate exhibits the characteristic property of forming a jelly (see Solution 2).

The solutions of cryptopia are more bitter than those of morphia. Cryptopia disappears slowly and silently in cold nitric and sulphuric acids, developing an orange colour with the former, and a splendid purple with the latter, and on heating the mixture the brightness of the purple gradually fades and gives way to dark rich sap green, which after prolonged heating changes to a very dark and opaque neutral tint. The purple colour is evanescent, the green very permanent. The purple, green, and neutral coloured solutions form, when mixed with a large bulk of water, each a brown solution of a different degree of intensity, and from which excess of ammonia precipitates a foxy brown, a dark rich brown, and a blackish brown amorphous precipitate respectively. With regard to the production of a purple colour, Messrs. Smith state 'that there is no difficulty in knowing if the cryptopia is mixed with thebaia. If the minutest particle gives a blue colour with sulphuric acid, it is pure; but if it gives the least tinge of purple, it still contains thebaia. To obtain the pure alkaloid, it must be precipitated from its watery solution by ammonia, and the precipitate, after washing and drying, is to be washed freely with ether or spirit.'

Finding that both specimens sent to me, and their deposits from water, æther, and the other solvents, all and without variation produced a rich purple with the monohydrated acid, and the simple blue colour only when the acid was slightly diluted, I applied the above process, but still found that the precipitate (1 grain), after being repeatedly washed with ℥ss. of alcohol and æther, furnished a purple colour of the same intensity as the original salt when mixed with the monohydrated acid, and a blue with only a more dilute acid, and that both the purple and blue solutions, on heating, presented the variation of colour above described; and, further, that the remainder of the precipitate left after boiling it with ʒjss. of alcohol still gave a rich purple colour. I presume, therefore, that the cryptopia used in the following experiments was free from thebaia—a point which I found very necessary to determine after witnessing the phenomena described in *Obs.* 40 *b*, p. 160.

Physiological Action.—I have administered cryptopia by the subcutaneous tissue only, and the following are the solutions which I have employed :—

Solution 1.—Cryptopia, 5 grains.
Acetic acid, 5 minims.
Water to measure 100 parts. Dissolve.

Solution 2.—Cryptopia, 5 grains.
Water, 40 minims.
Hydrochloric acid, 3 minims.
Alcohol to measure 100 minims.

Dissolve the cryptopia in the acid and water, and just as the solution is effected and begins to solidify into a pearl-like jelly, add the alcohol, and shake until the hydrochlorate is dissolved. ♏xx. of each of these solutions contain 1 grain of the alkaloid.

On the Dog.—The effect of cryptopia upon this animal is most remarkable. It throws him into a frenzy of delusion. He is aroused from a state of quietude to one of intense and uncontrollable excitement, and begins to act as if he were attacked by a swarm of savage bees and unable to effect his escape. He stands at bay, with the legs rigidly set upon the ground, and every muscle strained to the utmost, and the tail strongly curved on the perineum. Casting wild and anxious glances around him, he wheels from side to side, or darts forward a few paces and then as quickly retreats, with the head upraised and violently jerked backwards. Now and then, after much hesitation, and many attempts, he advances with a constrained rush to the other side of the room, and then begins to beat a forced retreat, jerking the head from side to side, slowly yielding ground, until at last, driven into a corner and thus secured in the rear, his rushing scrambling motions are for a time restrained, and he now contends, with remarkable vivacity and rapidity of motion, with those only of his fancied tormentors who approach from the front, and who, judging from the manner in which he squeezes himself into the corner with backward jerkings of the head, seem now to be actually flying in his face. These actions are accompanied by loud and rapid panting; the tongue is protruded, and saliva drops from the mouth. There is no indication of anger, nor indeed of fear: the dog is bent on escape, and resolute to effect it.

Obs. 40.—*Injected* ♏xx. *of Solution* 1 = 1 *grain* beneath the skin of the beagle (see *Obs.* 36, p. 152, &c.). Pulse 120; pupils $\frac{1}{6}$. He continued in his usual state until 20 minutes after the injection, when he began to move the head from side to side with an intent and animated look, as if under the influence of some optical illusion. He came up immediately when called. The pupils and pulse were unchanged; and as the above-mentioned symptoms passed off during the next few minutes, I doubted whether I ought to attribute any effect to the dose.

An hour after the first injection, I introduced ♏ xxxv. more of the solution = 1¾ *grain* beneath the skin. After 20 minutes, the symptoms first observed reappeared, and in such intensity as to leave no doubt of their reality. First he began to look intently and rapidly from side to side, then jumped suddenly round, and, after glancing eagerly about him, rushed forwards, but immediately stopped himself with a sudden jerk, and stood with the forelegs advanced and firmly set on the ground, ducking the head from side to side, now advancing with hesitating, constrained, and jerky motions; and thus the dog passed into the state above described, in which he continued *for* 1¼ *hour.* The intelligence was preserved throughout, and he noticed my calls by a slight wag of the tail and turn of the head. Such, however, was his agitation that I could only partially calm him by holding him firmly and caressing him, and so long as I continued to do so, he seemed relieved and comforted. As soon, however, as I loosed hold of him, he returned to his former state of wild excitement.

At the end of 2¼ *hours* from the first injection, he lay down for the first time, but still continued to pant. Pulse 160. Pupils widely dilated a few feet from the window.

During the next 40 *minutes* he lay still, and the panting gradually decreased. At the end of this time he seemed recovered. Tongue wet and clean. Pupils at light ¼″. Pulse

110 and of good power. He came up when called, and exhibited no more illusions or jerkings.

For the following hour he lay down quietly as if tired, but showed no tendency to sleep. At the end of this time, i.e. 4 hours after the first injection, he greedily devoured a plate of meat, and was afterwards lively, followed me about, and seemed quite recovered. During the action of the medicine he frequently passed a few drops of urine. There were no after-effects.

(*b*) After an interval of eleven weeks, I repeated the experiment upon this animal, injecting 2 grains of cryptopia dissolved in ♏xxx. of water by the aid of 2 drops of acetic acid. He continued quiet and evidently comfortable for 10 minutes, and then, erecting the ears, began to look attentively about him, now darting a glance under the table, and now fixing an eager look upon the carpet under his nose, as if he were watching a fly. Two minutes afterwards he was in full action as above described, moving the head rapidly from side to side, and swaying the body backwards and forwards upon the firmly set legs, in doubt whether to advance or retreat, and then at last wildly dashing forwards. 15 *minutes after the injection*, he began to pant, and during the next quarter of an hour continued in a state of the most frantic excitement, advancing and retreating from side to side, and wheeling round and round upon the hind legs, and at the same time retreating, now dashing recklessly against a piece of furniture, and now jumping on the tables and chairs to avoid his fancied pursuers. He was too much excited to pay any attention to my calls, but once, having lodged his rump firmly against my legs, he was a little quieted so long as I patted him. Soon, however, his excitement became extreme; the eyes were wild, the panting far too rapid to count, and the protruded tongue dropping saliva. When the excitement had attained its maximum, and as he was jerking the body to and fro as if doubtful where to rush, he was suddenly plucked backwards, and fell upon the right side in a strong tetanic convulsion with emprosthotonos, gasping for breath. So severe and prolonged was the spasm that I feared he would not survive it. It ceased in 15 seconds, and was followed by struggling, and it was a minute before the dog recovered his legs; and then for the next five minutes he was pulled about the room and rolled over and over in the most violent contortions. As soon as he had regained his legs, he was urged forwards, the hinder limbs being extended powerless behind him; then the body was quickly drawn up on the haunches, and the head and spine strongly curved forwards; after a few seconds, they were slowly but forcibly extended and curved in the opposite direction, and the animal was suddenly thrown over on the back. After a few struggles he recovered the legs again, and then went through the same evolutions as before, three or four times, after which these violent spasms ceased, and the animal stood still, panting 300 or 400 times a minute and slobbering. On catching my eye he greeted me for the first time—50 *minutes after the injection*—with a hearty wag of the tail, as if to say, 'I am recovered now.' A few minutes afterwards he vomited a quantity of undigested food and a little very viscid mucus. *For the next three hours* he continued quiet, but still under the influence of illusions of the sight. And having thus suddenly lost all hurry and excitement of motion, as he sat timid and nervous upon the haunches, or walked slowly about casting prying and fearful glances from side to side, occasionally panting a little and starting aside, his condition now greatly resembled that of a person in *delirium tremens*.

Three hours after the injection the illusions had passed away, and the dog seemed quite comfortable and in his usual condition. *During the next* 3 *hours* he was very quiet and dozed occasionally. He was unusually quiet throughout the next day. Urine and fæces were passed immediately before the injection, but none for 18 hours afterwards. At the 24th hour he vomited a second time, rejecting a little mucus. There were no other after-effects.

At the time of his greatest excitement the pupils were widely dilated, excepting in the direct light of day, when they contracted to $\frac{1}{4}''$. The heart's action was strong and rapid, but could not be counted for the panting. $2\frac{1}{2}$ hours after the injection the pulse was 140, regular, and of natural power; the respiration 48; and the skin, and particularly the ears, very warm.

On the Mouse.—Obs. 41. Injected ♏ v. of an acetous solution = $\frac{1}{16}$ *grain* beneath the skin of the mouse used in *Obs.* 37, p. 152, &c. He continued lively and apparently comfortable for 4 minutes, and then manifested great drowsiness, and closed the eyes. Respiration 160, regular.

After 1 *hour.* Had continued very sleepy, but occasionally awoke and pried about, and was now supporting himself on the haunches, the forepaws and snout near the ground, and the eyes nearly closed. Respiration 150, regular; and when called he put back the ears and worked the vibrissæ. Was easily disturbed, and then seemed quite awake for a short time, but soon fell off to sleep again. He continued in this condition—sleeping comfortably and composedly for an hour or two at a time, and then waking up for a few minutes and appearing moderately active—*for the next* 8 *hours.* He occasionally picked up a little food and nibbled it, but fell asleep as he held it in the paws. The somnolency continued during the greater part of the next day, after which he became as lively and active as usual. *At the* 5*th hour* the respiration was 140, regular.

Obs. 42.—Injected beneath the skin of another vigorous mouse ♏ v. of an acetous solution = $\frac{1}{10}$ of a grain. After a few minutes the breathing became panting, and between 300 and 400 a minute: and for the next half-hour the little animal appeared to be in a state of nervous excitement resembling that of the first stage in the dog, looking from side to side, and moving the body round and round, or sitting upon the haunches in a state of great attention, pricking the ears and turning the head at the slightest noise.

After 50 *minutes* the panting ceased. Respiration 200; and a few minutes afterwards he became dull and dozy. Respiration 160. The somnolency gradually increased during the next hour, and the respiration became slower and shallow.

After 2½ *hours* the respiration had fallen to 110, and was very shallow, and he was so torpid as to allow me to take him up. On pinching the skin he awoke and struggled, and on putting him down he ran a few paces pretty nimbly.

After 2½ *hours*, he had recovered from the deep sleep; the respiration was increased to 160, and deeper, and he appeared recovered. Moderate somnolency, however, continued throughout this and the two following days, during which time the respiration was 120. *At the* 8*th hour* he took a little food, but was too sleepy to eat more than a crumb or two.

Obs. 43.—Injected beneath the skin of another active adult female mouse ♏ v. of an acetous solution = $\frac{1}{24}$ *grain. After* 7 *minutes* she showed a little more excitement of movement than usual, and passed a little urine. Respiration excited, between 250 and 200. After this became quiet and a little dull, and allowed me to catch her without trouble ¾ *hour after* the first injection, when I *repeated the dose.* Drowsiness came on in a few seconds. After a minute she couched and closed the eyes, and the breathing, which before the second injection had fallen to the normal rate 160, was now 240, short and panting. The mouse continued to sleep soundly, but noticing calls, *for the next two hours*, and the breathing meanwhile gradually fell to 120, and became very shallow. At the end of this time the pulse was 280, the pupils unchanged, the temperature of the body very low. *Shortly afterwards* the breathing rapidly decreased and ceased, and the animal died without a movement, the pupils dilating as soon as the breathing stopped. 2½ *hours after* the first injection, and two minutes after the breathing had ceased, the body was opened. The lungs were of a salmon colour, and collapsed above and behind the heart. The right heart was distended with black fluid blood, the auricle contracting 120, and the ventricle 78 times a minute. The gall and urinary bladders were full.

Obs. 44.—$\frac{1}{30}$ of a grain in a small but active mouse was followed by rapid diminution of the respiratory movements, and death without movement 15 minutes after the injection.

The body was opened *two minutes after death.* The lungs were bloodless and collapsed. The left ventricle was contracting strongly and regularly 36 times, and the auricle 26 times a minute; two minutes afterwards they were empty and contracted, and had ceased to move. At first no motions were observed in the right ventricle, and the auricle was contracting faintly. On relieving it of a little blood, it contracted forcibly and regularly 58 times a minute, and an occasional contraction was revived in the right ventricle.

Twenty minutes after death, the right auricle was contracting regularly 68, and the right ventricle 14 times a minute.

Twenty-four minutes after death the contractions of the ventricle ceased. Contractions of the auricle could be excited *eighteen minutes later on.*

There was no undue vascularity of the brain.

On Man.—Obs. 45.—Subject, Samuel M. (see *Obs.* 38, p. 154). Pulse 72. Respiration 19.

(*a*) INJECTED ♏ XX. SOL. 2 = 1 GRAIN.

After 1 *hour.* Pulse 70. Respiration 19. Somnolency came on soon after the injection and continued, but had not yet amounted to sleep.

After 2 *hours.* Pulse 65, regular, of unchanged volume and power. Respiration 17, regular. Dozed continuously during the last hour without dreaming, and still felt very sleepy and comfortable. The tongue and pupils unchanged. Went home and continued very drowsy, retired to bed and slept soundly all night, and was, as usual after his opiates, sleepy throughout the next day. Urine free.

(*b*) INJECTED ♏ XXV. SOL. 2 = 1¼ GRAIN, six weeks after the previous dose. Pulse 74. Respiration 19–20. Pupil $\frac{1}{9}''$. Sleepiness came on in two minutes, and soon increased with yawning, so that in 10 minutes he 'could hardly keep his eyes open.' *After* ½ *an hour.* Pulse 74. Respiration 21.

After ¾ *hour.* Pulse 68. Respiration 19. Pupils and tongue unchanged. Continued very sleepy and looked very heavy.

After 1½ *hour.* Pulse 66. Respiration 19. Pupils $\frac{1}{8}''$. Continued very sleepy, and dozed most of the time.

After 2¼ *hours.* Pulse 66. Respiration 20. Pupils $\frac{1}{7}''$.

Had dozed and slept, and still continued heavy and drowsy, but the effect was decreasing.

Felt very comfortable and was very tranquil throughout. There was no dreaming nor feeling of stimulation. The pulse continued regular, and experienced no change in power or volume.

The effects were hypnosis, slight acceleration of the breathing, and dilatation of the pupils (see p. 185).

This patient experiences the hypnotic action of opium to the exclusion of all disagreeable effects.

Obs. 46.—Subject, John L. (see *Obs.* 34, p. 150). Pulse 72. Respiration 18.

(*a*) INJECTED ♏v. SOL. 2 = ¼ GRAIN.

After ½ *hour.* Pulse 70. Respiration 16. No somnolency.

After 1½ *hour.* Pulse 66. Respiration 20. Had dozed off twice.

After 2 *hours.* Pulse 66. Respiration 16. Had continued sleepy, and slept lightly for 10 minutes. Somnolency was the only effect, and this without any appearance of heaviness. The force and volume of the pulse, and the pupils, were unchanged throughout. He now walked home, and had half an hour's pleasant, dreamless sleep after arriving there. Dysuria followed.

(*b*) INJECTED ♏xx. SOL. 2 = 1 GRAIN, by two punctures, one in each arm, four days after the previous dose. Respiration 19. Pulse 78.

After 1 *hour.* Pulse 68, unchanged. Respiration 15. Had been feeling very comfortable, but not sleepy.

After 2 *hours.* Pulse 61, regular, a little fuller and stronger. Respiration 12. No change in tongue or pupils throughout. No feeling of sleepiness. Feels as if he had had a glass of spirits, with a diffused sensation of warmth. The face was cool, and there was no flushing or any change of appearance. Now walked home, and on his journey felt decidedly giddy for ¼ of an hour. Being bed-time, he soon retired to rest, and slept well without dreaming. In the morning he required the use of the catheter.

The dose did not produce the slightest somnolency on this

occasion. There were no after-effects. He has some chronic disorder which produces occasional retention of urine.

(*c*) INJECTED ♏xxiv. SOL. 2=1¼ GRAIN, a week after the previous dose. Pulse 66. Respiration 17.

After 2 *hours* 20 *minutes.* Pulse 64, regular, of good volume and power. Respiration 14. Had experienced a little drowsiness, passing into a doze for a few minutes at a time, and felt very calm and comfortable. Tongue and pupils unchanged. Skin cool; no outward appearance of any effect. Now walked home.

Narceine and cryptopia agree in having a very slight effect upon this patient.

Obs. 47.—Robert D., aged 16. Rheumatic fever. Insomnia, nausea. Vomited 5 hours previously. Pulse 88. Respiration 18-19. Pupils ⅛″.

(*a*) INJECTED ♏v. SOL. 2=¼ GRAIN.

After ¼ *hour.* Pulse 86. Respiration 20-21. No effect.

After 1 *hour.* Pulse 86, unchanged. Respiration 18. A little somnolency, and felt very comfortable.

After 2 *hours.* Pulse 77, not so full. Respiration 17. Pupils unchanged. Had continued dozy and comfortable. Obtained a short sleep afterwards. There were no after-effects.

(*b*) INJECTED ♏x. SOL. 1=¼ GRAIN, three days afterwards. Pulse 104. Respiration 17.

After ½ *an hour.* Pulse 91, regular. Respiration 18. Felt easy and comfortable.

After 1 *hour.* Pulse 85, regular, better volume and power. Respiration 17, shallow. Pupils unchanged. Tongue quite moist. Felt comfortable, but not sleepy. He obtained a few light dozes subsequently.

A tendency to sickness was present in this case, but it was not increased by the cryptopia.

Obs. 48.—Charles G., aged 30. Rheumatic fever. Pulse 126. Respiration 26.

INJECTED ♏v. SOL. 1=⅛ GRAIN.

After 1 *hour.* Pulse 118, regular, full, and bounding as before. Respiration 24. Pupils and tongue unchanged.

No somnolency or other effect. He obtained two hours' light sleep subsequently.

From the preceding observations it appears :—

1. That cryptopia, like morphia, has two distinct effects, a hypnotic, and an excitant of a most remarkable and exceptional kind, dependent partly upon an illusion of vision, and partly upon a tendency to convulsive action.

The transition from intellectual excitement to tetanic spasm in *Obs.* 40 *b.* was most unexpected; and since no tendency to involuntary action was observed in any other case—not even in the susceptible mouse—I am led to conclude that it was a consequence of strong mental excitement rather than the direct effect of the cryptopia upon the motor centres. In seeking an explanation of this phenomenon, I am reminded of having been once summoned to a man who had committed instantaneous suicide by cutting his throat; and, on entering the house, my first charge was a powerful middle-aged woman, who, upon sight of the deed, fell down before me in a severe epileptiform fit, the first and only attack of the kind she had had in her life. Whether the illusions of the dog were of a nature to produce such an effect is, of course, impossible to determine, but the convulsion, when it came, certainly appeared an appropriate climax to the intense mental excitement under which he laboured.

Apart from the inference derived from critical examination of the cryptopia used, other experiments upon this animal and upon the mouse (see THEBAIA) will show that the tetanising effect could not be due to contamination with thebaia.

Further acquaintance with cryptopia will perhaps more positively indicate that, as happens with morphia, individuals are affected in opposite ways by it, some being influenced entirely by its excitant action, while others are as exclusively affected by its hypnotic effects. Until this shall have been more clearly determined, we must be careful, I think, to avoid the use of cryptopia in those persons who are unpleasantly affected by morphia. I have as yet been unable, after careful inquiries, to detect the slightest tendency to illusory or other excitant effects in man.

2. The hypnotic effect is both considerable and protracted in those who are readily calmed by morphia. It is twice as active as meconine and narceine, and one-fourth as powerful as morphia.

3. Although no unpleasant effects have followed its use in man, further experience is required to show that, as a hypnotic, it possesses any advantage over morphia.

4. Its action upon the respiratory function is first stimulant, but subsequently depressent, and death is the direct consequence of this depressent effect.

5. The effect on the action of the heart is regulated by that on the breathing, and is therefore indirect. When the breathing is greatly accelerated, as in the dog, the pulse is proportionately stimulated; and when it is depressed, the pulse is lowered. Independently of the breathing, the heart is unaffected by the action of the drug.

6. In large doses cryptopia causes dilatation of the pupils.

V. CODEIA or CODEINE.—Opium contains from ¼ to nearly 1 per cent. of this alkaloid. That used in the following observations was manufactured by Messrs. Morson and Son. These are its

Characters. — Fine rhombic colourless friable crystals. Soluble in 25 parts of boiling water, and in twice that quantity at a temperature of 60° Fahr.; in less than 2 parts of either alcohol or chloroform; but requiring 40 parts of æther. Separates from water and alcohol in rhombic octahedra and prisms as perfect, brilliant, and refractive as those of triple phosphate. Similar crystals are also thrown down from a saturated alcoholic solution by the copious addition of water. If a large surface be exposed, the codeia separates from chloroform as a brilliant hyaline colourless varnish; and from æther under the same circumstances, partly in the same state, and partly in the form of bold crystals. More slowly evaporated, the deposit from æther is wholly crystalline.

The hydrochlorate crystallises in slender four-sided prisms. Both the alkaloid and this salt remain colourless in solution of perchloride of iron.

Disappears silently in monohydrated nitric acid, forming a pale greenish-orange-coloured solution. When the mixture is strongly heated, deutoxide of nitrogen is evolved with slight effervescence.

Soluble in warm sulphuric acid without change of colour, but when strongly heated the solution assumes a faint brownish tinge, rapidly passing, as the temperature increases, into a neutral tint, and then becoming thick and black and evolving sulphurous acid with slight effervescence. Now mixed with 100 times its bulk of water, a voluminous bluish-black deposit is formed, which becomes dingy green when the mixture is supersaturated with ammonia or potash. When the codeia is heated with a mixture of monohydrated nitric and sulphuric acids, a bright pale orange-coloured solution, like that from the action of nitric acid alone, results; and on strongly heating it, red fumes continue to be quietly evolved, while the clear mixture loses colour.

Physiological Action.—The statements of the early observers[1] respecting the action of codeia are so much at variance, arising probably from the imperfect separation of the various principles of opium, that little reliance can be placed on them. Had these experimenters been careful to identify the substances they employed, their observations would have lost nothing by lapse of time; but as it is we find it necessary to go over the ground afresh in order to satisfy ourselves of the action of a medicine which was discovered about thirty-four years ago. The only modern observer I am acquainted with, to whom I can apply for help, is M. Bernard; but a statement which I have already quoted (see p. 141) engenders a doubt in my mind whether the codeia employed by him corresponds with that I have described above.

His observations[2] on the action of codeia are as follows:—Five centigrammes = 0·77 of a grain of hydrochlorate of codeine, injected beneath the skin of a young dog of moderate size, is sufficient to send it to sleep, but the sleep is not so

[1] See Kunkel, Barbier, and Gregory, Journ. de Chim. Méd. vols. ix. and x., and Magendie, Formulaire, 8me éd. p. 87.

[2] Bulletin général de Thérapeutique, t. lxvii.

profound as that caused by an equal or a smaller dose of morphia, and the animal is easily awakened by pinching the extremities or by noise. It is especially on waking that the effects of codeine are distinguished from those of morphine. Animals to whom an equal dose of codeine has been given wake up without fright and without paralysis of the hinder extremities, and in their natural temper; neither do they present the intellectual disturbances which follow the use of morphia. He then relates an experiment in which 0·77 grn. of hydrochlorate of morphia and hydrochlorate of codeia were respectively administered to two dogs, 'et ils dormirent tranquilles à peu près trois ou quatre heures' (pp. 195, 196).

As to poisonous properties, codeine comes next to thebaine, and is much more dangerous than morphia. The contrary opinion exists among physicians, who prescribe codeine in larger doses than morphia.

Injected into the veins, hydrochlorate of codeine is much more poisonous than hydrochlorate of morphia. Nearly 30 grains of the latter so used, failed to cause death (p. 199).

The following observations on the dog do not accord with these statements. The bitch employed in *Obs.* 5 was so susceptible of the hypnotic action of morphia that half a grain invariably induced prolonged narcotism, and yet double and quadruple the quantity of codeia failed to produce decided drowsiness at any time.

In order to obtain a decided hypnotic effect, it will be seen that it is necessary to give codeia in much larger doses than morphia; but in man these are far too small to produce toxical effects. The fact, therefore, that a smaller quantity of codeia is required to destroy life is of no real importance in reference to its medicinal use.

I now proceed to detail the results of my own observations:—

On the Dog.—Obs. 49.—(*a*) Injected 1·07 *grain*, dissolved in a little water faintly acidulated with hydrochloric acid, beneath the skin of the bitch (see p. 106). Pulse 120. Pupils $\frac{1}{10}$″ at light.

After twenty minutes the eyelids were red and swollen, and she rubbed them with the forepaws.

After ¾ of an hour. Pulse 80, regular. Respiration 18, occasionally panting. Nose dry; swelling and redness of the eyelids nearly gone.

After 1¼ hour. Pulse 64, regular, of good volume and power. Respiration 18. Experienced nausea half an hour before, and at this time there was a rumbling of air in the intestines.

After 2 hours. Pulse 72, regular. Respiration 18, regular; continued to feel a little nausea.

After 3 hours. Pulse 104, having the respiratory character described at p. 107. Respiration 18, occasionally interrupted by a sigh. Much intestinal rumbling.

After 6 hours. Pulse 76, accelerated 3 or 4 beats by every long inspiration.

She remained very quiet and timid, as if feeling poorly, until the *9th hour*, when she had returned to her usual condition. As yet no excretions had been passed.

The effect of the medicine was quieting. For the first three hours there was nausea, and during part of this time a little tendency to sleep, but she did not close her eyes throughout. The pupils were unaffected, the secretions of the mouth increased.

(*b*) Five weeks afterwards 2 *grains*, dissolved in a little alcohol and water, were injected.

After ½ hour the pulse was 60, and regular.

After 1½ hour. Pulse 72. Respiration 16; both natural. The animal was watched for the next ten hours, and her general condition throughout was that described above, but there was no apparent nausea, nor any tendency to sleep. In neither experiment was there the slightest indication of excitement, either of the mental faculties or of motion.

Obs. 50.—Injected beneath the skin of the beagle (see p. 159, &c.) 3 *grains* of codeia dissolved in ♏xxxv. alcohol and water.

At the *end of an hour* there was no change. I therefore injected 3 *grains more* in ♏xl. of alcohol and water. Within a few minutes he began to whine and often change posture, as if very uncomfortable. The bowels acted repeatedly.

Two hours after the first injection. Respiration irregular, wholly abdominal; the expiration consisting in a prolonged collapse of the abdominal muscles, and accompanied by a subdued whine ending with a slight grunt. Heart's action weak and irregular. Pulse 65, respiratory—two quick beats towards the end of each inspiration. Pupils contracted to $\frac{1}{8}''$ in a light which previously allowed of their dilatation to $\frac{1}{4}''$, but opened wide away from the light.

After 3½ *hours* the respiration had fallen to 8; but now there were slight thoracic movements. The pulse 84, very feeble—towards the end of expiration 60, at the end of inspiration about 120. Pupils natural. The dog lay down most of the time; he walked well, but feebly, and after leaving the room for a short while, he came up and greeted me on my return.

After 6 *hours.* Pulse 88, a little more regular. Respiration 12. Continued in the same restless whining state, not lying more than two or three minutes in the same place.

After 8 *hours.* Slowly recovering. Respiration 16, but still very shallow and almost wholly abdominal. Pulse 88, stronger and more regular, but still respiratory. Passed a little urine. He now walked downstairs slowly and with a little hesitation. Refused to eat. A few hours later on, he seemed pretty comfortable. He passed a very quiet night, and the next day was in his usual condition.

Distress, apparently from the thoracic derangement, and restlessness, were the prominent symptoms from beginning to end, and throughout there was no inclination for sleep. The intelligence was unaffected. The animal walked well, but feebly, and came up with a wag of the tail as often as I called him, but his uneasiness was such that he got up and changed his place every few minutes. As he lay down the flaccid legs were slightly jerked with faint spasms, and this was observed up to the 7th hour. At the 2nd hour the muscles of the chest occasionally quivered with a much stronger spasm.

Derangement of the vagus, chiefly manifested in depression of the respiratory function, and consequent depression of the action of the heart, and a latent tendency to convulsive action, represent the chief effects of codeia upon the dog.

On the Mouse.—Obs. 51. Injected $\frac{1}{20}$ of a grain of uncombined codeia beneath the skin of the mouse used in the experiment recorded at p. 9.

After 17 *minutes* the animal began to assume the contour of the hog; the tail was curled stiffly upwards, and the hind legs as stiffly extended backwards, the body being every now and then suddenly impelled forwards—the effect, so far, exactly resembling that of morphia (see p. 114).

During the next 28 *minutes* the forced motions occurred only occasionally, the animal remaining in the intervals in a fixed and cramped condition, the thoracic and cervical vertebræ being strongly arched, and the head and shoulders raised from the ground by the deep lumbar curve. As he thus couched on the depressed haunches, an arm was occasionally raised and shaken with rigid spasm.

The respiration meanwhile was very irregular, becoming faint and vibratile, and every few seconds almost imperceptible. The movements appeared to be almost exclusively diaphragmatic.

At the forty-sixth minute the forward impetus was converted into a spasmodic jerk, and without the slightest warning the mouse was advanced forwards by a succession of little jumps, the body being maintained meanwhile in a state of cramp.

At the forty-eighth minute the little animal was thrown perpendicularly upwards, and came down again upon the side in a strong convulsion, the spine, limbs, and tail being alternately flexed and extended.

As soon as these movements ceased the body and hind legs were rigidly extended in a straight line, and the respiratory movements became rapidly slower. The fit ceased after about five seconds, when the mouse regained the couching position and sat in an attitude of constrained expectancy; the respiration being 100, irregular and shallow, but yet laboured.

During the next eight minutes the animal preserved his intelligence and seemed free from distress, turning the head from side to side and sniffing as often as he came to a state of rest, for he still continued to be urged round and round by a succession of little jerks, the spine meanwhile, and also during the short intervals of rest, being maintained in the state of cramp described at p. 114, and there was in consequence the same stiff lagging action and backward extension of the hind legs.

At the fifty-sixth minute the animal was seized with a second convulsion beginning with emprosthotonos, the nose, tail, and legs being drawn together in vibratile spasm. This forward cramp relaxed in a second, and excepting the head, which continued to be bent strongly upon the chest, the mouse was now fully extended, and the hinder parts of the body drawn backwards by curvature of the lumbar portion of the spine. The muscles of the trunk meantime were strongly cramped, and the respiratory movements were suspended and the urine evacuated. The spasm having relaxed, the mouse recovered the legs, but no sooner had he done so than he was danced up and down, and at last thrown a foot perpendicularly upwards. Coming down upon the back, he rolled over on the side in another tetanic spasm, the body was extended as before, and the breathing was again arrested, and when the violent cramp gave way the animal was dead and flaccid.

The chest was opened 9 *minutes after death*, when the body was still flaccid. The lungs were bloodless and collapsed above and behind the heart; the right auricle alone exhibited a faint occasional contraction. The left heart was empty, pale, and contracted, the right was distended with black fluid blood, and the coronary veins were turgid. On relieving the auricle of a little blood, its contractions were revived, and ten minutes after the chest was opened they numbered 195 a minute, and were quite regular.

Twenty-five minutes later on, i.e. 44 minutes after the death of the animal, and when the blood had dried upon its surface, the auricle still contracted forcibly and pretty regularly 33 times a minute. The stomach was firmly contracted upon a little food.

Obs. 52.—Injected $\frac{1}{14}$ *of a grain* beneath the skin of a vigorous adult male mouse.

During *the first* 25 *minutes* he continued very active, chiefly endeavouring to jump out of the glass-shade in which he was confined. Respiration 160, but irregular.

At the 25*th minute,* and when he was sitting still, the cramping flexure of the spine, with extension of the hind legs and curvature of the tail, came on, and the animal was now suddenly jerked forwards every 10 or 15 seconds. The frequency of these impulses increased, and in the course of 2 or 3 minutes the mouse was carried round and round the limits of his cell by a succession of rapid jerks. This appeared to be unaccompanied by any particular distress, and when the motion ceased the animal looked about him and worked the vibrissæ. After a short interval the impulse again commenced; the head was jerked up, and the body thrown forwards and advanced in a state of rigid spasm, as if upon tightly stretched vibrating wires. Even the skin was implicated in the spasm, and the fur became rough. *At the* 30*th minute* he was thrown over upon the side in a strong convulsion; the phenomena described in the previous observation were repeated, and the breathing was finally arrested in a second tetanic spasm 37 *minutes after the injection.* The chest was opened 2 *minutes after death.* The lungs were bloodless and collapsed. The right heart greatly distended, the left contracted and nearly empty. Both auricles contracting, the left very faintly 50 times, and the right 200 times a minute. A slight contraction was occasionally observed at the base of the left ventricle.

After relieving the right auricle of a little blood, the contractions fell at once to 120 a minute, and became regular and much stronger. The stomach was contracted; the bladder empty and contracted; the vessels of the brain a little full.

On Man.—I have given codeia *by mouth* in the following form:—

℞ Codeiæ gr. xij.
Acidi hydrochlorici ♏ x.
Aquæ ℥iij. Solve.

in doses varying from 1 to 4 grains.

(*a*) In one patient, George W., aged 17, 2 *grains*, taken every other morning an hour before breakfast, caused slight contraction of the pupils, a feeling of stiffness in the integument over the forehead, and occasionally a troublesome itching of the skin, chiefly affecting the forehead and shoulders.

The same dose taken towards bedtime sometimes caused slight drowsiness, a contraction of the pupil from $\frac{1}{5}''$ to $\frac{1}{8}''$ and an acceleration of the pulse a few beats. The medicine had no marked influence on the epileptic fits for which it was taken.

(*b*) Another patient, William W., aged 19, took the same mixture in increasing doses. No effects followed until the dose reached 4 *grains.* Half an hour afterwards there was a sudden increase in the temperature of the skin, and, to use his own words, 'he broke out into a feverish heat.' The pulse was slightly accelerated, and the pupils contracted, and there was considerable giddiness, obliging him to sit down for an hour. After continuing about two hours, these symptoms passed off. On one occasion he took food at the end of three hours, but within an hour vomited it, and for some hours afterwards was troubled with rumbling and griping pain in the belly. On another occasion the symptoms were repeated, but the nausea did not end in vomiting, and it passed off after two loose stools.

There was no inclination for sleep on either occasion.

Given by the *subcutaneous tissue*, somnolency is a more prominent effect, but it only occurs in certain individuals.

Obs. 53.—Samuel M. (see *Obs.* 45, p. 164, &c.) Pulse 74. Pupils at light $\frac{1}{9}''$, aside $\frac{1}{6}''$.

(*a*) Injected 1 *grain of codeia* dissolved in ♏xxiv. of alcohol and water.

After 20 *minutes.* Pulse unchanged. Felt a little heavy.

After 1 *hour.* Pulse 68. Pupils $\frac{1}{10}''$, aside $\frac{1}{8}''$. Was drowsy.

After 2 *hours.* Pulse 64, unchanged. Pupils contracted as before. Somnolency passed off. A little heaviness continued for the following hour. There were no after-effects.

The somnolency on this occasion was insufficient to produce sleep.

(*b*) Injected at 7.45 P.M. $1\frac{1}{2}$ *grain* in equal parts of alcohol and water, in all ♏ xxij. Respiration 19. Urine acid, sp. gr. 1026·4. Took a chop and tea at 6 P.M. Skin sweating. Temperature 80° Fahr.

Somnolency came on in $\frac{1}{4}$ of an hour.

After 1 *hour.* Pulse accelerated 15 beats, fuller and

stronger. Pupils $\frac{1}{10}$, sideways $\frac{1}{8}$; was drowsy, but presented no appearance of heaviness.

After $2\frac{1}{4}$ *hours.* Pulse accelerated 10 beats of the same volume and power as before the injection. Respiration 18. Pupils as at first hour. Anterior part of the tongue and hard palate a little dry. Had been very drowsy during the last hour, and still continued heavy, gapish, and sleepy. The looks betokened heaviness, and the membranes of the eye were slightly injected. Skin moist and cool. No nausea. Passed, after a prolonged effort, nearly ℥ij. of urine sp. gr. 1024, otherwise resembling that passed before the injection. It is probable that the bladder was not completely emptied. Walked home, still feeling a little drowsy. Slept well. No after-effects.

(*c*) Injected 2 grains, dissolved in ♏ xxiv. of alcohol and water, into the right arm. Pulse 74. Tongue wet and clean.

After 1 *hour*, pulse accelerated 8 beats, unchanged. Pupils $\frac{1}{11}$, sideways $\frac{1}{8}$. Somnolency came on after a quarter of an hour, and continued, but as yet he had not slept.

After 2 *hours*, acceleration of pulse continued. It was now slightly increased in volume and power. Had slept for half an hour. Felt a little squeamish.

After 3 *hours*, pulse accelerated 4 beats, of the same volume and power as before the injection. Pupils contracted as before, the sclerotic a little injected. Anterior part of the tongue dry. The nausea continued for an hour, and was accompanied by general warmth and itching of the skin, especially between the shoulders. Still continued to feel drowsy. The somnolency passed off as he walked home. There were no after-effects.

The doses above mentioned were subsequently given in combination with atropia. (See Action of Opium and Belladonna.)

Obs. 54.—John F., aged 57, a strong ablebodied man, suffering from sciatica. Injected 1 grain, dissolved in ♏ x. water acidulated with acetic acid, by two punctures. The only effects were, an acceleration of the pulse 15 beats, with an increase in the volume and power, and a contraction of the pupil from $\frac{1}{7}$ to $\frac{1}{8}$. There was no somnolency during the $2\frac{1}{2}$ hours that he remained under observation.

Obs. 55.—Frederick C., aged 55, an ablebodied but weakly man. Pulse 72. Pupils $\frac{1}{9}$. Injected ♏ x. of the solution used in *Obs.* 54 = $\frac{1}{4}$ grain. After an hour the pulse was 66 unchanged, no appreciable contraction of the pupils. There was neither giddiness nor somnolency, nor did these or any other effects follow.

The $\frac{1}{12}$ *of a grain of acetate of morphia*, used in the same way, caused contraction of the pupil from $\frac{1}{9}$ to $\frac{1}{11}$, giddiness as if he had been drinking spirits, and a little somnolency; the pulse at the end of an hour was 62 and unchanged.

It appears, from the facts above stated, that the action of codeia on man closely agrees with that of morphia. Like this substance, it possesses hypnotic and excitant properties. In those who are susceptible of the hypnotic action of opium, it induces somnolency when given by the areolar tissue, in doses of 1 or 2 grains. The effects, however, are much more transient than those of the other somniferous principles of opium; this was very marked in the case of Samuel M., in whom the effects of morphia, narceine, &c. were always considerably prolonged. (See *Obs.* 53, p. 176, &c.)

The excitant properties of the drug are indicated in all, by the stimulant action upon the heart and motor centres, resulting in acceleration of the pulse, contraction of the pupils, and derangement of the vagus. Given by the skin, 2 grains are equivalent at most, and in those who are easily influenced by a soporific, to $\frac{1}{4}$ of a grain of morphia.

By the stomach larger doses are required, but it appears that these may be often objectionable, on account of their tendency to cause gastro-intestinal disturbance.

Codeia is prescribed in the French Codex, in the form of syrup, in doses varying from the 0·15 to the 0·6 of a grain. As a medicine it appears to have no advantage over morphia, but rather the contrary. It cannot, therefore, be recommended as a useful or desirable addition to our Materia Medica.

VI. THEBAIA or PARAMORPHIA.—For specimens of this alkaloid I am indebted to Messrs. Morson and Son, and Messrs. T. and H. Smith.

Characters. — Minute, colourless, rectangular plates or prisms, soluble in about 45 parts of alcohol. More soluble in æther, and still more soluble in chloroform, and separating from these solvents in silvery crystals of the original form. Forms with cold sulphuric acid a deep orange-red solution, soon losing the orange tinge, and becoming blood-red. Heated, the mixture rapidly decolorises, and then begins to assume a neutral tint, which deepens until the thick fluid is nearly black, as in the third stage of meconine and cryptopia. Thrown into water and treated with ammonia, a similar amorphous brown precipitate is formed (see p. 157).

Physiological action.—The tetanic action of this alkaloid is well known. The following will serve as illustrations:—

On the Dog.—*Obs.* 56.—(*a*) Injected 1 *grain*, dissolved in ♏xlviij. of alcohol, by two punctures beneath the skin of the brown bitch (see *Obs.* 49, p. 170).

Besides a little panting within the first hour, there were no appreciable effects. The pulse, pupils, and tongue remained unchanged. After 4½ hours, she hesitated to follow me down some steep stairs, and on taking her up I noticed that the muscles generally were rigid. She ran and jumped well, however, and took food at this time, and was afterwards naturally lively, and there were no after-effects.

(*b*) The next day I injected 2 *grains* dissolved in ♏xxiv. water by the aid of 2 drops of acetic acid. During the first half-hour there was a little panting at intervals.

After 35 *minutes.* No change in pulse or pupils. She lay quietly in the usual condition, jumped up when called, and wagged the tail, but walked rather stiffly with the hind-legs.

After 45 *minutes* she got up, but immediately fell over on the side in a convulsion, which instantly passed into rigid and universal spasm; the body being arched backwards, the chest and abdomen fixed, and the legs stiffly extended before and behind; the ears were erect, the eyelids drawn backwards, the eyeballs fixed, and the pupils unchanged;

the mouth was open, and the tongue protruded; the heart's action greatly accelerated, the systoles numbering between 200 and 300. Such was the intensity of the rigor that the stiffened and motionless body could be held straight out by one of the hind-legs.

At the end of a minute the action of the heart had nearly ceased, and the tongue was purple. After keeping up artificial respiration for six minutes, the pulse was restored full and strong, and numbering 200, and the tongue recovered its rosy hue; and at the end of this time the breathing was also restored, and consisted of rapid but regular panting movements. The animal appeared now to have recovered; but after two or three minutes, the whole body was again fixed with intense cramp, and the respiratory and all other movements were effaced, and again the heart had wellnigh stopped, and the tongue had become purple. The breathing was again restored by strong and rapid compressions of the chest-wall, and the dog was so far recovered as to be able to stand on the rigidly-extended legs, and to walk very stiffly a few paces; the respiration panting, as after a hard run, and the pulse 300.

In the course of 2 or 3 minutes she again fell over on the side, in another opisthotonous spasm. After keeping up artificial respiration for five minutes, the breathing was restored. The relaxation of the spasm was but of short duration; two or three rigid jerks of the limbs preceded another violent cramp like the former. The respiration was again suspended, and the pulsations of the heart became rapidly slower; the mouth was now closed and the jaws locked, and the chest so rigidly fixed that it was with difficulty that I could compress it, and cause a little air to pass through the nostrils. The chest became extremely collapsed, and the heart's action ceased to be felt; the eyes were widely open still, and the pupils dilated; then the spasm ceased, and left the lifeless body flaccid, $\frac{3}{4}$ of an hour after the administration of the poison. Excepting at the times when the breathing was arrested, consciousness was preserved, and in the intervals the sensibility was greatly increased: upon the approach of a finger the animal immediately closed the eyes, and the legs were instantly withdrawn from the lightest touch. The external temperature was 80° Fahr.; rigor mortis came on very soon, and was strong.

The body was not examined until five hours after death. The lungs were collapsed behind the heart, non-crepitant, and of the colour and consistence of fœtal lung. The right heart was greatly distended with very dark venous blood; the left heart was contracted, and nearly empty. The stomach, intestines, and bladder were empty and contracted.

Obs. 57.—♏ix. of the solution used in the last experiment, =¾ *of a grain*, was injected beneath the skin of the beagle (see *Obs.* 50, p. 171, &c.), the bowels and bladder having been freely relieved half an hour previously. *A quarter of an hour after the injection* he passed a large quantity of fæces again, together with urine, and immediately afterwards vomited, and then lay down and became quiet, occasionally hanging out the tongue and panting.

After half an hour, he seemed distressed and lay on one side, with quick panting breathing—the mouth and eyes closed, and the head thrown back; the disengaged hind-leg was occasionally twitched, and the toes and claws were frequently extended; now and then there was an audible attempt to swallow, as if there was some irritation of the œsophagus. When called he got up and came to me, and then went back and threw himself down with the muzzle on the floor, and as he thus lay the head was occasionally twitched from side to side. He was restless, and often changed posture. The breathing was panting, 120 and regular; the pulse 160 regular, rather weak.

After 1 *hour*, continued in the same state, but seemed rather sleepy, lying with the head raised up, and the eyes half-closed; the muscles of the neck, the eyelids, the nose, and the legs, one or other, were almost constantly twitched; and once the whole body was simultaneously twitched and raised a few inches from the ground with a sudden jerk. Respiration in a succession of pants, which gradually decreased in rapidity and force until again renewed. Still appeared dozy, but got up when anyone came into the room.

After 1¼ *hour*, appeared to have lost all distress, and

continued to lie down quietly. Respiration 24, regular. Pulse 104, regular, occasionally accelerated a beat at the end of inspiration. The muscles, and especially the half-closed eyelids, were still affected with slight twitches, and during inspiration the hind-legs and lower part of the abdomen were jerked inwards.

After $2\frac{1}{2}$ *hours* seemed quite recovered, and was moderately lively and playful. Pulse 120, regular. Nose a little dry. Lapped water.

After $3\frac{1}{2}$ *hours*, recovery perfect; was full of animation and ate a large quantity of meat voraciously. There were no after-effects.

On the Mouse.—*Obs.* 58.—A little of the acetous solution ($=\frac{1}{12}$ *of a grain*) was injected beneath the skin of an adult active mouse. The animal was dead and flaccid *three minutes after*, death resulting from arrest of the breathing, but without outward indication of convulsion or spasm. Rigor mortis was strong after a few minutes.

Obs. 59.—Injected a quantity of the same solution ($=\frac{1}{36}$ *of a grain*) beneath the skin of another full-sized vigorous mouse.

He continued, as usual, prying and walking about actively for *the next four minutes*, when he was seized with a violent rigor. The limbs were thrown into the most rapid vibratile action which it is possible to conceive, whereby the little animal was jumped upwards; and by the occasional but invisible contact of the extended toes with the ground, he was kept nearly half a minute suspended in the air. Then he fell over on the side, and while the head was forcibly bent downwards to the chest, the body was extended and fixed with cramp; the respiratory movements were arrested, and the eyelids closed. When the spasm relaxed the eyelids again separated, but the animal was flaccid and dead.

The chest was opened within two minutes, and, as in all the other animals killed by the constituents of opium, the lungs were collapsed and bloodless, the right heart distended with black blood, and the left empty and contracted. Contractions still affected the right cavities, and continued for some time.

It thus appears that thebaia acts almost exclusively on

the motor centres, inducing in them that highest degree of excitement which results in cramp, and which is only fatal to life because it arrests the respiratory movements. Directly, the poison has no action on the heart; and so long as the excitant action is moderate, only acceleration of the breathing and some distress from over-excitement of the vagus, together with general muscular twitches, result.

In *Obs.* 57 a hypnotic effect was certainly observable, but it was very faint.

Unlike the soporific constituents of opium, the action of thebaia is comparatively transient. The dog in *Obs.* 57 had completely recovered from its influence after 3½ hours; but the effect of an equivalent dose of morphia, meconine, &c., upon the same animal, continues for many hours. In this respect, as well as some others, thebaia agrees with codeia.

Having completed my examination of the active constituents of opium individually, and regardless for the time of more general considerations, deduced such conclusions respecting the action of each as seemed to be required by the facts, I will now devote a little space to the brief study of these substances collectively. We shall thus be able to judge how far the partial conclusions already formed accord with those which may be inferred from a more comprehensive view of the subject.

The constituents of opium resemble the members of a family group who, having no feature or character that is not common to the whole, are as individuals alone characterised by the manner in which these common features are blended, or by the prominence which some one obtains over the others—we fail to recognise in any one of them a character that is not common to all. In our survey of the action of these constituents we have been able to detect only two effects —a soporific, and an excitant. In morphia we have seen these associated, equally or disproportionately, according to the quality of the nervous system subjected to its action. We have concluded that narceine and meconine are devoid of any excitant action, and this is doubtless true of the facts adduced; but if our observations on morphia had been limited to doses which produce a soporific effect no greater than that which followed the largest doses of narceine and meconine given in those observations, we should have concluded, with as much reason, that morphia was devoid of excitant properties. In reference to narceine we must not lose sight of the fact that its characters, like those of silica, are masked by insolubility, and that it is on this account incapable of that rapidity and intensity of action which the more soluble constituents are free to exercise. The feebler action of these substances, indeed, appears to be directly proportionate to their insolubility. Note 1 of the Appendix refers to an observation which goes far to prove that when given in large doses, meconine is really capable of producing convulsive action. *Physiologically*, narceine and meconine may be regarded as insoluble forms of morphia.

At present, therefore, I think we must hesitate to conclude that narceine and meconine are devoid of an action which is so eminently characteristic of the other members of the group. Their hypnotic effect, moreover, is comparatively speaking so feeble, that if our observations had been confined to one class of individuals, we might have concluded that they were devoid of this property also.

Facts referred to below, respecting the action of cryptopia on the dog and mouse respectively, give still greater force to these considerations. Turning to thebaia, we find its soporific effect is so very slight, in comparison of its excitant action, as to be wholly disregarded by some observers. On comparing, however, *Obs.* 57, 29, and 37, we find that $\frac{3}{4}$ of a grain of thebaia exercises a more decided hypnotic action than 5 grains of narceine or 2 grains of meconine. Still, it is true that in this member of the group the common features are so unequally developed that one-half of them—the hypnotic action—is very easily overlooked.

Codeia is the natural associate of both morphia and thebaia, and occupying a mid-place between them, the three are linked together in a natural order of succession. As an excitant, codeia is stronger than morphia and weaker than thebaia; as a hypnotic, it is stronger than thebaia and weaker than morphia. This relationship receives full illustration in the action of the three substances on the mouse.

Cryptopia is in every way a remarkable body, and if we did not know its connections, we should hardly recognise them by its effects; and yet an attentive consideration of its action will show that it has no property which is not common to the other members of the group to which it belongs.

Contraction of the pupils is so constant and decided a result of the action of opium upon man, that the discovery of an active dilating power in one of its constituents is extremely interesting. I have shown that morphia dilates the pupil in the horse; and from my observations of its effect upon the sympathetic and spinal nervous systems, I have concluded that the contraction or dilatation of the pupil is a mere accidental result of the shorter or longer distance between the centre of the third nerve and that of the ganglion which supplies sympathetic fibres to the iris, and the iris itself. The dilating effect of cryptopia still further illustrates the insignificance of the contraction of the pupil as a mark of a difference of effect in the action of opium on the several parts of the nervous system. The individual in whom the dilating action of cryptopia was noticed, invariably experienced speedy and long-continued contraction of the pupil under the influence of either morphia or codeia. As a further illustration of this effect, and of the action of a still larger dose of cryptopia than that under which the phenomenon was first observed, I will add the following.

Obs. 60, on the same individual, Samuel M. A week after the last dose (see p. 164), 1½ *grain,* dissolved, by the aid of ♏ jss. acetic acid, in ♏ xxij. of water, was injected into the arm. Pulse 66. Respiration 18. Pupils at light $\frac{1}{9}$; sideways $\frac{1}{6}$. Somnolency came on at the fifth minute.

After ¾ *hour.* Pulse 64, unchanged. Respiration 19, regular. Pupils very slightly dilated. Had continued to feel very sleepy and comfortable. Yawned now and then.

After 1¼ *hour.* Pulse 64, unchanged. Respiration 18–19, regular. Pupils at light, $\frac{1}{7}$ nearly; sideways, $\frac{1}{6}$ nearly. Had been sleeping tranquilly, now and then muttering a little.

After 2 *hours,* was still sleepy, but the effect was passing off. Pulse 64, a trifle fuller and stronger (?). Respiration 20, regular. Pupils at light, $\frac{1}{6}$; sideways, $\frac{1}{4}$.

After 2¾ *hours*, somnolency nearly passed off. Pulse 60, regular, of the same volume and power as before the injection. Respiration 21, regular. Pupils remained dilated, as at 2 hours. Felt very comfortable, and still somewhat sleepy; there was no heat of skin, flushing of face, or injection of the conjunctiva throughout, nor a trace of twitching. A stimulant effect upon the respiratory movements was observable as on previous occasions. The hypnotic effect was equal to about ¼ of a grain of acetate of morphia administered the same way. Somnolency continued throughout the following day.

The excitant action of cryptopia is a feature of much interest. Generally, it serves to illustrate the connection which subsists between cerebral and motor excitement, between mere intellectual excitement and convulsion. I have concluded that the extraordinary excitement of motion (see *Obs.* 40, p 159), was the result of illusions; but there is much in the character of the actions that resembles chorea, and perhaps it would be nearer the truth to say that cryptopia produces general chorea, affecting both the intelligence and the movements.

There is a remarkable uniformity in the character of the mental excitement and that of the actions of the body. Without loss of intelligence the animal is wholly absorbed by his illusions, and is urged by them into movements which are so far voluntary, that when the illusions are dispelled, by diverting the attention, the movements are for the time arrested.

Illusions are a frequent result of the action of opium on an excitable brain, and the waking dream of the dog is a very closely-related phenomenon. At its height the mental excitement produced by cryptopia almost amounts to the positive delirium which in some individuals follows a large dose of morphia (see p. 102). The muscular movements, on the other hand, which at first are under the control of the animal, rise to such a degree of intensity that they overpass the limit of voluntary action, and become convulsive and even tetanic, when there is no difference between the action of cryptopia and thebaia. Cryptopia also teaches us another important fact—viz., that the same medicine may produce exactly opposite effects in different animals. The mouse, we have seen, is exquisitely sensitive to the excitant action, not only of thebaia and codeia, but of morphia itself, which indeed, in moderate doses, fails to produce any other effect. Yet, of all the constituents of opium, there is none which produces a more ready, decided, and simple hypnotic effect on this animal than cryptopia. I have given it to five different mice; three of the animals died—one after 15 minutes, a second after 2 hours, and a third after 2¼ hours—and two recovered. In only one of these was there any tendency to convulsive movement observed, and this was indicated by slight occasional twitches only of the head and tail during the last ten minutes of life.

To sum up briefly what has been said, two effects only are observable in the action of opium, and these are traceable, in various degrees of development, in each of its constituents. And, taken as a whole, these bodies form a natural series connected in the following order: morphia and its feeble associates meconine and narceine, cryptopia, codeia, and thebaia—the first and last member representing the common characters, hypnosis and convulsion in their highest degree of development.

The action of opium being thus reduced to its simplest terms—hypnosis and convulsion, we have yet to consider the connection which throughout we have seen to subsist between these two apparently opposite effects.

That which first strongly arrests the attention is the invariable association of the two; and, secondly, their apparent convertibility the one into the other; thirdly, we have seen in cryptopia that a variation of the one effect is attended by a corresponding variation in the other.

An attentive consideration of these facts will lead, I think, to the conclusion that the action of opium and each of its constituents is simple, and that the effects which flow from this action are simple or compound, according to the quality of the nervous system in different individuals; or in the same individual, according as one portion of the nervous system is more readily impressed than the other. What, then, is this simple action? One of intense stimulation of the whole nervous system, I believe. The maximum effect upon the brain is a paralysis of the conducting fibres between the organs of sense and the brain on the one hand, and between the ideational centres of the brain itself on the other.

The impressions excited in the intellectual part of the brain being, like the rapid intermittences of a powerful galvanic current, sufficiently powerful to paralyse the conductors, narcotism is the result. In lesser degrees of cerebral excitement currents are allowed to pass, but they are too feeble to excite the imagination, and hypnosis is the simple effect.

In some individuals the impulses are irregular, the whole brain is thrown into disorder, and every variety of stupor, delirium, and delusion may result.

Applying the same explanation to the effects of the drug upon the motor centres, *mutatis mutandis*, precisely the same phenomena result. In their maximum degree of excitement paralysis ensues, and accompanies the corresponding condition of the cerebrum—namely, narcotism.

A lesser degree of excitement is manifested in the convulsive twitches and cramp which affect the muscles of the trunk, and in the restlessness of the half-paralysed extremities; and this condition, it is to be observed, is associated with more or less active delirium or stupor, indicating a similar obstruction of the ideational conductors.

The weakest degree of excitement of all is in the tetanus produced by thebaia; the excitations produced by this substance are insufficient to paralyse or even obstruct the conductors, and they pass off to the most distant parts of the body as rapidly as they are generated. This condition of the motor centres is also attended by a corresponding condition of the brain, the intelligence being perfectly preserved, and the sensory impressions intensely acute—the animal shrinking from the slightest touch, or motion of the air.

We are thus unexpectedly brought face to face with the conclusion, that whereas all the constituents of opium are endowed with the same excitant action, the particular one which appears to possess the greatest share is that which is really possessed of least, and that its fatal effects are merely accidental, and the result of the greater facility with which the weaker impulses generated by its action are carried to the muscles of the chest in common with those of the rest of the body. If this be the true view of the action of thebaia, it should follow that morphia is its appropriate antidote, because, according to the theory, it should intensify its excitant effects to such a degree that the impressions could no longer pass to the muscles. The combined action of thebaia and morphia is illustrated in the following:—

Obs. 61.—Injected ¾ of a grain of the thebaia used in *Obs.* 57, p. 181, mixed with ½ a grain of acetate of morphia, and dissolved, by the aid of a drop of acetic acid, beneath the skin of the same beagle. Vomiting and defecation commenced after three minutes, and recurred at intervals, with retching and tenesmus for the next half-hour, until only frothy mucus from the stomach, and bright orange-coloured jellylike mucus from the bowels, were voided.

After 5 minutes, and at short intervals, there was slight panting, which increased during the next 24 minutes, when it finally ceased. The respirations were 48, gradually decreasing during the next hour to 32, and then, as the effects of the thebaia passed off, rapidly falling to 12. Up to the end of *the first hour* there were no perceptible thoracic movements; the breathing was irregular, and apparently wholly diaphragmatic, consisting sometimes of short easy pants of the flanks, but chiefly of slow, resisting, unequal, and irregular contractions, accompanied by a prolonged audible expiration, often attended by a faint subdued whine. The inspirations were equally laboured, and sometimes accompanied by a whistling through the nostrils. Muscular twitchings commenced at *the 24th minute*, and continued until 2½ hours after the injection, when—having previously decreased in degree and frequency—they ceased entirely. The respirations fell at the same time to 12 a minute, and the dog became more restless. The pulse throughout ranged from 70 to 80; it was weak and regular so long as the respirations were quick, but manifested the respiratory character as soon as they became slow.

With regard to the main point of the enquiry—the effect

of morphia in restraining the convulsive action of thebaia—it appeared, generally, that the action of the latter took place just as if no morphia had been given; the only difference between the effect in this experiment and that recorded at p. 181 being, that the somnolency during the first 2½ hours was greater on this last occasion. On a little closer observation, however, it will appear that the excitant effect on the respiration was considerably shortened, and the muscular twitchings were certainly less severe; they continued, however, as long in the one case as the other.

As to the influence of the thebaia on the action of the morphia, on referring to *Obs.* 13, p. 112, it will be seen that the same dose of morphia alone was followed by decided hypnotic effects. But in the present experiment there was certainly a stronger tendency to sleep during the 2½ hours that the influence of the thebaia remained; and but for the twitchings during the first part of the time, and the restlessness which increased as the effect of the thebaia wore off subsequently, the animal would have slept soundly in the intervals, when the eyes were closed, and the nodding head sunk from time to time upon the floor.

I conclude from these experiments that morphia does, to a slight degree, diminish the convulsive action of thebaia, and that thebaia, on the other hand, increases the hypnotic effect of morphia — both of which conclusions show that there is some truth in the theory which I have advanced above. I think, however, that the facts adduced prove, that this theory is insufficient to account altogether for the different effects, individually and relatively, of the several constituents of opium, and it will be necessary to seek some additional explanation. The best that I can offer is, that each constituent excites impressions of a certain degree of intensity and rapidity, and that these are to some extent independent of those excited by the others in the same nervous centres. But this question will be considered a little further on. To return to the general effects of opium.

In the stage of narcotism and paralysis, the spinal cord, as usual, escapes for a time, and only manifests a slowly increasing depression of function.

Slight hindrances to the transmission of impressions in the brain, or in the motor system generally, producing a little delirium, cramp, or twitching, are of little importance; but when the same difficulties affect the conductors of motor impressions to the respiratory apparatus, they become at once a source of extreme distress, or of actual danger (see p. 123).

All the thoracic disturbances which result from the use of opium may be traced to these effects upon the vagus, the phrenic and the other respiratory nerves. Their conductivity is deranged and diminished; strong impulses pass irregularly, and with temporary intermissions; and the muscular parts are maintained in a condition of slight cramp, or there is a continual alternation of partial paralysis and cramp—the intensity of the latter being low, so as to be easily overcome in a measure by the endeavour to relieve the urgent feelings of distress.

The respiratory function being thus directly depressed, blood accumulates in the right heart, and its contractions are enfeebled. Sufficient evidence has been adduced to show that the heart, after the prolonged action of opium, is merely in a condition of restraint, the stimulant action upon the sympathetic being sufficient, so long as there is no great impediment to the flow of blood through the lungs, to maintain it in a state of vigorous contraction. (See *Obs.* 4, pp. 103, 119, &c.)

Having now freely discussed the general action of opium, we may again be allowed to enquire how it is that morphia and other constituents of opium produce such opposite effects upon different individuals.

Nerve-force, if it be not mere electricity, is regulated by the same laws. The nerves, we know, are simply conductors, and conduction is effected by molecular changes, which doubtless vary in their rapidity and polar intensity in different individuals.

There are some in whom the molecular changes—or, in other words, the conduction—is extremely rapid. These people think and act with great rapidity; they are easily impressed, and as easily exhausted.

There are others, in whom the conduction is effected more slowly. These people are not easily impressed; they are capable of excitement as intense as that manifested in the former class, but its accession is more gradual; a greater force is required to produce an equal effect in the two classes of individuals.

In the latter class, stronger impressions are required to excite and sustain the molecular changes, while at the same time there is a greater capacity of endurance.

In the first class, morphia rapidly arouses these molecular changes, until the polarity becoming too intense, the intermittences are irregular, and the conductors become exhausted, as is manifested by delirium, cramp, or convulsive movements.

In the other class the stimulation is readily borne, and after moderate doses the individual sleeps tranquilly, and the rate and force of the pulse are increased.

If this view of the effects of morphia and the action of the nerves be correct, we have a ready explanation of the fact, that the same medicine produces convulsion in one individual, and tranquil sleep in another.

We have seen that cryptopia throws one animal into convulsions, and acts as a pure hypnotic to another; it is not unreasonable therefore to suppose that there may be nervous systems which are able, in like manner, to convert a large portion, if not all, of the impressions excited by thebaia into soporific effects.

The action of the several alkaloids of opium in combination with belladonna, and the treatment of poisoning by opium, will receive full consideration in Chapter VI.

ATROPA BELLADONNA

CHAPTER V.

Of late years, and since its active principle has come into general use, belladonna has engaged a large share of attention; and so much that is at variance has been written concerning its physiological action, that in treating of the subject there is risk of falling into controversy. In presenting a statement of my own observations, I shall avoid all unnecessary departure from matters of fact, and shall only refer to the conclusions of other investigators when I find my own observations at variance with them.

My labours have been directed to ascertain the effects of belladonna on the body in medicinal doses, with a view to its rational employment in disease. Its influence in poisonous doses will engage our attention only incidentally, or by way of illustration. I shall commence with a description of the effects of the medicine on the horse, the dog, and on man in increasing doses; in the second place, I shall endeavour to interpret the several phenomena presented during its action; and, thirdly, I shall try to apply the knowledge thus obtained to the relief of those morbid conditions to which it seems adapted.

Physiological Action.—In the following observations the same sample of sulphate of atropia was invariably employed. Its action is so fully illustrated that it is unnecessary to give its physical and chemical characters.

On the Horse.—The effects of belladonna on this animal chiefly consist in cardiac and mental excitement. The following observations illustrate the action of increasing doses of sulphate of atropia given by the subcutaneous tissue. The grey horse and brown colt referred to are the subjects of *Obs.* 1–3, p. 100 *et seq.*

Obs. 62.—Effects of $\frac{1}{12}$ *of a grain* (*a*) on the grey horse. Pulse 30 to 36.

After 35 *minutes* an acceleration of the pulse 10 beats, considerable dilatation of the pupils, slight dryness of the anterior part of the tongue.

After 1 *hour* the pulse, stronger and fuller, attained a maximum acceleration of 28 beats; only a very narrow margin of iris visible; tongue and mouth generally dryish. The pupils returned nearly to their original size fifteen minutes afterwards.

After 1¼ *hour* the maximum acceleration of the pulse and the dryness of the mouth still continued.

The symptoms had entirely passed off 2 *hours after* the dose, except an increase of the pulse 6 beats, which still existed at the *end of the third hour.*

(*b*) The effect on the pulse of the brown colt was slighter. Pulse 36 to 40.

At the end of 35 *minutes* the acceleration was 4 beats; at the *end of* 1 *hour,* 10 beats; and a few minutes afterwards the maximum 16 was reached, and continued for an hour. Dilatation of the pupils commenced at the thirty-fifth minute, attained the maximum at the end of an hour, but soon yielded; 1¼ *hour after* the injection, the pupils had nearly returned to their usual size.

At the *end of the third hour* the pulse was still accelerated 8 beats, but the other effects disappeared at the end of the second hour.

(*c*) In a chestnut pony (pulse 32), the maximum acceleration occurred *at the end of* 1¼ *hour,* and amounted to 26 beats. *At the end of* 2 *hours* the increase was 22 beats, and *at the end of* 3 *hours,* 14. The maximum dilatation of the pupils and dryness of the mouth were observed at the end of the first hour; twenty minutes later on, both these effects had disappeared; but an hour afterwards, the pupils were again slightly dilated, and the mouth was a little dry.

The general effect was calming, and affected all three animals alike, and they were dull and quiet. There was no previous restlessness, no subsequent sleepiness, nor any decided effect upon the excretions.

Obs. 63.—Effects of ⅛ *of a grain.*

(*a*) On the grey horse:—

After 17 *minutes.* An acceleration of the pulse 24 beats.

After 35 *minutes.* A maximum increase of 34 beats, the mouth dry, pupils fully dilated.

After 2½ *hours.* Pulse full and compressible, accelerated 20 beats; mouth quite moist; iridial margin not quite so narrow.

After 5½ *hours.* Pulse reduced to its normal rate (36), but fuller and softer than before the injection. Mouth clean and wet. Was eating.

(*b*) In the brown colt, at the times above stated, the acceleration was 18, 40, 20, and 12 beats respectively. *At* 5½ *hours* the only remaining effect was the increase of the pulse. As occurred with the other animal, the symptoms —complete dilatation of the pupils and dryness of the mouth—were fully developed 35 *minutes after* the injection. At 2½ *hours* the pupils had nearly returned to their ordinary dimensions.

(*c*) In a third animal the maximum acceleration of the pulse amounted to 44 beats. *At the end of* 5½ *hours* the only remaining effect was an increase of the pulse 12 beats, as in the colt.

The general effect upon these three animals was the same. From the tenth to the twenty-fifth minute there was slight restlessness, caused apparently by dryness of the mouth, for the tongue was frequently rolled about. As soon as the symptoms were fully developed, the animals became dull and quiet, and remained so for about four hours. Each animal staled copiously twice or thrice during the 5½ hours.

Obs. 64.—Effects of ¼ *of a grain* on the brown colt. Had just dunged, and the reaction of the intestinal mucus was neutral.

After 12 *minutes.* Pulse 56, having attained a maximum acceleration of 18 beats. Pupils slightly dilated. Gaped much, and rolled the tongue in an uneasy manner about the dry mouth.

After ½ an hour. Pulse still 56, slightly irregular. Pupils almost fully dilated. Tongue and lips completely dry; the latter glazed, the former rough and cracked.

At the end of an hour the pupils were fully dilated.

The maximum effects upon the pulse, pupils, and mouth continued unabated to the end of the *third hour.*

At the end of the sixth hour the pulse was still 56; the pupils were only moderately dilated, and the mouth was moist.

At the eighteenth hour the pulse was small and soft, and accelerated 4 beats. Had voided one large mass of dung, rather dark-coloured, and moister than natural from the presence of an increased quantity of mucus, the reaction of which was acid. There was one separate mass of this mucus the size of a florin. Appetite good. The general effect was quieting, and the animal remained still and dull.

Obs. 65.—Effects of ⅓ *a grain* on the grey horse. Pulse 30. Had just passed dung of an acid reaction.

At the twelfth minute the pupils suddenly and completely dilated as I was examining them; the pulse at the same time obtained a maximum acceleration of 42 beats; the mouth was drying.

After 20 *minutes* the pulse was soft and full, and four beats less (68). The mouth was only dryish.

After an hour the pulse was still 68, of diminished volume, and very soft; the lips, mouth, and tongue quite dry. Had gaped occasionally, and remained very quiet and dull.

After 3 *hours.* The pulse 68, and compressible; the mouth parched, and there was considerable nervousness.

After 6 *hours.* Pulse still 68, but so weak and compressible that it was counted with difficulty. Pupils fully dilated, and a bright red glare was reflected from the fundus of the eye; conjunctivæ of a bright crimson colour; mouth moistened with a creamy secretion; tongue white. For an hour previously the horse had been restless, and now there was a moderate amount of delirium; a touch made the muscles quiver, and he was startled by the least noise, and jerked the head backwards on moving, as if he misjudged distances;

and when startled he became restless. During *the next four hours* the effect gradually wore off. At the *eighteenth hour* the pulse was 36—six beats more than before the injection—very soft and weak. Mouth wet and healthy. Pupils contracted as before the injection. Had passed a softer and more copious stool than usual, and the reaction was rather strongly acid. Appetite good, but the animal was tired and gapish.

Urine was frequently voided during the action of this and the previous dose.

Obs. 66.—The effect of *two grains* on the grey horse. *After* 15 *minutes* pulse accelerated 35 beats, but weaker. Pupils fully dilated; mouth completely dry. Yawned and remained very quiet.

After 20 *minutes* the pulse attained its maximum acceleration of 37 beats; restlessness and nervousness set in, and the membranes of the eye were strongly injected.

After one hour the muscles were very tremulous, the restlessness was considerable, and the animal was evidently partially blind, and apparently under the influence of illusions; for he not only misjudged distances, but with the ears pricked forwards, and the head raised and retracted, he continually jerked it backwards.

After 1¾ *hour* there was occasional hiccup, twitching of the intercostal muscles, and quivering of the pannuscorium. The restlessness increased at intervals, and the animal started at the least sound or touch; but when the head was gently held, it gave him support and confidence, and he became quieter. Pulse as at 20 minutes. He continued in this condition for *the next fourteen hours*, towards the close of which time the restlessness and other symptoms abated, but the pupils remained fully dilated for the next twenty-four hours.

During the action of the medicine urine was frequently voided.

By the *mouth*, ℥j. = 480 grains of either the vacuum extract, the powdered leaves, or root; or 4 grains of sulphate of atropia given in a pint of water, produced no more effect than the $\frac{1}{12}$ of a grain of the latter introduced by the skin.

Conclusions.—In the horse, belladonna causes—

1. Powerful stimulation of the heart, the maximum effects following doses which are insufficient to produce nervousness. In large doses the maximum acceleration is often less, and the force of the heart's action and tone of the bloodvessels is always diminished.

2. The action of moderate doses on the cerebro-spinal nervous system results in a general quieting effect. Large doses cause undue sensibility to external impressions, a slight noise or a touch causing a start or a tremor; the hearing becomes excessively acute, vision is obscured, but the impressions conveyed by the eyes are acutely perceived.

Wakefulness and slight restlessness are the results of the prolonged action of large doses. Injection of the membranes of the brain, as indicated by the effect on the sclerotic and conjunctivæ, accompany this condition.

3. The urinary function is excited, and generally the quantity of urine is increased. The mucous secretions of the alimentary canal and the bile are slightly increased. The skin appears to be unaffected.

4. No appreciable effects on the respiration are observable.

On the Dog.—Intense stimulation of the heart, accompanied by dilatation of the pupil and dryness of the mouth, is the most prominent effect of the action of belladonna on this animal.

Obs. 67.—(*a*) Effects of $\frac{1}{96}$ *of a grain* on the bitch (see *Obs.* 56, p. 179, &c.). Pulse 120. Pupils $\frac{1}{8}''$ at the light. Secretion of the mouth and conjunctiva freely alkaline. Urine acid.

After 14 *minutes.* Pulse 300, regular and strong; cardiac systoles felt through the chest-walls with unusual distinctness. Pupils $\frac{1}{8}$, completely dilated towards darker side of room. Nose, lips, mouth, and tongue all dry; the little moisture on the floor of the mouth alkaline. *After half an hour,* pulse 260; *after an hour,* 230. Mouth not quite so dry; secretion alkaline. Voided a little alkaline urine. *After* 1¼ *hour,* pulse 168; mouth moist; begged for food, and ate it readily. *After* 2¼ *hours,* pulse 130, regular and strong, as at 14 minutes. Pupils retained their maximum dilatation of $\frac{1}{8}''$.

From the time of injection until after the effects had passed off, the playfulness and intelligence had not abated for a second, and an ordinary observer would have recognised no difference in the animal.

(*b*) $\frac{1}{48}$ *of a grain* was followed by the same effects, together with a slight diminution of activity and playfulness; and she continued, when left to herself, to walk slowly around the room, and examine every object in her way in a prying manner, as if things appeared somewhat different from usual to her. The acceleration of the pulse was not greater than that caused by the smaller dose; but the dryness of the mouth and the dilatation of the pupils were increased, and all the symptoms continued longer. At the end of *three hours* the pulse was 160, the pupils $\frac{1}{5}''$, and the mouth moist. *After* 3½ *hours*, pulse 136, pupils $\frac{1}{8}$, and the animal was quite playful.

(*c*) $\frac{1}{24}$ *of a grain* reproduced the same symptoms, and with no greater acceleration of the pulse than the $\frac{1}{96}$ of a grain. There was increased dulness and disinclination for exertion, a little tottering from weakness of the hind-legs, and an inclination for sleep. During the first 1½ hour urine was voided three times.

Obs. 68.—The effect of $\frac{1}{32}$ *of a grain* on the terrier (p. 111) differed very little from that of the $\frac{1}{24}$ on the bitch. *After* 2½ *hours* the pulse was 168. Respiration 28.

He slept lightly and comfortably between the *third and fourth hour*, at the end of which time the pulse was 150, the respiration 28. Mouth parched, and the pupils completely dilated.

Obs. 69.—(*a*) The effect of $\frac{1}{6}$ *of a grain* upon the bitch (urine alkaline) was as follows:—

After 7 *minutes*, pulse 224, regular. *After* 9 *minutes*, pupils completely dilated and fixed; nose, mouth, and tongue quite dry. *After* 27 *minutes*, had lost much of her playfulness, and began prying slowly about the room, and now and then stumbled from unsteadiness of the hind-legs. *After* 1 *hour*, continued in the same state, consciousness unimpaired; was partially blind and went about cautiously, walked awkwardly, and hesitated to jump off a chair. Pulse between 300 and 400; respiration 22, regular. She now lay down for the first time, and closed the eyes and dozed for ten minutes; and then got up and walked about as before, smelling her way rather than seeing it.

After 1½ *hour.* The heart beating with the noise and rapidity of a winnowing-machine, about 400 times a minute, each beat distinctly perceptible through the chest-wall. Walked stiffly and clumsily; when placed upon a shifting ground, such as a cushioned chair, she reeled very much, and could hardly keep her footing; and when called she walked straight into the air, and tumbled down. When called came in the direction of my voice, hesitatingly and with groping movement, as if advancing in the dark.

After 3 *hours*, pulse as before, strong, regular, 250; respiration 18, interrupted at intervals by a long-drawn sigh. *After* 5 *hours*, pulse 200, quite regular and strong. *After* 5½ *hours*, mouth moistening. *After* 6 *hours*, pulse still 200 to 190; mouth moist; had recovered some activity, and came up when called, but when placed on the table walked off into the air, and fell down. *After* 7 *hours* she took food, had recovered her activity and sight, and was quite frolicsome. The pulse was 160, the pupils still fully dilated and fixed, and the fundus reflected a green light. *After* 11 *hours*, pulse 140 to 150, regular, and of good power; pupils, before a gas-lamp, surrounded by a very narrow margin of iris. Was very lively and well. There were no after-effects.

The general effect was quieting; there was no distress at any time, but once she manifested a little irascibility on being disturbed. The motions were slow and free from excitement; she rarely lay down, and during the eleven hours took four naps, only one of which lasted fifteen minutes.

Urine was passed at 1¼ hour, 1½ hour, 2¼ hours after the injection. That voided at 1½ hour was acid; the others were alkaline, and deposited triple phosphate on standing a few hours. Two drops of that first voided dilated the pupil within an hour from $\frac{1}{10}$ to ¼″, and after eighteen hours it still dilated to ⅛. The other urines readily dilated the pupil.

(*b*) ¼ *of a grain* produced the same effects on the heart, the pupils, and mouth.

After half an hour she seemed almost blind. *After an hour* lay down for a short time, but did not sleep; and during the following eight hours she continued to pry about slowly and cautiously—often, with a clear space around her, jerking the head back as if she had unexpectedly come upon some object, and then actually walking against the wall or an article of furniture.

The intelligence was unaffected throughout; and when called she pricked the ears, wagged the tail, and began to walk slowly and hesitatingly in the direction of the voice; and on meeting her she was able, after repeated exertions, to raise herself upon the hind-legs, and put the paws on my knees. 3½ *hours after* the injection the pulse was 210; *at the eighth hour*, 120, when she had recovered the sight, was quite active and lively, and took a little food. During the twelve hours following the injection, she lay down only five or six times for a few minutes, and sometimes dozed, but did not sleep at any time five minutes continuously. Small quantities of urine were frequently passed. The next day she was as frolicsome as ever.

Such are the effects of belladonna on the dog. They correspond exactly with those which accompany its action on the horse, with these differences:—

1. In comparison of its size, the dog bears a much larger dose than the horse.
2. In the dog the influence on the heart is more strongly marked, while the cerebral effects are much less decided.

If the effect of ½ a grain on the horse and ¼ grain on the dog be compared, it will be seen that in the former case the pulse is little more than doubled, while in the latter it is trebled. The effect, however, is more prolonged in the horse. At the end of five hours the maximum acceleration, minus only 4 beats, was maintained; whereas in the dog, the pulse had decreased at least 150 from its maximum acceleration at the end of 3½ hours, when it was not quite double its usual rate. In the horse there was decided restlessness and nervous delirium, and the increased vascularity of the brain was further indicated by the injection of the membranes of the eye. In the dog the excitement amounted to mere disinclination to remain quiet, and there was no outward indication of any increased vascularity of the brain, the conjunctivæ remaining quite pale. Lastly, the action was not accompanied or followed by any indication of those symptoms of exhaustion, both of the heart and the brain, which were declared during the last hour or two of the action of the medicine and subsequently in the horse.

3. There were no appreciable effects upon either the respiration or the functions of the alimentary canal.

4. The effect on the kidneys was decidedly diuretic, and atropia continued to be eliminated with the urine as long as the action of the medicine continued. As in *Obs.* 69, the presence of the alkaloid can be readily proved by dropping a little of the urine into the eye. The following plan may be substituted :—Soak up the urine with sufficient bibulous paper to thoroughly damp it, crumple it together, and put it upon a filter and wash with a drachm or so of chloroform, which will carry out the atropia; let the chloroform evaporate, and then dissolve the remaining stain in two or three drops of water, and place one in the eye.

On Man.—(*a*) For *subcutaneous use*, a solution of 1 grain of sulphate of atropia in ℥j. of water is most suitable, as there is less liability of using too much than with a stronger solution; and the medium dose, ♏ v. to ♏ viij., is a convenient quantity.

Obs. 70.—If $\frac{1}{160}$ *to* $\frac{1}{120}$ *of a grain* = ♏ iij. to ♏ iv. of the solution, be injected beneath the skin of a person in health, we shall observe, after 10 or 20 minutes, an acceleration of the pulse, and generally a slight increase in volume and power. If the pulse was previously slow and feeble, or intermitting, the change will be very decided. The acceleration will generally amount to 20 beats a minute; it will take place suddenly, and attain its maximum within one or two minutes. After being maintained for half an hour, a gradual decline takes place, and the heart soon returns to its usual state, and continues to beat as quickly and powerfully as before. Just as the pulse rises, a slight giddiness is often perceptible. Usually these will be the whole of the symptoms; but in weak and delicate adults, a feeling of dryness of the mouth and throat, and, at the end of an hour or two, a slight dilatation of the pupil in a subdued light, will be superadded.

Obs. 71.—When $\frac{1}{96}$ *of a grain* = ♏ v. of the solution, is used, the acceleration of the pulse will usually amount to 25 beats; the anterior part of the tongue and hard palate will be generally dry; and about the tip of the tongue, the dryness will often be so complete as to render this part parched, rough, and brown. At the end of *two hours*, the pupils will have dilated from $\frac{1}{10}''$ to $\frac{1}{8}''$.

Obs. 72.—After the use of $\frac{1}{60}$ of a grain = ♏ viij. solution, the acceleration of the pulse will be found to range between 20 and 60 beats in different individuals. The rise is attended by considerable giddiness and waviness of the vision. The patient walks cautiously, and with an inclination to unsteadiness. *After* 20 *or* 40 *minutes*, he will complain, with some huskiness of voice, of great dryness of the throat and mouth; and the anterior part of the tongue, or the whole of the dorsum, excepting a wide margin, will be found dry, brown, and rough. The hard and, in many persons, the soft palate also will be perfectly dry and glazed. There will be more or less somnolency, and sometimes a little flushing of the face. The dilatation of the pupils will amount to $\frac{1}{7}''$ or $\frac{1}{6}''$, according as they measured the $\frac{1}{10}$ or $\frac{1}{8}$ previously.

Obs. 73.—The effects of $\frac{1}{48}$ *of a grain* = ♏ x. solution (a full medicinal dose) are as follows:—After 10 or 15 minutes, an acceleration of the pulse from 20 to 70 beats; no apparent change in volume, but a decided increase in the force of the cardiac contractions and of the arterial tone; a general diffusion of warmth, a slight throbbing or heaving sensation in the carotids, and a feeling of pressure under the parietal bones; giddiness, heaviness, drowsiness, or actual sleep, with a great tendency to dreamy delirium, and, in women, slight occasional startings; complete dryness of the tongue, roof of the mouth, and soft palate, extending more or less down the pharynx and larynx, rendering the voice husky, and often inducing dry cough and difficulty of deglutition; a parched state of the lips, occasional dryness of the mucous membranes of the nose and eye, and increasing dilatation of the pupils.

After continuing about *two hours*, the dryness of the mouth is suddenly relieved by the appearance of a viscid acid secretion of an offensive odour, like the sweat of the feet. The mouth becomes foul and clammy, and a bitter coppery taste is complained of.

As moisture thus returns to the mouth, the pulse is observed to fall, and it now rapidly resumes its ordinary rate and character. At this time, the pupils have reached their maximum dilatation, and measure about $\frac{1}{3}''$; but when exposed to the brightest light, they will still contract to $\frac{1}{4}$, $\frac{1}{6}$, or even $\frac{1}{8}''$, according to their original size.

During the action of the medicine there will be a slight elevation of the temperature of the surface, rarely exceeding 1°, and a still slighter and less appreciable rise of the internal temperature of the body. No difference will be observed in the rate of respiration, except (as may happen in a nervous woman) a little emotional excitement on the sudden accession of the giddiness.

The breathing will be as tranquil as before the injection. The patient occasionally heaves a deep sigh, and still oftener takes a prolonged yawn, as he sits still in a dull, apathetic, or drowsy condition.

After the pulse has resumed its ordinary rate, and the mouth has moistened, the giddiness and drowsiness pass off, and the patient appears tolerably lively and brisk in mind and body. But he will himself continue to feel for some hours longer such languor of body and mind as will render him disinclined for, or even incapable of, active bodily or mental exertion. A little dimness of vision also remains, and the patient is unable to thread a needle, or even to read.

Obs. 74.—The $\frac{1}{40}$ *of a grain* = ♏ xij. of the solution, reproduces, in the young and robust, the effects last described a little intensified and prolonged. The pulse, which before the injection numbered 64, will be raised in twenty minutes to 128, or just doubled, and an increase in volume and force will be still observable.

The mucous layer of the tongue will become completely dry, brown, and hard; the hard and soft palates, arches of the fauces and uvula, and back of the pharynx, dry and glazed, so that the moveable parts are wrinkled as often as the muscular tissue contracts. The rate of breathing remains the same as before the injection. The maximum acceleration of the pulse is sustained for about half an hour, during which time the rate of breathing is undisturbed; then the pulse begins slowly to decline, and at the end of $2\frac{1}{2}$ *hours* from the time of injection, it will have fallen to 74. The dryness persists to some extent for many hours; and if the patient sleeps, he is troubled by dreams, and at intervals disturbed by a start. A fancied noise is a common cause of awakening, and at these times the patient generally manifests a little delirium.

Obs. 75.—The $\frac{1}{32}$ of a grain = ♏ xv. of the solution, produces symptoms of much the same intensity as the $\frac{1}{40}$, but the cerebral effects will be slightly increased; and if the patient be weakened by disease, or be unusually susceptible of the action of the medicine, instead of sleep there will be a little meddlesome delirium, and he will require attention to prevent him from getting out of bed. He will have little or no inclination for sleep, and will probably be busily influenced by pleasing illusions and delusions, meddling with everything in his way, picking at and handling imaginary objects in the air, and accompanying his acts by muttering and smiling, or with loud chattering, interrupted by subdued laughter.

A person less susceptible of the action of the drug will get a very good amount of sleep, disturbed at intervals by vivid dreams.

I have occasionally injected the $\frac{1}{20}$ *of a grain* = ♏ xx. of the solution; and, agreeably with what I have observed after the use of doses larger than the $\frac{1}{30}$, the effect upon the pulse has been less apparent than after an ordinary full dose of the $\frac{1}{48}$ of a grain.

The following were the effects of $\frac{1}{20}$ *of a grain* = ♏ xx. of the solution :—

Obs. 76.—Samuel M. (see p. 176). Pulse 78. Respiration 20. Tongue clean and moist. Pupils $\frac{1}{9}''$.

After 20 *minutes*, pulse 110, unchanged in volume and power; respiration 20. The hard palate and the anterior part of the tongue and the soft palate dry, but not parched or glazed. Pupils $\frac{1}{7}$. Felt sleepy and a little giddy, but walked steadily. *After* 1 *hour*, pulse 108, unchanged. Respiration 20. Dorsum of tongue dry and parched, and the entire roof of the mouth and velum palati dry and glazed. Pupils $\frac{1}{6}$. Continued giddy, but walked steadily, though slowly and cautiously; was greatly inclined for sleep, but had not slept. Gaped very often, and said he should soon be asleep if he were in bed. There was no tendency to delirium, and perfect freedom from nervousness and restlessness. Skin naturally cool and moist.

After 2 *hours*, pulse 98, diminished in volume and power, but still quite regular. Respiration 20, regular. Mouth as parched as before; voice husky. Pupils between $\frac{1}{6}$ and $\frac{1}{5}$; slight injection of the conjunctival membrane. Had slept, and continued very drowsy. Was entirely free from nervous symptoms, and stated that he was quite able to walk home —a distance of a mile. He did so at this time; but when he reached the house he could not put the key into the door, 'because he felt so stupid and shaky in the hand,' and had to seek assistance. Went to bed, and slept heavily all night. The throat and mouth were very dry in the morning; but this passed off after breakfast, and there were no after-effects.

Such are the phenomena which attend the action of belladonna when given in medicinal doses; and it will never be necessary, I believe, to induce symptoms of greater intensity than those I have described. If larger doses be given, there will be superadded a distressing fluttering sensation in the cardiac region, slight delirium (such as I have described under *Obs.* 75), exquisite sensibility of hearing, and frequent illusions of this sense also; staggering, or complete inability to walk; insomnia, restlessness, and frequently great nervous agitation of mind and body (see below). As far as my observations extend, *nausea* and *headache*, either during the action of the medicine or afterwards, are rare and exceptional consequences of the *subcutaneous* use of atropia.

The desire for food returns soon after the operation of the medicine; but during its action, insalivation and deglutition are almost or altogether impossible.

Dysuria, or more or less complete retention of urine for two or three hours, or longer, invariably follows the action of a full medicinal dose of the drug.

(*b*) *Administration by the Alimentary Canal.*—Precisely the same symptoms, occurring in the same order, follow the administration of belladonna, or its active principle by the alimentary canal—the stomach or rectum. But when given by the stomach in full doses, nausea and headache sometimes follow.

Obs. 77.—The operation of ♏ xxx. *of a Succus Belladonnæ*, prepared by Messrs. J. Bell and Co., was noted in six adults male and female. Belladonna action was fully developed in all within an hour. The pulse in one was accelerated only 10 beats; in another 20 beats; in a third 26; in two others 40 beats; and in the sixth (a youth of twenty), the cardiac systoles were more than doubled, the pulse rising from 60 to 140 beats.

Associated with this acceleration of the pulse, the other effects of belladonna were well developed; but in none of the patients was there any observable increase in the *respiratory movements*. The individual in whom the acceleration of the pulse amounted to 80 beats did not, throughout the 45 minutes during which the maximum acceleration continued, outwardly manifest or express the slightest excitement. The respirations never exceeded 18, and at the time when the cardiac excitement first reached its acme, and afterwards, the inspirations numbered 15 or 16, and were natural and easy.

In order to complete the view of the action of belladonna on man, I cite the following two cases:—

(*a*) A woman, aged 30, swallowed with suicidal intent a spirituous solution of atropia = 1⅓ *grain of the sulphate.*

After ½ *hour* she experienced nausea and dimness of sight; was unable to feel the arms and legs, and being alarmed called for help.

After 1 *hour*, pulse 150, small and weak; face was flushed; conjunctivæ vividly injected; only a circular line of iris visible; was nearly blind, a thick cloud obstructed vision, and images were confused with a reddish tinge; there was tendency to sleep, chilliness, and cramp, and tingling of the extremities.

After 2½ *hours*, vomiting, readily induced by drinking warm fluids.

After 4½ *hours*, violent delirium, and restlessness; tenesmus, and frequent desire to pass urine.

After 10 *hours*, pulse 110; the patient had become calm, but could not sleep although inclined. During the *next* 12 *hours* she obtained but little sleep; she was able, however, to rise in the morning, when the pulse was frequent, small, and irregular, and she complained of lassitude; the pupils were still widely dilated, and vision confused. Recovery was now rapid.[1]

(*b*) An adult male took about 1 *grain of sulphate of atropia* in solution before breakfast, and 15 hours after food.

After 1 *hour*, pulse very quick and of good volume; pupils completely dilated, eyes restless; was generally restless and unmanageable, refusing to answer, to swallow, or to be examined; appeared profoundly intoxicated.

After 2 *hours*, pulse very weak; hands cold; dragged the legs when compelled to walk; was kept in forced exercise from the second to the sixth hour.

After 8 *hours*, insomnia, incoherent quarrelling, loss of memory, partial paralysis of the arms and legs. Continued wakeful and delirious during the *next* 4 *hours*, and passed very little urine, and throughout the succeeding night the hearing and sight were morbidly sensitive.

After 48 *hours*, pulse 108; tongue dry and furred; skin hot; was quite rational and nearly recovered. Catheterism was required during the next four days.[2]

[1] M. Roux. Journal de Chimie médicale, 1860, p. 529.
[2] H. Leach. Medical Times and Gazette, 1865, ii. 34.

Having so far ascertained the effects of belladonna in the healthy body, it will be well to consider certain conditions which affect its action:—

1. *Children*, as Dr. Fuller ('Medico-Chirurg. Trans.' vol. xlii.) has observed, are remarkably insusceptible of the action of belladonna. But this is more apparent than real. Children in this respect resemble the lower animals, and while acceleration of the pulse, dilatation of the pupils, and dryness of the mouth, are most readily induced in them, cerebro-spinal effects,—giddiness, drowsiness, illusions of the senses, and unsteadiness of gait,—are only developed after very large doses.

2. In the subject of *Obs.* 25 (p. 129), *pregnancy* appeared to diminish the activity of atropia. I injected it many times during the fifth and sixth months of gestation, in doses varying from $\frac{1}{96}$ to $\frac{1}{34}$ of a grain. The last-named dose was required to produce effects as great as those which follow the use of $\frac{1}{48}$ of a grain in most persons. The chief effect of the action of the drug in this case was somnolency, and after the larger doses she often slept continuously for half an hour at a time. Nothing, however, can be inferred from the action of any drug upon a single individual, and the exceptional results in this case may have been due to other causes than pregnancy.

3. *The weak, and those of excitable temperament*, are more readily and powerfully influenced than the strong. The $\frac{1}{96}$ of a grain of the atropia salt will produce as great an effect on a delicate nervous woman as $\frac{1}{60}$ of a grain upon a man of average strength. As a rule $\frac{1}{60}$ of a grain in a man, and $\frac{1}{80}$ in a woman, are followed by equal effects.

4. *Condition of the Renal Function.*—With reference to this point I have observed two facts: first, that in kidney disease, where the secretion of urine is diminished, or only very moderate in quantity, the effects of belladonna are both readily induced, and are considerably prolonged; and secondly, that in persons in whom the kidneys are usually active, the action of the drug is less powerful.

In corroboration of the first of these statements, I may refer to *Case* 2, Table I., in which belladonna effects continued during a considerable portion of the night, and caused some little anxiety to the wife.

As an illustration of the second statement, I give the following:—

Obs. 78.—John F., a grey-haired man aged 60, of weakly health, and having a complete *arcus senilis*, excreted, on the average, from 3 to 4 ounces of urine an hour. Table I. No. 6.

The $\frac{1}{40}$ *of a grain* by the skin, produced on two occasions an acceleration of the pulse 40 and 20 beats respectively; but the action of the medicine was rapid and of short duration, the pulse attaining its initial rate in each case two hours after the injection of the remedy. Besides this effect on the pulse, dilatation of the pupil from $\frac{1}{10}''$ to $\frac{1}{8}$, and a slight and transient feeling of dryness and sleepiness, there were no symptoms. During the two hours occupied by the action of the medicine, ℥viij. of urine were secreted on both occasions, and a few drops of either were sufficient to cause considerable dilatation of the pupil. At the time I used the atropia, and previously, the patient was taking 20 grains of bicarbonate and 15 grains of nitrate of potash thrice a day, and on both occasions a dose was taken two hours before the injection. It was necessary, therefore, to determine whether the diminished effects were not due to excess of alkali in the blood. This was done by substituting ♏xx. dilute sulphuric acid, and ℥j. of decoction of yellow bark, for the alkaline medicine. He continued its use for two months, and after intervals of a week, a fortnight, and three weeks from the time he commenced taking it, I repeated the injection of atropia, using on two occasions $\frac{1}{40}$ of a grain as before, and on the last $\frac{1}{30}$ of a grain. On each occasion he took a double dose of the medicine (acid sulph. dil. ♏xl. &c.) two hours before the injection.

When the $\frac{1}{40}$ of a grain was used, the effects were no stronger than when he was taking the alkali; and the $\frac{1}{30}$ of a grain produced only slight giddiness and drowsiness in addition. Its effects, in fact, were just such as would have followed the use of the $\frac{1}{60}$ of a grain in most persons. On all occasions, whether taking acid or alkali, the moisture on the tongue was either neutral, or faintly alkaline; and the urine, naturally *acid before* the injection, continued so during the early part of the action of the medicine. On all occasions but two, the secretion was strongly alkaline at the close of the action of the atropia. The disagreeable odour of the breath was only very slightly developed, the tongue was never even dryish, and only on two occasions did the secretion give an acid reaction.

In this individual the kidneys were unusually active, and to this circumstance, rather than to excess of either alkali or acid in the blood, we must attribute the comparative impotency of the drug. These remarks apply almost equally to Charles V. (See No..1, Table I.) Taking the whole of these facts together, I think we may safely conclude that the effects of belladonna are in some degree proportionate to the activity of the renal function, the action of the drug being much more marked when the quantity of urine is small than when it is abundant. This fact will to some extent serve to explain the comparative immunity of children.

5. *Alkalies.*—Dr. Garrod ('Medico-Chirurg. Trans.,' vol. xli. p. 53) has shown that atropia is decomposed, in contact with caustic potash and soda, in the course of two or three hours. These bodies have no power, however, of annulling or even diminishing the action of belladonna within the body, as the following observations show.

Having determined the action of given doses of belladonna juice and atropia, upon eight individuals, male and female, I repeated the dose some days afterwards, with the addition of from 1 to 2 grains of caustic potash and soda, ♏ ij. to ♏ iij. of saturated solution of ammonia, and from ʒvj. to ℥ij. of lime-water, respectively, at the time the medicine was taken.

In every case the belladonna effects were reproduced as speedily, were developed as fully, and continued as long, as when the medicine was given without the alkali.

With regard to *caustic ammonia*, my observations do not agree with Dr. Garrod's statement that this alkali does not destroy atropia. I find that if a third of the following mixture—belladonna juice, or an equivalent quantity of sol. atropia, ʒjss. spirit of wine, ʒiij. saturated solution of ammonia, ♏jx. and water ʒjss.—be given at intervals of a week, the first dose, taken on the day when it was mixed, will produce effects of undiminished intensity; the second will be followed by much feebler effects, or no symptoms at all; and the last dose will produce no effect whatever. I have made this observation in three different individuals with uniform results.

I have further ascertained that *lime-water*, in the quantities above mentioned, so acts as completely to annul the action of belladonna after a few hours.

The *carbonates of the alkalies*, as Dr. Garrod has observed, have no decomposing influence upon atropia.

Neither have *alkaline salts*, such as phosphate of soda.

5. *Acids*, as above shown, have no particular influence on the action of belladonna.

When administered by the stomach, I have occasionally seen the action of belladonna exceptionally postponed in certain cases for two hours, when the pulse has suddenly increased sixty beats, and the other symptoms have followed with corresponding intensity. This occurred on one occasion after the simultaneous administration of xxiv. grains of phosphate of soda, and ʒss. of succus belladonnæ.

Elimination.—Atropia passes undiminished and unchanged through the blood, and the kidneys are active in its elimination from the minute that it enters the circulation until it is entirely removed from the body. After a full medicinal dose, between two and three hours are required for this purpose.

Availing myself of its dilating action upon the pupil, I have repeatedly demonstrated the presence of atropia in the urinary secretion of different individuals, 18, 19, and 20 minutes after the subcutaneous injection of the $\frac{1}{48}$, and even the $\frac{1}{96}$ of a grain, of sulphate of atropia, and have ascertained its existence in urine secreted $2\frac{1}{2}$ hours after a larger dose. This fact is readily demonstrated by placing one or two drops of the urine between the eyelids, at intervals of ten or twenty minutes, for two or three hours. Twelve drops of eight ounces of urine secreted during the action of the $\frac{1}{48}$ of a grain of the salt, are sufficient to dilate the human pupil from $\frac{1}{10}$ to $\frac{1}{6}$ of an inch in diameter, and maintain it thus dilated for 8 or 10 hours. It may be desirable to separate the atropia from the urine, and this can be done very readily by shaking it with a quantity of chloroform equal to a sixth of its bulk, separating the chloroform, and allowing it to evaporate spontaneously. The remaining stain is dissolved in a few drops of water, and a drop placed between the eye-lids. That the $\frac{1}{96}$ of a grain of sulphate of atropia—a quantity insufficient to destroy an infant—may thus easily be detected in the urine, is a fact of practical importance in a medico-legal point of view, while to the physiologist it is one of great interest. That the drug passes undiminished through the system is evident from the fact that the atropia urine dilates the pupil as readily, and to as great an extent, as an equal bulk of water to which the dose of atropia has been added.

I conclude, therefore, that the fullest medicinal doses are wholly removed by the kidneys alone.

Belladonna indeed is, in the truest sense of the word, a diuretic, and more powerful perhaps than any other that we possess. After excessive doses in both man and the lower animals, frequent emission of urine is a marked symptom. In medicinal doses the diuretic effect is often masked by retention of the urine; but if that which is excreted during the operation of belladonna and a few hours afterwards be examined, an increase either in the specific gravity, or of the quantity, will be observed. In the latter case the specific gravity will, of course, be proportionally diminished. Analysis will show an increased elimination of all the solid constituents, excepting generally the chlorine, which, on account of the increase of the other constituents, appears to be diminished. The urea is always increased, and often to a considerable extent; but the effects of the drug are most manifest in the increase of the phosphates and sulphates, which are sometimes doubled.

If the action of the medicine take place during a period of fasting, and the maximum acceleration of the pulse be great, and sustained for an hour, the urine will resemble that voided after the digestion of a hearty meal, in the richness of its solid constituents.

TABLE I.—THE URINE BEFORE AND

Name, disease, diet, and treatment.	Urine before the administration of Atropia.
1. Charles V., æt 32. *Lumbago.* 8·30 A.M. Tea, bread and butter. 11 A.M. $\frac{1}{48}$ grain *Atropiæ Sulph.* by skin.	A. Urine secreted between 8·30 and 11 A.M. ℥vj. sp. gr. 1016, freely alkaline: phosphatic opalescence on boiling. 1000 grain measures contained:— Chlorine, 3·71; Urea, 14·66; Phosphates and Sulphates, 9·41 grains.
2. T.F.R. æt. 28. *Chronic albuminuria.* 9 P.M. Feb. 3, a light supper. 7·30 P.M., Feb. 4, $\frac{1}{48}$ gr. *Atrop. Sulph.* by skin	A. Urine secreted between 11·45 P.M. and 8 A.M. Feb. 4. ℥x. sp. gr. 1015·6. 1000 gr. meas. contained: Grs. Chlorine . . 2·62 Urea . . . 13·33 Phosphates and Sulphates . 7·2 Uric acid, normal. B. Urine secreted between 3 and 6 P.M. Hearty dinner at 3 P.M. ℥vij. sp. gr. 1022·8. 1000 gr. meas. contained:— Grs. Chlorine . . 4·92 Urea . . . 20·0 Phosphates and Sulphates . 9·71 Uric acid, normal. C. Urine secreted between 6 and 7·30 P.M. A glass of ale at 6·15 P.M. ℥iiʒvj. Sp. gr. 1023·2. 1000 gr. meas. contained:— Grs. Chlorine . . 4·64 Urea . . . 22·33 Phosphates and Sulphates . 10·0 Uric acid, an excess.
3. Michael E., æt. 48. *Sciatica.* 10 A.M. Bread and butter, tea, and two eggs. 11·45 A.M. $\frac{1}{48}$ gr. *Atrop. Sulph.* by skin.	A. Urine secreted between 10 and 11·45 A.M. ℥ii. sp. gr. 1021, acid clear. 1000 grain measures contained:— Chlorine, 6·07; Urea, 23·33; and Sulphates and Phosphates, 7 grains.
4. John J. H., æt 19. *Colic.* 5 P.M. Tea, bread and butter. 6·28 P.M. $\frac{1}{24}$ gr. *Atrop. Sulph.* in ℥j. aquæ by mouth.	A. Urine passed at 6.27 P.M. ℥ij. ʒvj. sp. gr. 1016·8, faintly acid. 1000 grain measures contained:— Chlorine, 3·21; Urea, 16·66; Sulphates and Phosphates, 9·8 grains.
5. Alfred L., æt. 21. *Gastralgia.* 5 P.M. Tea, bread and butter. 7·15 P.M. $\frac{1}{24}$ gr. *Atrop. Sulph.* in ℥j. aquæ by mouth.	A. Urine secreted between 6.15 and 7.14 P.M. ℥jss. sp. gr. 1010, very faintly acid.
6. John F., æt. 60. *Sciatica.* Before breakfast, at 10·43 A.M., $\frac{1}{48}$ gr. *Atrop. Sulph.* by skin.	A. Urine passed at 10.42 A.M., ℥xijss. sp. gr. 1008·4, acid, bright on boiling.
Before breakfast, at 8 A.M. gr. xx. Potas. bicarb. and Pot. nit. 10·20 A.M. $\frac{1}{48}$ gr. *Atrop. Sulph.* by skin.	C. Urine passed at 10.19 A.M., sp. gr. 1008, freely acid, whitewashes the vessel with lithates, bright on boiling.
8·15 A.M. Tea, bread and butter. 9 A.M. Dose of potash as above. 10·30 A.M. $\frac{1}{40}$ gr. *Atrop. Sulph.*	F. Urine passed at 10·29 A.M. ℥xj. ʒij. sp. gr. 1006·8, acid, bright on boiling.
8·15 A.M. ℥j. bacon, toast and tea. 9 A.M. ♏︎ xl. Acid. Sulph. dil. 10·38 A.M. $\frac{1}{40}$ gr. *Atrop. Sulph.*	H. Urine passed at 10.37 A.M. ℥xj. sp. gr. 1006·4, acid, bright on boiling.

FTER THE ACTION OF ATROPIA.

Urine after the administration of Atropia.		Remarks.
. Urine 3 hours after Atropia, ℥xvii. sp. gr. 1010, freely alkaline, phosphatic opalescence on boiling. 1000 grain measures contained :— Chlorine, 1·95; Urea, 6·33; Phosphates and Sulphates, 8·13 grains.		Much difficulty in voiding urine B. Incomplete retention for 12 hours after.
D. Urine passed at 9·45, 2½ hours after the Atropia, and when its action was declining, ℥iiiss. sp. gr. 1022·8. 1000 gr. measures contained :— Grs. Chlorine . . . 4·50 Urea . . . . 23·33 Sulphates and Phosphates . . . 11·06 Uric acid, an excess.	E. Urine secreted between 9·45 P.M. and 9 A.M. next day. A little soup, bread, and ℥iv. light beer at 10·30 P.M. ℥xivss. sp. gr. 1023·6. 1000 gr. measures contained :— Grs. Chlorine . . . 3·33 Urea . . . . 28·33 Sulphates and Phosphates . . . 13·95 Uric acid, an excess.	All the urines had the same general characters—freely acid, bright, free from crystalline or organic deposit. The influence of the Atropia continued all night.
B. Urine 2 hrs. after atropia, ℥iv. ʒii. sp. gr. 1010, pale, faintly acid. 1000 gr. meas. cont.: Grs. Chlorine . . . 2·62 Urea . . . . 10·66 Sulphates and Phosphates. . . . 3·40	C. Urine 2½ hours after Atropia, ʒii. sp. gr. ·1015. 1000 gr. measures contained :— Grs. Chlorine . . . 3·76 Urea . . . . 14·00 Sulphates and Phosphates . . . 5·10	12 drops of B. dilated the pupil from $\frac{1}{10}$ to $\frac{1}{6}$, and kept it so for 7 hours.
B. Urine 2½ hours after the Atropia, ʒvii.ʒvj. sp. gr. 1018, alkaline, phosphatic opalescence on boiling, and separation of much triple phosphate after 2 hours. 1000 grain measures contained :— Chlorine, 3·03; Urea, 16·66; Phosphates and Sulphates, 11·6; Uric acid, 0·081 grain.		Urine B. dilated the pupil.
B. Urine passed 2 hours after the Atropia, ʒvss. sp. gr. 1013, freely alkaline, phosphatic deposit on boiling. Separation of much triple phosphate on standing two or three hours.		Urine B. dilated the pupil.
B. Urine 2 hours after the Atropia and still fasting, sp. gr. 1001·6, alkaline, phosphatic cloud on heating. The urine was retained, and it was only after much straining that a few drachms could be obtained.		Urine B, D, E, G, I, and K, dilated the pupil readily. No difference was observed between the dilating power of D and E, or I and K.
D. Urine passed 18 minutes after Atropia, ʒvj. sp. gr. 1016. Deposits uric acid.	E. Urine 2 hours after Atropia, alkaline, strong phosphatic opalescence on boiling.	
G. Urine removed by catheter 2¼ hours after Atropia, ʒviii.ʒvj. sp. gr. 1012·2, freely alkaline. Phosphatic opalescence on boiling.		The urine was always retained after the action of the medicine, and it was only after prolonged straining that a few drachms could be obtained.
I. Urine passed 19 minutes after Atropia, ℥j. sp. gr. 1009·2, bright on boiling.	K. Urine passed 2 hours after Atropia, freely alkaline, phosphatic opalescence on boiling.	

Taking a general survey of Table I., which illustrates the influence of belladonna on the renal function, we find, in *Case* 1, that the specific gravity of the atropia urine is diminished about a third, while the quantity of the secretion for equal times is more than doubled; and it will be found, on calculation, that the chlorine and urea are slightly increased, while the phosphates and sulphates are doubled.

In *Case* 2, it is to be observed that only one light meal was taken between 3 P.M. and 9 A.M. the next day, and that urine D was secreted during a time of fasting; and yet it is even richer in solid constituents than urine C, which was secreted towards the end of the digestion of a hearty dinner of mixed diet. Urine E is still richer in solid constituents, which is partly due to the food taken, and partly to the operation of the belladonna, which continued through a considerable portion of the night, and produced a little dreamy delirium and broken sleep.

In *Case* 3, the urine secreted in equal times, before and after the administration of belladonna, was nearly trebled in the latter case; while the chlorine and urea, in equal parts of the urine, were only diminished one half, thus giving in the whole a considerable increase of all the solid constituents.

The same facts will appear on examination of the other cases. The fact which stands out before the others is the invariable and considerable increase in the phosphates and sulphates. The atropia urine is more frequently alkaline than acid, and triple phosphate often separates in the course of two or three hours in considerable quantity; and when the fluid is heated, it almost invariably deposits a cloud of phosphates.

The uric acid does not appear to suffer any diminution, and it is, on the contrary, often increased.

Further illustrations of these statements are given in subsequent pages.

Having completed my description of the phenomena which attend the operation of belladonna, and discovered some of its results, I now pass on to a more general consideration of the subject, with the view of ascertaining, as far as possible, the nature and mode of an action which in every aspect is as wonderful as any in nature. An infinitesimal quantity of atropia—a mere atom—as soon as it enters the blood, originates an action which is closely allied to, if it be not identical with, that which induces the circulatory and nervous phenomena accompanying meningitis, enteric, or typhus fevers; and as the alkaloid gradually passes out of the body, and is finally eliminated, undiminished and unchanged, we see these great functional disturbances decrease *pari passu*, until the body is restored to its natural condition. Such an action is strictly comparable with that of sunlight on a mixture of chlorine and hydrogen, or of spongy platinum on hydrogen. Atropia determines an action as powerful and almost as rapid as either of these agents, and like them it is only the determining cause—it undergoes no change itself.

Action on the Sympathetic Nervous System and the Circulation.—How far the operation of belladonna illustrates the causation of the febrile condition is a question of the highest interest. If we take the simplest view of its action, it is that of direct and powerful stimulation of the sympathetic nervous system; and, indeed, in children and many of the lower animals, this is so far the chief effect that, in moderate doses, it may be regarded as the only one (see *Obs.* 67 (*a*), p. 198). During the operation of medicinal doses the heart contracts with increased vigour, the arteries increase in tone and volume, the capillary system fully participates in this general excitation of the circulation, and a diffusion of warmth is felt throughout the body.

The stimulant effect is so intense that, if the dose be excessive, signs of exhaustion are soon manifest. The maximum effect is observed after moderate doses only; generally $\frac{1}{96}$ of a grain of sulphate of atropia, used subcutaneously, will be sufficient to produce it; and, as a rule, the dose should not exceed $\frac{1}{48}$ of a grain, if we want to induce stimulant effects alone. The influence of large doses in diminishing the force and activity of the heart is illustrated in *Obs.* 65, p. 196. The pure stimulant effect on the heart is well seen in the following case:—

Obs. 79.—With a view more of ascertaining the influence of belladonna in progressive failure of the heart's action in inanition than of hoping for a permanent good result, I injected the $\frac{1}{240}$ *of a grain* of sulphate of atropia into the arm of an infant ten weeks old, at a time when, excepting a few beats now and then, the pulse was imperceptible at the wrist, and the cardiac systoles only 80. Within *four minutes* the pulse rose to 100, and each beat was quite perceptible at the wrist. In *eight minutes* it had increased to 110, and was quite regular and distinct. The stimulant action continued for the next three hours, and at the end of this time the pulse was 100, of good volume, and of sufficient force to bear moderate compression. The respiration remained unaltered, and the pupils dilated from $\frac{1}{10}''$ to $\frac{1}{7}''$. The pallid skin of the extremities was suffused with a dark crimson flush. The stimulant effect upon the pulse continued to within half an hour of the death of the child, *five hours and a half* after the injection of the atropia.

The effect on the circulation is readily studied in the frog. Mr. Wharton Jones long ago called attention[1] to the fact that the application of a solution of atropia to the web of the frog's foot caused contraction of the artery, and he considers that 'the effect of this arterial constriction is to produce a congestion of the capillaries and the venous radicles,'[2]—the congestion being a consequence of the arterial constriction.

My own observations accord with these statements only so far as a moderate contraction of the artery is concerned. Congestion, I believe, only results when the whole of the bloodvessels of the part are more or less paralysed.

Dr. Meuriot's observations[3] on this point are as follows:—'When a few drops of solution of sulphate of atropia are placed upon the inter-digital membrane of the frog, the circulation is instantly accelerated, so that the globules can no longer be distinguished or followed. The web is at the same time injected, new capillaries become visible, and the circulation is established in many vessels from which it was previously absent. From the commencement, a diminution in

[1] Guy's Hospital Reports, second series, vol. vii. p. 21.
[2] Medical Times and Gazette (1857), vol. i. pp. 27–79.
[3] De la Méthode physiologique à l'Étude de la Belladonna (1868), pp. 37–40.

the calibre of the arteries, sometimes a third or even the half, may be proved by means of the micrometer;' but 'we have never,' he says, 'obtained the complete obliteration that some authors say they have observed. The diminution of the calibre does not persist long; the circulation is maintained with rapidity for several hours, provided the dose of atropia employed be small. If it be large, hyperæmia of all the vessels results; little by little the veins and capillaries become engorged; the circulation becomes slow, and is at last completely arrested. The stasis commences in the capillaries and veins, and the circulation always continues in the artery for some time after it has completely ceased in the veins.'

Dr. Meuriot's observations are reliable, because he has adopted the proper means to make them so—that of keeping the animal under observation for several hours together. It is a plan which I have myself always adopted, and it will be seen from the following experiments that our observations closely agree:—

Obs. 80.—Having divided the spinal marrow of a frog in the occipital region five hours previously, I placed the animal on the frog-plate, and examined the circulation in a particular division of the web for 2½ hours, measuring the size of the bloodvessels, and noting such variations as followed ordinary disturbances. The slightest reflex movement of the limb caused for several seconds an almost complete obliteration of the main artery; a few minutes, however, sufficed to restore it to its original calibre.

At the end of this time I injected $\frac{1}{480}$ *of a grain* of sulphate of atropia in ♏ij. water, beneath the skin of the abdomen. *After* 7 *minutes* the rapidity of the circulation was much increased. During the next 3 *hours* the circulation continued regular and extremely rapid; the artery at first slightly contracted, so as to lie barely within the lines of the micrometer. After an hour this slight contraction was no longer observed, and throughout there was no change in the calibre of the veins or capillaries. The respiration and the pupils remained unchanged.

Obs. 81.—After 12 *hours*, and having previously watched

the main artery between the micrometer-lines for ¾ of an hour, I injected $\frac{1}{240}$ *of a grain* of the sulphate of atropia. *After* 10 *minutes* the artery had slightly contracted, so as to lie just within the micrometer-lines; the size of the capillaries closely corresponded with the short axis of the red corpuscles, and the walls of the vessels appeared to be sometimes contracting upon them in their onward passage, and generally the capillary circulation was a little slower. *After* 20 *minutes*, the artery was contracted $\frac{1}{6}$, and the rapidity of the current so far diminished that the stream was seen to be composed of corpuscular particles. In the venous capillaries two corpuscles flowed readily side by side, and the circulation in the veins was so slow that individual corpuscles could be clearly recognised. *After* 30 *minutes*, the artery was contracted ¼, the current so slow that the individual corpuscles could be seen, the arterial capillaries sufficiently dilated to allow the corpuscles to pass with their long axes nearly transverse, and the blood moving everywhere, but slowly.

After 1 *hour* the circulation was slow and uniform, but now the rate in the artery was such that individual corpuscles could not be seen.

After 1¼ *hour* the circulation had recovered its original rapidity, being such that the artery appeared as a mere red streak, and the blood in the largest veins an indistinctly corpuscular stream. *At the fifth hour* a little retardation was again noticed, but from this time, and *during* the next 30 *hours*, the circulation was uniform and excessively rapid; the artery being indicated by a faint red vibrating streak, and the rate of the blood in the largest veins such that the corpuscular elements could no longer be distinguished; the capillaries and the veins were themselves slightly contracted. At the twelfth hour the artery was contracted ⅓, and at the twenty-fourth hour it was reduced to half the original size. The walls of the capillaries appeared tense, and to be exerting pressure on the red corpuscles as they passed along.

During the next two or three days the circulation continued unusually rapid. During the early part of the action the pupils dilated from $\frac{1}{10}''$ to ⅛. Throughout the respiration was unchanged, and the pigmentary corpuscles of the skin were retracted.

It appears from these experiments that a moderate dose of atropia produces a tonic and slightly contracted condition of the whole of the circulatory tubes, accompanied by increased force and rapidity of the heart's action. The blood is equally distributed, and the circulation in any given part is so tight and rapid that it really contains a little less blood than when in a quiescent state, and the tissue is consequently a little paler. But the quantity which passes through it in a given time is greatly increased. After the use of moderate doses these effects are maintained throughout.

After larger doses the stimulant effect is manifested by a still greater contraction of the arteries, and, if the dose be excessive, this is associated with diminished activity of the circulation, which gradually approaches stasis. In *Obs.* 81 the dose was almost sufficient to produce this result, and it was only after some hours, and when a portion of the alkaloid had been eliminated, that the stimulant effects were declared.

These results strictly accord with the effects of atropia on man. After moderate doses the whole circulation is increased in force and rapidity. The tone of the larger arteries is good, and, if the circulation was previously slow, we find that they are usually increased in volume as well as in tone. After larger doses we observe no further increase in the force and rapidity of the circulation, and usually a decided decrease in the volume of the smaller arteries. If the dose be still further increased, we shall observe only a moderate acceleration of the pulse, a diminution of the size of the artery, and a positive decrease in the force of the pulsations. When the dose is excessive, the artery will often be found dilated, and its coats flaccid, and collapsing under the slightest pressure.

The effect *on the heart* itself is obvious to the touch. Pulsations, which before a dose of atropia are only faintly felt through the chest-wall, afterwards become each one very strong, distinct, and still regular, and no artificial contrivance is needed to demonstrate increased pressure of the arterial current. Nor, after excessive doses, is the hæmometer required to prove loss of power in the cardiac contractions and diminished arterial pressure. When the heart is fully under the influence of the excitant action, it is remarkable how rarely cardiac sensations are complained of. A little fluttering in the region of the heart, simultaneous with the development of the maximum acceleration of the pulse in three or four patients, has, in my own practice, constituted the whole of the sensational effects. In the lower animals there is not the least evidence of distress when the rapidity of the action of the heart is such that it is best described as vibratile.

The *local action of atropia* is illustrated in the following experiment:—

Obs. 82.—Using the same frog, and after an interval of eight days from the last experiment, placed a large drop of solution of sulphate of atropia (1 grain in ℥j. of water) upon a part of the web in which the circulation had been carefully examined, and the dimensions of the bloodvessels ascertained for $\frac{3}{4}$ of an hour previously. The circulation was uniform and rapid, and the pigment-corpuscles radiated. *After half an hour*, the main artery was contracted $\frac{1}{8}$, the rapidity of the circulation increased, and the pigment-corpuscles retracted. *After* 1 *hour* no further change had occurred. The web was now wetted with a solution of 1 grain of sulphate of atropia in ♏ xl. of water, a drop of which was also placed on the eye.

For the *next* 2 *hours* no change was observable, the contraction of the artery and the rapidity of the circulation remained as at half an hour after the first application. The pupil was unchanged. *Fourteen hours* after, the artery was still a little contracted, the rapidity of the circulation moderate, and the pigment-corpuscles radiated. *At this time* another drop of the stronger solution was placed upon the same portion of the web.

After *an hour* the rapidity of the circulation was considerably increased in all parts of the web. The main artery was *dilated* to its original calibre, and the pigment-cells were retracted. During the *next* 6 *hours* no further change took place, but at the end of this time the pigment-corpuscles were again fully radiated. The subcutaneous injection of atropia at this time caused slight contraction of the artery, acceleration of the current, and retraction of the pigment-corpuscles.

It appears from the foregoing observations that atropia, when administered generally or topically, or both ways simultaneously, fails to reduce the artery to less than half its original dimensions; and that this contraction cannot be regarded as the cause of the congestion and stasis which occasionally follows the local application of the alkaloid, for the maximum degree of arterial contraction is compatible with great rapidity of the circulation. It further appears that strong doses have little or no contractile action. In no case have I ever witnessed a greater contraction than occurred in *Obs.* 81.

With regard to congestion and stasis, it is remarkable that very weak solutions of atropia will sometimes produce these effects. I have myself once or twice experienced slight congestion of the entire conjunctiva, with dryness of the membrane and dull aching pain in the eyeball, lasting for several hours, after the use of a very weak solution of atropia. On one occasion this condition followed the instillation of 12 drops of a solution of 1 part of sulphate of atropia in 400,000 parts of water. It was accompanied by a dilatation of the pupil from $\frac{1}{9}''$ to $\frac{1}{7}$. I have also experienced the same effects on two occasions after the use of dilute solutions of hyoscyamine. These effects are independent of any direct irritant action; they are rare, and the conditions under which they occur appear to be very obscure.

Action on the Skin and Mammary Secretion.—Simultaneously with that general diffusion of warmth which accompanies the equal distribution of the blood throughout the body, a scarlet suffusion of the skin is often observable in young children (see *Obs.* 79) and those who have a delicate skin. This injection of the cutaneous capillaries has been described as a 'scarlatinous rash.' Generally it is nothing more than a temporary blush, but in rare cases, and in persons who are liable to vascular irritation of the skin, the redness remains, and its disappearance is attended with slight roughness and desquamation. As far as my experience goes, belladonna has no particular action on the skin, in either exciting or repressing its secretions. Its general effects on the circulation predispose to sweating, and this should not be lost sight of when we wish to excite the action of the skin.

Dr. Meuriot (*opere citato*), including a probability (p. 58) in his conclusions, states :—'L'atropine s'élimine par les reins, par toutes les muqueuses, et parfois par la peau chez l'homme' (p. 139). Whether this occurs after fatal doses, and when the kidneys are paralysed, remains to be proved. I am free to assert that in medicinal doses, and when the kidneys are moderately active, atropine is exclusively eliminated by the kidneys.

While the skin is undoubtedly predisposed for sudorific action by moderate doses of belladonna, we shall find that its effects on the mammary secretion are depressent where, as in the early days of lactation, there is undue excitement of the glands. The effect of belladonna in repressing hyperæmia and its consequences will be considered hereafter, and I will only observe here that this influence is most unmistakably displayed in allaying and removing excessive functional activity of the mammary glands.

Action on the Mucous Membrane and Glandular System generally.—Theoretically we might conclude that an increased flow of blood through the parenchymatous glands would result in an increase of their secretions. With the kidney this is actually the case; but in the mouth we have positive evidence of the arrest of secretion. While there is no lack of moisture on the finger-ends and skin generally, or in any other part of the body, and while the whole circulation is endowed with increased tone and activity, those parts of the mouth which are adjacent to the median plane are so completely parched that they fail to impart the least moisture to a bit of bibulous paper or sugar kept in contact with them for several minutes. It is observable that this dryness is greatest along the median line and on either side of it, and that after moderate doses it extends only a short way outwards. Dryness of the lips, the buccal mucous membrane, and the pillars of the fauces, only occurs after large doses; but a very moderate dose is required to render the central part of *the tongue* dry and parched from back to front, and the hard and soft palates and back of the œsophagus as dry and glazed as a piece of paper. If we examine the parts last mentioned, we shall find them dark red and congested, and there will be no difficulty in recognising a turgid vessel here and there. It is plain, therefore, that the absence of moisture is not due to occlusion of the bloodvessels; we have in fact a condition which exactly resembles that accompanying the typhous state. The bloodvessels of the part are congested and the blood is arrested.

It is very remarkable that stasis should result in any part during the operation of a dose of belladonna, which, as we have seen, produces general activity of the whole circulation, with increased tone in the bloodvessels and greater force and rapidity of the heart's action. But we must remember that the tongue is a most delicate indicator of vascular as distinguished from simple cardiac excitement, and that it frequently becomes dry in slight febrile affections in which there is no evidence of congestion in any other part of the body. The disproportionate share which the tongue takes in the general vascular disturbance may be due to several causes, and these all have their origin probably in the peculiarity of anatomical structure which results from bilateral development, and generally, we may account for the ready production of congestion in the parts above mentioned by the fact that the intercommunication of the bloodvessels across the median plane is less free than in the other parts of the body.

While dryness is the invariable result of the use of belladonna in health, it is remarkable that the reverse effect occasionally follows its use in disease. A quarter of an hour after the injection of a medicinal dose of atropia beneath the skin of a patient suffering from fever, I have several times observed the tongue, which for days before had been parched, contracted, and hard, swell out again and become moist for a time. (See Effects of Belladonna in Acute Disease.)

The dryness very often extends to the mucous membrane of the lower passage of the nares. I cannot say that I have ever noticed positive dryness of the conjunctival membrane during the action of belladonna, but the injection of the $\frac{1}{8}$ of a grain produced ophthalmia of several days' duration in one of the dogs employed in my experiments.

The foregoing considerations lead me to conclude that more or less complete arrest of secretion in the mouth during the action of a moderate dose of belladonna is a local and exceptional effect and one that does not extend to the glandular system generally.

While the mouth continues in the dry condition above described, *the salivary glands* appear quiescent, and the morbid state of the tongue and palate renders it difficult, if not impossible, to excite a flow of saliva by gustatory impressions; but they readily pour out abundance of secretion when an appropriate stimulus reaches them. In a patient who was suffering from severe neuralgia of face associated with profuse salivation, the secretion was in no degree diminished when, as occasionally happened, a severe paroxysm of pain came on during the action of a full dose of atropia. In another case—that of a healthy woman, aged 53, the subject of chronic idiopathic ptyalism—I injected sulphate of atropia in doses varying from $\frac{1}{60}$ to $\frac{1}{48}$ of a grain, without producing decided temporary or permanent effect on the disease.

The influence of medicinal doses on the intestinal secretion is not very marked, but when given by the mouth and in large doses belladonna frequently causes nausea, and in poisonous doses vomiting (see cases of poisoning in a subsequent table), and sometimes diarrhœa.

The following observations lead me to regard belladonna as a cholagogue:—*Firstly*, we have seen an increase in the colouring matter of the bile, as well as in the quantity of intestinal moisture and mucus, in the horse (see *Obs.* 64, p. 196). *Secondly*, I have carefully watched the effect of belladonna in several cases of jaundice, and have failed to observe any diminution of the bile in the urine which has been secreted during the operation of a full dose of atropia used subcutaneously, or in that voided from time to time when the patients were taking repeated doses of belladonna by the mouth daily. *Thirdly*, we may obtain positive evidence of an increase in the secretion of bile in animals who have died after the prolonged action of the drug. The following are instances:—

Obs. 83.—(*a*) An active male mouse was scantily fed and prevented access to water for 24 hours, and at the end of this time $\frac{1}{20}$ *of a grain of sulphate of atropia* in ♏ iij. of water was injected beneath the skin. Respiration 160.

After 5 *minutes*, the animal was in a sleepy semi-torpid state, in which he continued for 10 hours, when he recovered much of his activity and took a few crumbs of bread and butter. A few minutes after the injection the pupils were fully dilated, and the respiration 130.

At the 3*rd hour* the respiration had fallen to 112, but it was regular and accompanied by good expansion of the chest. Pulse 220, regular.

At the 5*th hour* the respiration and pulse had fallen to the minimum, the former being 81 and shallow; and the latter 200 and feeble. The feet and legs were cold and purplish. At this time the animal was very torpid, he could hardly stand, and made very feeble resistance when handled.

At the 6*th hour* the torpor was less, the respiration increased to 92 and deeper, and the pulse 280.

Between *the* 7*th and* 9*th hours* urine was passed for the first time, at first a little, and subsequently a large quantity.

The mouse at the end of the 9*th hour* was tolerably active when disturbed. Respiration 108; cardiac contractions 264, small and weak.

At the 13*th hour* I left the little animal still dull and sleepy, and found him next morning dead and stiff, killed probably by cold, and exhaustion from the effects of the medicine. He had lived during 13 hours of the action of the belladonna, and survived its direct effects, and no intestinal secretions were passed during the time he was under observation. The ventricles were empty and firmly contracted; the right auricle full of dark blood. The lungs were free from congestion. The blood was chiefly contained in the large abdominal veins. The brain, spinal cord, and kidneys were free from congestion; the urinary bladder empty and contracted. The stomach and intestines pale. Several coils of the large intestine were moderately distended with thin yellow fæcal matter, and the gall bladder was full of rich bile and more capacious than I have usually observed it.

(*b*) The same dose in another fasting animal caused considerable excitement for the first three hours, during which time the mouse was in active motion, couching for a short doze only occasionally. Narcotic effects then appeared and continued, under the influence of two other doses ($\frac{1}{30}$ and $\frac{1}{80}$ of a grain), until the 36th hour, when the animal became torpid and cold, the skin dusky, the respiration slow and shallower, and the heart's action too feeble to be heard. He continued, however, in this condition for 12 hours more, lying on the side and making a little movement when disturbed, and died without convulsive movements.

During the first 15 hours of the action of the poison the heart's action was strong and regular, sounding like that of the dog, but the vibrations were twice as rapid and at least 600 a minute. The respiration meantime was equally vigorous and rapid, the pants numbering from 300 to 240 a minute.

At the end of the 15th hour the senses of hearing and smell were remarkably acute; common sensibility seemed to be slightly diminished, and when the little animal was thoroughly aroused from his heavy sleep he was as active as ever.

At the end of the 24th hour, and 10 hours after the second dose, the respiration was 128 and good. During the 12 hours preceding death, when the animal was lying on the side in a state of semi-torpor, the respiration continued regular, and numbered from 100 to 72.

Death occurred 56 hours at least after food or water was taken. During the time the animal was under the influence of the poison, urine was frequently voided, and about the 7th hour a large dark relaxed motion was passed. Rigor mortis strong. The intestines both large and small were relaxed and contained a considerable quantity of fluid bilious matter; the gall bladder was distended; the urinary bladder half full. The nervous system and the other viscera were in the condition mentioned in (*a*).

In both of these cases there was a considerable increase of the biliary secretion during long periods of fasting. Whether atropia is eliminated to any extent by the liver does not appear. From my observations on the quantity eliminated by the kidneys I am inclined to think that it is not, and that the increase of the bile is simply due to the increased activity of the circulation, in which the liver must participate.

Dr. Meuriot (p. 58) concludes from an analysis by Dr. Marcet,[1] that atropia is eliminated by the intestine; but on turning to the analysis in question (vol. ii. p. 612) I find that Dr. Marcet's statements expressly refer to a mixture of urine and fæces. At present, then, we have no proof of the elimination of atropia by the mucous membrane. From what we know of the vicarious action of the mucous membrane in the elimination of other poisons there is, however, a strong probability of a similar action in the elimination of atropia when the dose is sufficient to exhaust the renal function.

Action on the Involuntary Contractile Tissue.—Belladonna has been deemed efficacious in obstinate constipation;[2] in incontinence of fæcal matters;[3] in incontinence of urine; and in the expulsion of biliary and renal calculi.

Its influence in these cases is usually attributed to the property which it is assumed to possess 'of increasing the peristaltic contraction of the intestines, by virtue of which it easily overcomes obstinate constipations, and which very often goes so far as to cause diarrhœa.'[4]

Now I think it will be conceded that diarrhœa is no evidence of increased peristaltic action, but rather of the reverse condition.

Increased peristaltic action often follows the subcutaneous use of morphia, and the intestines are emptied without the production of diarrhœa (see *Obs.* 13, p. 112), which may, however, subsequently arise from causes independent of the contraction of muscular fibre.

The modus operandi in the above-mentioned conditions

[1] The Lancet, 1859.
[2] Debreyne. Des vertus thérap. de la Bellad. Paris, 1852.
[3] MM. Bercioux et A. Richard. Gazette hebdomadaire. 1858.
[4] Meuriot, op. cit. p. 102.

appears to be as follows:—Belladonna allays pain, and thus removes irritation and the attendant spasm which is at once a cause of pain and an obstructing agent. And further, the drug relaxes the circular muscular fibre and in the case of a calculus thus facilitates its passage. In incontinence arising from spasm the expulsive efforts are moderated. I take the condition of the bladder as sufficient evidence of the correctness of this view. Retention of urine almost invariably occurs during the action of a full dose of belladonna, and dysuria very often follows. We may encourage a patient to make prolonged efforts to pass urine when fully under the influence of the drug, and he will either fail altogether or only pass a few drachms, and this not in little jerks which indicate spasm, but in weak driblets. Indeed the absence of spasm is readily determined. If a full-sized flexible catheter be introduced under these circumstances, it meets with no opposition, but slips readily into the bladder, and the urine flows as sluggishly as from the bladder of a patient afflicted with paraplegia. I have several times had occasion to use the catheter in order to complete an observation. When the viscus has contained quantities varying from ℥ vij. to ℥ xix. its contractile force has only been such that, while the handle of the instrument (No. 11) has been held horizontally with the orifice lodged upon the margin of a glass jar 3½ inches across, the stream has fallen in a short curve a space of 4 inches before it has come in contact with the opposite side of the vessel.

The frequent micturition which is observed after poisonous doses, and sometimes after medicinal ones, is the result of repeated calls to empty a distended and weakened bladder. Faint spasms (p. 240) may occasionally happen.

It appears pretty conclusive, then, that belladonna relaxes the hollow viscera, and it is to this effect that we must attribute its antispasmodic, as well as its expulsive, action. Experimental enquiry shows that the circular fibres of the hollow viscera are under the control of the spinal cord, and there is reason for believing that the longitudinal fibres are equally under the influence of the sympathetic nervous system. We can easily understand, then, how the action of belladonna, by relaxing the circular fibres and at the same time tightening the longitudinal, may be effectual in getting rid of a biliary or renal calculus, or an intestinal concretion, and this without the strong expulsive efforts which accompany contractions of the circular fibres. For the same reason no particular intestinal contractions would occur in the absence of such fulcra for the contractions of the longitudinal fibres.

These remarks are equally applicable to the bladder, the contractile fibres of which, as disease of the spinal cord teaches us, receive their stimulus through the spinal nerves.

I am glad to find that one of our best authorities[1] on the practical use of medicines agrees with me in this view of the action of belladonna. He says :—'Externally used I believe belladonna to be the best remedy we possess against tenesmus, whether the womb, the anus, the urethra, the nipple, or the eyelid be the seat of the forcing action. I have applied cotton-wool soaked in a solution of sulphate of atropia, to the neck of the womb, to quell the forcing pains of uterine tenesmus' (p. 98).

Action on the Pupils.—Next to its influence on the circulation, the most prominent effect of the action of belladonna consists in dilatation of the pupils. We may watch them in the horse, opening rapidly within a quarter of an hour after the subcutaneous use of atropia. It is no passive relaxation of the radiating fibres of the iris; on the contrary, there is evidence of direct antagonism between the occluding and dilating fibres. Just at the time when the dilatation of the pupils was beginning, after a dose of belladonna, I have called in a patient from a subdued light, and placed him at a distance of three or four feet from an ordinary gas-lamp. On examining the pupils, I have now and then been much surprised to find them decidedly smaller than they were under the same circumstances before the injection. This contraction has persisted for several minutes, when all at once the pupil has given way and become broadly dilated. It would seem from this observation as if the third nerve had been roused

[1] Dr. Tilt on Uterine Therapeutics, 3rd edit.

to unusual exertion just at the time when the increasing influence of the sympathetic began to be first felt, and that the sudden stimulus of light had called forth its opposing energy to such a degree that for a few minutes it was able to repress the rising force of the sympathetic, which a little later on would become overpowering.

The cause of the dilatation is variously interpreted. Meuriot says:—'At the present day experimental physiology has acquired the irrevocable fact, that after the action of atropia the 3rd pair do not respond to electrical excitation. None of the other nerves of the iris are similarly affected, and the sympathetic and trigeminal preserve their excitability. Atropine paralyses the common motor oculi in its peripheral extremities, and to this cause the loss of accommodation is referable' (p. 129). He admits that the more obvious effects of this paralysis, such as ptosis, loss of movement of the eyeball, and strabismus, are very rare, and states that 'in the majority of cases the interocular ramifications are alone implicated,' because 'the ciliary branches are bathed in a fluid in which it is easy to discover the presence of the poison' (p. 130).

Disregarding the somewhat unscientific postulates that the humours of the eye contain more atropia than the blood and common fluids, and that the effect of the poison on the conductors is more powerful than on the nerve centre, I readily admit that when a patient is comatose from the effects of belladonna, the 3rd nerve is completely paralysed, but I see no reasons for admitting that the *third* nerve is more readily paralysed under the influence of belladonna than any other motor nerve in the body. The effects of a poison which has a direct paralysing action on the third nerve have been fully considered at p. 8. Impairment of accommodating power, it has been seen, is the first of these effects, then follow ptosis and strabismus, and last of all, and in only a slight degree as compared with the result of belladonna action, dilatation of the pupil.

But let us carefully ascertain the effects of a mydriatic.

Obs. 84.—Having injected $\frac{1}{12}$ of a grain of sulphate of hyoscyamine beneath the skin of the subject of *Obs.* 20,

p. 126, the visual phenomena were carefully observed. The medicine produced powerful soporific effects without causing the least intellectual disturbance. At the end of 2½ hours the dilatation of the pupil was such that it contracted to ¼" only at the bright light of day. On endeavouring to read the paper at the usual distance, he was unable to decipher even the head lines. There was a thick haze before the eyes, he could not distinguish his finger-nails, and the letters appeared run together to form an uniform dark surface. But when he held the paper at full arm's length and at the same time inclined the head backwards—making a distance of 36 inches—he was able to read the smallest type of the 'Times' newspaper fluently. He recognised the lightning conductor (p. 127), as a film projecting from the side of the chimney pot. By means of a double convex lens or a pin hole in a card, he was able to read any type at a distance of three inches from the eye.

The condition of the vision just described appears to be exactly that which is experienced by the lower animals under the same influence. When the dog is called from a distance he turns round and looks directly at the face of the person who has called him, as if he recognised him clearly, but when he attempts to meet him, he seems to be enveloped in darkness. He hesitates at every step, runs his nose against every object in his way, and, unless assisted and encouraged by frequent calls, becomes distracted and gives up the endeavour. If he should succeed in approaching his master, he stops at a distance of one or two feet from him, then raising himself on the hind legs in the act of putting the paws upon his master's knees unexpectedly falls short of his legs altogether.

Such are the well-known effects of atropia on the vision. They are quite distinct from those produced by conium. Atropia causes dilatation of the pupil and presbyopia or more correctly hypermetropia, a defect that may be removed by the use of double convex glasses. But conium is followed by distinct weakness and sluggishness of the accommodating action, attended with every degree of overlapping of the images, and ultimately double vision; and these effects are accompanied by the outward signs of paralysis of the third nerve, viz., ptosis, strabismus, immobility of the eyeball, and dilatation of the pupil. It is further to be observed that dilatation of the pupil is neither the first, nor the most marked effect of the action of conium, and this is in accordance with what we observe in chronic disease of the third nerve. In such cases ptosis and strabismus are much more prominent symptoms than dilatation of the pupil. The two conditions then produced by belladonna and conium respectively are so different that we cannot reasonably refer both to the same cause; and as there can be no doubt that paralysis of the third nerve is the cause of the visual effects produced by conium, we must seek some other explanation of the atropia phenomena.[1] Leaving out of consideration any effect on the retina, dilatation of the pupil and hypermetropia constitute the whole of the effects of such doses of belladonna as may be taken without danger to life, and I venture to suggest that these are simply cause and effect. There is much difference of opinion as to the mode in which the accommodation of the eye for near and distant vision is effected. The explanation generally accepted is that which has been given by H. Müller and Bowman.

Mr. Bowman supposes that the lens is advanced by the action of the ciliary muscle, the contractions of which draw the ciliary processes forwards and outwards. But Helmholtz has shown that the adaptation of the eye is accompanied by a change in the figure of the lens, the antero-posterior diameter being increased, and the anterior surface rendered more convex, while the posterior surface undergoes no apparent change either in figure or in place. Now, traction upon the circumference of the lens in a forward and outward direction, if it had any effect at all, would cause a flattening of the anterior surface; the reverse of what occurs in the accommodation of the eye to near objects. It is plain, therefore, that we must find some other explanation of the manner in which accommodation is effected. And first it

[1] The best proof of the undiminished activity of the third nerve, is the *contraction* of the pupil which may be witnessed during the sleep of one under the influence of a full dose of atropia. (See p. 250.)

appears that the agent must be a *compressing* one, for as the lens is firmly fixed in the vitreous humor and is not free to move alone, no traction-force which does not compress the eyeball wholly or in part, can alter the position, or to any extent the form, of the lens.

A little consideration, I think, will show that the iris itself is the chief agent in compressing the lens, and effecting the accommodation of the eye.

Let us regard the anatomical connection of the part for a minute.

Towards its circumference the iris separates into two planes, an anterior and a posterior.

The anterior passes vertically upwards to be attached to the slightly incurved line of junction between the cornea and the sclerotic; the only part of the globe, it is to be observed, that can be drawn inwards by the action of the internal muscles of the eyeball. The posterior, diverging a little from the anterior, passes upwards and a little backwards over the anterior half of the ciliary body, where it comes forwards between the lens and the iris.

Very shortly after they come in contact the radiating fibres of the iris become continuous with the outer surface of the ciliary body. Traced further upwards and backwards, they become lost in the ciliary muscle itself which covers over the posterior half of the ciliary body. A muscular plane then extends from the choroid, just behind the ciliary body, to the pupillary margin. In its passage forwards, where it lies between the incurving ciliary processes and the corresponding part of the sclerotic, it becomes increased in quantity, and fills up the space left by the separation of the two, and forming an attachment to the margin of the sclerotic thus constitutes the so-called ciliary muscle. The contractile tissue is thence continued downwards over the anterior part of the ciliary body as the converging fibres of the iris (the dilator of the pupil), the inner extremities of which are connected with the sphincter.

The interruption between the fibres constituting the ciliary muscle, and those forming the circumferential parts of the iris, and the attachment of the former to the sclerotic, is an arrangement which is observed in other parts,[1] the fresh points of attachment giving greater power to the contraction of the muscle.

Such being the arrangement of the whole of the muscular tissue contained within the globe, we come now to ascertain the effects of its contraction.

In the first place we may safely assume that the muscular tissue of the iris is constantly maintained in a state of moderate tension, and that dilatation and contraction of the pupils is merely the result of a slight preponderance of one set of fibres over the other.

It necessarily follows that forcible contractions of the pupil tighten the whole system of radiated fibres, and that pressure is thus exerted upon the whole of the ciliary body, from its posterior limits behind, forwards and inwards in the direction of the lens, upon the margin of which the pressure is transmitted. Since, however, the central parts of the lens can alone offer substantial resistance, the margin is compressed, and the figure of the body altered accordingly.

Thus adaptation may be effected to some extent by the mere contraction of the pupil and when the radiating fibres are in a passive state. When, however, the sphincter, no matter what be the size of the pupil within moderate limits, forms a fixed line for the active contraction of the radiated fibres, the pressure on the ciliary body, and through it on the margin of the lens, is increased. If the pupil contract under these circumstances, the pressure becomes very considerable, and the anterior convexity of the lens is proportionately increased.

This pressure, which if directly transmitted to the lens by the irregular surface formed by the ciliary processes, would disturb the regularity of the refraction, is most equally distributed by the fluid contained in the canal of Petit; and thus the perfection of the refracting medium is maintained under every degree of pressure. The attachment of the ciliary muscle to the edge of the sclerotic is no doubt a contrivance for regulating the pressure upon the lens and preventing it from becoming at any time excessive.

[1] See a paper by Mr. Ellis, Royal Soc. Trans., 1856.

But further, the anterior plane of the iris is inserted into the slightly inflexed line of junction between the sclerotic and cornea, and contractions of the iris must, therefore, constrict the eyeball at this line and compress the contents to some extent. As results of this action, the convexity of the cornea is slightly increased, and there is a correspondingly slight advance of the central parts of the vitreous humor and the lens. The slight compression occurring simultaneously with the contraction of the radiating fibres, serves to steady the lens at the moment it undergoes compression.

The ciliary body being completely under the control of the radiating system of fibres, not only effaces at the time of its compression that portion of the lens in which spherical aberration chiefly occurs, but being simultaneously drawn forwards and inwards so as to overlap the front of the lens to a greater extent than in distant vision, still more completely obviates this defect by excluding the circumferential rays.

The effects above described are regulated by the contractions of the pupil, the degree of which in the production of a given amount of accommodation varies with the individual. Thus, as I have before stated, contraction of the pupil and accommodation are simply cause and effect.

The instant that the strain of near vision ceases, the pupil dilates, the engorgement of the ciliary vessels is relieved, the lens returns by its own elasticity to its original form, and the vision becomes presbyopic.

Such I believe is the intelligible and perfect contrivance by which the accommodation of the eye to near vision is effected. The action of mydriatics is the reverse of that which we have been considering. The stimulant effect of belladonna on the sympathetic simply overpowers the influence of the 3rd nerve. The sphincter gradually yields to the stronger contractions of the radiating fibres and the pupil opens. As this occurs, the action of the circular fibres is progressively weakened, and when the sphincter is fairly overpowered, it no longer affords an unyielding line to the contracting fibres, and these having now no fixed support for *inward and convergent* action are thus rendered incapable of compressing the lens.

When the dilatation is complete the hypermetropic effect is probably increased by the traction of the ciliary body forwards and outwards towards the lines of attachment of ciliary muscle and the anterior plane of the iris, which, while it cannot disturb the position of the lens removes from its edge even that pressure which is due to mere turgidity of the ciliary body. It thus appears that dilatation of the pupil and hypermetropia may exist independently of paralysis of the 3rd nerve.

Action on the Cerebro-spinal Nervous System.—The general effects of belladonna resemble those of opium in being excitant and hypnotic, but in their particular character they differ greatly from those produced by opium. The excitement does not produce cramp; and convulsion is an after-effect and not a result of the direct action of belladonna. So also the soporific effect is much less marked; it never, in man, amounts to narcotism, and the coma which supervenes during the last hours, after a poisonous dose, is, like the convulsions which occasionally happen, a remote consequence, and not a direct effect of the action of the drug.

It is only after moderate medicinal doses that we witness soporific effects (see *Obs.* 73, p. 203). If the dose be duly proportioned to the individual, the soporific effect is often very considerable, and in some persons as great as that produced by morphia. The effect, moreover, is quite independent of the alleviation of pain.

The influence of $\frac{1}{2}$ a grain of acetate of morphia on Samuel M. and John W. is recorded in *Obs.* 19 and 20, pp. 125 and 126. The hypnotic effect of $\frac{1}{40}$ of a grain of sulphate of atropia (given by the skin) on these patients was, to say the least, equally powerful. The latter patient, after remaining for 2 hours under observation in a very drowsy lethargic condition, has gone home, a distance of 4 miles, laid down, and slept soundly and continuously for 3 hours. (See also ACTION OF BELLADONNA IN ACUTE DISEASE, p. 250.)

After larger doses, insomnia and delirium arise, and poisonous doses prolong these effects for many hours, and the patient gradually passes into the comatose state. As occurs with opium, insomnia with illusions, or delusions, or both, occasionally follows a medicinal dose of belladonna (see *Obs.* 75, p. 205); the particular effect, as happens with opium, being determined by the condition of the patient or the peculiarity of nervous constitution.

The cerebral action of excessive doses of the two drugs differs chiefly, if not entirely, in the intensity of the soporific effects. Both cause delirium, but sleep converts that produced by opium into a dream, while the insomnia which accompanies the belladonna action allows of its active manifestation. In the dog and the horse the soporific effects of belladonna are not so powerful as in man, but in the mouse they are as potent as those induced by any of the principles of opium. In some the hypnosis amounts to narcotism, but the animal may generally be more or less completely aroused. Some observers[1] regard the insomnia as a consequence of the activity of the circulation; but the most intense excitement of the heart is compatible with heavy sleep or profound narcotism. (See *Obs.* on the dog, &c., COMBINED ACTION OF OPIUM AND BELLADONNA; and also *Obs.* 83 (*b*), p. 228).

The influence of belladonna on the *sensory centres* agrees very closely with that of opium. More or less anæsthesia results from the action of both drugs, but the effect of belladonna is less marked in diminishing natural sensation than in relieving neuralgia. Its influence on morbid sensation is not merely palliative, it is remedial. The beneficial effect is partly attributable to a direct action on the nerve tissue, and partly to the influence of the drug in removing hyperæmia, from which the neuralgia often results.

The action on the motor centres and the spinal cord is comparatively slight. The *corpora striata* participate both in the hypnotic and in the excitant effects. Giddiness and muscular weakness, from inability for exertion, rather than from real loss of motor power, accompany the hypnotic effect; while restlessness and insomnia as invariably occur when the hypnosis is overruled by the excitant action. Restlessness is a marked result of the action of belladonna in excessive doses. The subsequent exhaustion is proportionate to the previous excitement, but actual paralysis, like coma,

[1] Hammond on Wakefulness: Philadelphia, 1866. Meuriot. op. cit. p. 108.

is a result of exhaustion, and not an effect of the direct action of the drug.

The spinal cord is least of all affected by the action of belladonna. Some observers conclude that the reflex function is diminished, others that it is increased. The truth appears to be that it is only when the voluntary control over the muscular movements is weakened, that any effect on the cord is observable. When the movements become feeble and tremulous, it is then that jactitation appears and feeble convulsive movements are apt to supervene. It is not that the reflex activity of the cord is increased, but that the voluntary control is weakened.

The action of poisonous doses of belladonna is usually very protracted. In warm-blooded animals it will often continue for many hours (see *Obs.* 83 (*b*)); and in the cold-blooded frog it persists for several days. During the last few hours of the life of the mouse, and after it has been reduced to a torpid (comatose) condition, reflex movements can be readily induced by pinching the skin, or pricking the rump portion of the tail, or the soles of the feet; and sometimes the little animal will experience a general spasm, which only differs from that produced by codeia or thebaia (see p. 173) by the comparative slowness and feebleness of the movements.

The action of atropia leaves the frog in an excessively nervous state; the least disturbance causes great agitation, with increase of the respiratory movements, and a touch often throws the animal into a tetanic convulsion. (See Appendix.)

That the reflex activity of the cord is diminished by contraction of the vessels of the pia mater is a view in which we can no more agree than with that which attributes the dilatation of the pupil to a similar condition of the vessels of the choroid. We are satisfied that belladonna is unable to produce these mechanical effects (see p. 223).

We have seen that the effect on the 3rd nerve is neither depressent nor excitant to any appreciable extent; and if we examine any other of the nerves proceeding from the cranio-spinal axis we shall be led to the same conclusion.

Those which administer to the respiratory function are of the first importance and require separate consideration. With regard to the respiratory movements, it is generally considered that belladonna increases them. My own observations lead me to conclude that belladonna has no direct influence on the respiration (see *Obs.* 77, p. 207). In man, I have only once seen the maximum rate of respiration maintained during a state of rest (see *Obs.* 76, p. 206). In all other cases the breathing has assumed the diminished rate proper to repose. Large doses which cause general restlessness and nervousness, may indirectly excite the respiratory movements, but even this, according to my experience, is rare. Once I have noticed a persistent acceleration in the mouse (*Obs.* 83 (*b*) p. 228). I have occasionally witnessed, during the rise of the belladonna symptoms, a little transient panting in the dog; but these effects are too rare and inconstant to be attributed to a direct action on the respiratory movements. My enquiry, however, has reference to the action of medicinal doses, and of these I can say unreservedly that they have no influence whatever on the respiration.

On the Vagus.—Dr. Meuriot[1] concludes that atropia is a *vasculo-cardiac* poison, and that it accelerates the action of the heart by paralysing the peripheral branches of the pneumo-gastric nerves; and that while small doses accelerate the respiratory movements by exciting the respiratory centres, large doses diminish the respiratory movements by paralysing the extremities of the vagus.

The facts from which he derives these conclusions are the following:—

1. Weak doses of atropine accelerate the heart and increase the pressure of blood in the arteries. Poisonous doses have the reverse effect.

2. In animals in whom the vagi had been previously cut, atropia causes no acceleration of the pulse, but the contrary, together with a diminution of the pressure.

3. In animals in whom the vagi were divided when they were already under the influence of atropia, no acceleration of the heart followed, and only an increase of the pressure;

[1] Op. citato, pp. 138 and 73.

but afterwards there was a decrease both of the pulse and of the pressure.

Now let us stop to enquire for a minute what are the effects of the section of the pneumo-gastric nerves? First the pulse attains its maximum acceleration for the individual; 'the heart is in a continuous thrill, and never empties itself, so that the whole systemic circulation is unduly distended; the thoracic cavity is in a state of permanent distension in consequence of the excessive action of the inspiratory muscles; the veins in the chest are so distended that they admit of no further amplification during the inspiratory effort.' [1]

Section of the vagi then is tantamount to removing the great forces of free active dilatation of the right heart, and the suction power of the inspiratory act, in advancing the flow of the blood, and to placing at the same time a loose ligature around the pulmonary artery. We are all familiar with the consequences. It is not that the heart is cut off from the influence of inhibitory nerves, as M. Sée, MM. E. and M. Cyon and others would have us believe; but the organ is naturally excited to the utmost, and the pressure within the left ventricle is greatly increased by the engorgement of the pulmonary circulation. The heart is exercised to the full extent of its power, and neither belladonna nor any other agent can urge it beyond. Nay, further, an artificial stimulus under these circumstances, if it have any influence at all, acts only as a tonic and apparently as a sedative in reducing the irritability of the organ. Now according to Dr. Meuriot's statements, belladonna actually does this, and if the pneumo-gastric nerves be divided before or during the action of atropia, the result is the same, the heart's action is a little quieted, and the arterial pressure is decreased.

This is precisely the effect of belladonna in disease. When the pulse is at its maximum in fever, and when the lungs are perfectly free, the subcutaneous injection of atropia in either small or large doses fails to increase it, and generally

[1] 'Influence of the Movements of Respiration on the Circulation of the Blood,' by J. Burdon Sanderson, M.D. Phil. Trans., vol. 157, p. 591. 1867.

causes a diminution of its frequency. (See THERAPEUTICAL USE OF BELLADONNA, p. 246).

Belladonna, we conclude then, has no paralysing action on the vagus nerves, as exhibited in either the respiratory movements or in the action of the heart. Further confirmation of this conclusion will be derived from a consideration of the effects of the combined action of belladonna and opium in a subsequent chapter.

That large doses of belladonna ultimately decrease the rapidity and force of the heart's action, I have already admitted; but this arises from no special action on the vasculo-cardiac system, but from overstimulation, or exhaustion of the sympathetic nervous system.

I have compared the action of atropia on the nervous centres to that of spongy platinum on hydrogen, and the simile admits of being fully and consistently applied. The most decisive result of the action of belladonna is a very considerable increase of the phosphates, and the whole of the accompanying phenomena may be attributed to excessive stimulation of the nerve-centres attended by increased oxydation. This effect is determined by the presence of atropia, just as the union of oxygen and hydrogen is determined by the presence of spongy platinum.

The essential action of atropia, therefore, appears to be hyperoxydation of nerve-tissue. If the supply of oxygen were proportionately increased the process would probably be sustained until nutrition failed; but the respiratory function is only rarely and exceptionally increased, the blood suffers deoxydation, the nervous system manifests general depression and symptoms resembling those of poisoning by carbonic oxide or carbonic acid supervene. The temperature falls and the skin becomes cold and dusky, the respiratory movements become slower and shallower, the heart's action weaker, and torpor soon ends in coma.

The influence of the respiration in sustaining the action of the drug is well seen in contrasting its effects on the subjects of *Obs.* 83 p. 227. To all appearance the animals were equal in development and vigour, and the circumstances under which they were placed were also equal.

CHAPTER VI.

THERAPEUTICAL USE of BELLADONNA.—We come now to the practical application of the facts which the study of the physiological action of belladonna has disclosed, and it is obvious that our notions of the therapeutical powers of the plant must not be confined to its anti-spasmodic and anodyne properties; and that in the medicinal use of the drug its deliriant effects should rarely or never be induced.

Belladonna must be regarded in the *first* place as a direct and powerful stimulant to the sympathetic nervous system, or in other words, to the heart and bloodvessels; *secondly*, it is a potent diuretic; *thirdly*, by virtue of a direct action on the nerve-centres, and of its stimulant effect on the circulation, it is an oxydising agent; *fourthly*, it possesses powerful anodyne and hypnotic properties; and *fifthly*, as a result of its action on the sympathetic and sensory centres, it is a most valuable antispasmodic. First as a *vasculo-cardiac stimulant.* It is remarkable that this, the primary and essential result of the operation of belladonna, should have been so long overlooked. So simple and immediate is the influence of this plant in exciting the action of the heart, and so powerful in increasing and sustaining the force and rapidity of the circulation, that none but its own natural allies datura and hyoscyamus at all approach it; and in the directness and simplicity of its action it is superior to either of these. Simply then as a general diffusible stimulant, belladonna surpasses all other drugs, whether derived from the animal, vegetable, or mineral kingdom. In all conditions and diseases, therefore, in which there is depression of the sympathetic nerve-force, such as syncope from asthenia or shock; in the collapse of cholera; in failure of the heart's action from chloroform or other cardiac paralysers, the subcutaneous use of atropia is the most appropriate and hopeful means of resuscitation.

In using it as such, we must bear in mind the fact which has so often forced itself on the attention in the course of our enquiry, that the stimulant action of the drug is soon superseded when the dose is excessive by depressent effects. In other words, the power of the drug to exhaust is in direct proportion to its power to stimulate. Given with a view of exciting or sustaining the heart's action, the dose will range from the $\frac{1}{100}$ to the $\frac{1}{60}$ of a grain, and it should never exceed the $\frac{1}{40}$ (see p. 205–6, &c.).

While cases such as the above will call for the occasional use of belladonna, my experience of its beneficial action in *acute disease*, in hyperæmia and stasis from impaired power or disordered action of the sympathetic, either general or local, leads me to believe that it has not yet attained to its legitimate place as a therapeutical agent, and to anticipate that its sphere of usefulness will be acknowledged before long to be co-extensive with that of acute disease itself.

The similarity of the general phenomena which attend the operation of belladonna and those which accompany pneumonia, enteritis, the development of pus in any of the tissues or organs of the body, &c., has already arrested the attention (p. 217). We know that these local diseases are the result of hyperæmia and stasis, and of exsudations, and their transformations which are taking place to relieve the congestion of the bloodvessels of the part. But what of the pyrexia? why the general vascular excitement because the bed of the nail or the tonsil is inflamed? We call it symptomatic or *sympathetic*, and so it really is. One part of the sympathetic nerve is irritated, and the whole system is aroused against a local offender. The pyrexia then is not the disease but the remedy. Nature is lavish in this as in all else; she developes the curative means abundantly, and leaves us to control them. Thus we dissociate pyrexia from morbid action; but if it be left uncontrolled, exhaustion ensues, and the remedial action becomes a part of the morbid process, just as occurs with belladonna itself when used injudiciously. Thus in applying belladonna to the treatment of acute disease, we are not blindly led by an unscientific dogma, but simply follow nature. We divert the sympathetic from the original cause of its irritation by the introduction of another and more general influence which can be readily adjusted and controlled.

Under the influence of this more general action we see the local irritation and pain abate as soon as the hyperæmia is relieved; and the products of the inflammatory process are rapidly removed as the circulation through the part is freely established.

A clear view of the action of belladonna on the healthy tissues is readily obtained by such an experiment as that recorded at p. 220. The blood is equally distributed throughout the body, we see the ultimate ramification of the arteries contracted to about $\frac{3}{4}$ of their original dimensions, there is increased tone with slight contraction in the capillaries themselves, and the part contains only a moderate amount of blood, while the rapidity of the circulation is such that it receives in a given time twice the quantity of blood that it does in health.

We may as readily satisfy ourselves of the influence of the drug in removing congestion and stasis. Thus, if $\frac{1}{200}$ of a grain of sulphate of atropia be injected under the skin of a frog in which some cardiac paralyser has previously produced a condition of stasis in the web, we shall soon see the oscillating current begin to take a forward course, and in a short time the flow will be re-established, the dilated vessels will recover their original dimensions, the circulation will proceed with unwonted tone and vigour, and for many hours a slight contraction of the bloodvessels may be observed.

Passing from these observations to a consideration of the action of belladonna in inflammation, we are naturally led to expect that its influence would be both powerful and beneficial. And such indeed is the case, provided that the medicine be timely and judiciously used.

The action of belladonna in febrile diseases is frequently attended with results which are not only unexpected, but exactly the opposite of what is observed in health. Thus it may happen, if we give a full dose of atropia to a patient with a pulse of 120 and higher, a dry and hard tongue, and pupils measuring the $\frac{1}{6}''$, that after 10, 20, or 30 minutes, when the action of the belladonna is fully developed, the pulse will be decreased (*Cases* 5, 10), the tongue be moist (*Cases* 6, 8), and the pupils contracted (*Case* 5). Two similar effects, the one arising from a local irritation, and the other from the presence of belladonna, like spreading circles on a smooth sheet of water, interfere and neutralise each other. The coincidence of the two actions and a corresponding augmentation of the effects may be possible, but this I have never witnessed. Diminution of the pulse with increase of force, moistening of the tongue and unchange of the pupil, may often be observed; contraction of the pupil is a rarer effect.

It appears, therefore, that the stimulant action of belladonna is converted in great measure in febrile diseases into a tonic and sedative influence.

With this brief introduction, I will now record some of the results of my experience of the medicinal use of belladonna in acute disease.

1. **Pneumonia.**—If it were true that belladonna paralysed the vagus, its influence in this disease would be the reverse of beneficial. If on the other hand it is to be regarded as a simple stimulant to the sympathetic nervous system, it is of all remedies the most hopeful. The results of its use prove that the latter is the more correct view of its action. In watching the following severe cases I have been astonished to witness the rapidity with which the grave symptoms have subsided, and perfect convalescence has been established.

Case 1.—Jane R., aged 14, admitted into the London Fever Hospital on the *fifth day*. P. 120. Respiration 54, shallow. Much pyrexia and cough. Lower half of the left back quite dull with faint naked tubular breathing towards the spine, and absence of breath-sounds elsewhere. Diminished resonance of and fine moist crepitation in the rest of the lungs. Diffuse crepitation in the right lung behind. Expectoration scanty, viscid, and rusty-coloured.

℞ ♏ xx. *Tincturæ belladonnæ cum* ℥ ss. *Spiritûs ammoniæ compositæ*: quartis horis.

Tenth day. Had continued to improve rapidly. Pulse 65.

Respiration 25, with good heaving of the chest. No cough or expectoration. The left back perfectly resonant, and the respiration full and clear, excepting below the left axilla, where there was a little pleuritic rubbing, and pain on taking an extraordinary inspiration. Tongue moist and clean. Skin cool. A desire for food.

Twelfth day. Convalescent, bright, and cheerful. Pulse 86. Respiration 19. Lung-sounds everywhere clear and strong. Tongue clean, dryish anteriorly. Pupils $\frac{1}{3}$ away from the light. Omitted the belladonna.

Case 2.—Michael T., aged 30. Admitted on the *3rd day.* Pulse 118. Respiration 40, shallow. Much pyrexia. The left back, from the spine of the scapula downwards, completely dull and silent; above this part fine crepitation and tubular breathing. The right back resonant, but there was diffuse moist crepitation. Cough very troublesome; expectoration frothy, viscid, and tawny.

♏ xx. *Tincturæ belladonnæ quartis horis.*

Twelfth day.—Had progressively improved, and now there was a total absence of febrile action. Pulse 60, of good volume and power. Respiration 25, regular. Pupils $\frac{1}{4}''$. Tongue clean and pink, but dryish and rough anteriorly; complained that the medicine caused great thirst, and that he could not see to read. Left side everywhere resonant, but less so than the right. Air entered everywhere with a clear, loud, vesicular sound, excepting beneath the spine of the scapula, where the breathing was naked and only tubular and the voice bronchial; and at the extreme base, where there was fine crepitation and occasionally a little mucous râle, but the air entered well. Breathing good in the front with a little snore or crackle at the end of inspiration. Right lung clear everywhere. No expectoration; a very little dry cough.

Fourteenth day. Felt well, but complained of the dryness of the mouth and thirst. Pulse 50. Respiration 25. Pupils $\frac{1}{4}''$. Left chest everywhere resonant, and the air entered freely but feebly with a natural vesicular sound, excepting in a limited portion about 2 inches square beneath the spine of the scapula; and in the axilla. In the former situation the respiration was still tubular, and in the latter there was diminished resonance with obscure breathing accompanied by a faint and distant crackling. There was no cough. Appetite good, and wished to get up.

Sixteenth day. Pulse 55. Respiration 25. Morbid area behind reduced to a little spot which could be covered by the end of the stethoscope. In this there was tracheal breathing still. There was but little change in the axillary region.

Twentieth day. Resonance and breath-sounds everywhere normal. Convalescent. Omitted the belladonna.

Case 3.—In the following case mercury, opium, and belladonna were used simultaneously. In advanced stages of the disease the combination is a most valuable one:—

Georgina G., aged 15, admitted into the London Fever Hospital on the *seventh day* of the disease. Pulse 120. Respiration 40. Pupils $\frac{1}{7}$″. Tongue furred and moist, red at the tip and edges. Severe pyrexia. Face flushed and anxious. The right back quite dull throughout; breathing inaudible in the lower half, in the upper half tubular. Cough severe, with rusty-coloured pneumonic sputum.

Ordered *ammonia and senega*, constant *sinapisms* to the back, and ℥iv. of *wine.*

Ninth day. Pulse 120. Respiration 50. Physical signs unchanged. Skin very hot. Face deeply flushed. Slept little and was restless.

Calomel $\frac{1}{4}$ grain and *opium* $\frac{1}{6}$ grain, every two hours. ♏xv. *Tincturæ belladonnæ* every four hours, and a *blister* to the right back.

Eleventh day. Pulse 96. Respiration 35. Pupils $\frac{1}{4}$. Cough and expectoration less. Slept well. Skin moderately cool.

Thirteenth day. Pulse 84. Respiration 25, no cough. Air entered all parts of the lung audibly. Tongue cleaner. *One pill every* 12 *hours.*

Fifteenth day. Pulse 65, regular. Respiration 20, natural, very little cough; expectoration ceased. The right back everywhere resonant, and almost naturally so. Air entered every part, but there was fine crepitation throughout the lower lobe. The blister was healed. Tongue clean and moist.

Skin cool, flushing gone. Gums unaffected by the mercury. Appetite returning. To have fish.

Eighteenth day. Progressively improved. Pulse 76. Respiration 14. Pupils $\frac{1}{3}''$, inability to read. Tongue clean and moist. Gums slightly swollen, faint mercurial fœtor. Skin cool. Appetite good. Right back everywhere naturally resonant. Respiration everywhere clear and natural. Omitted the pill. Full diet.

Twenty-second day. Bright and cheerful and quite convalescent. The lungs were perfectly healthy, and the functions quite natural. Pulse 70, of good volume and power. Pupils $\frac{1}{3}$. Mercurial effects disappeared. Appetite good. Omitted the belladonna.

II. Enteric Fever.—I have used belladonna in this disease with the best results. As my space will not allow me to record cases, and as the number in which I have used the remedy is too small to furnish statistical evidence, I must content myself with indicating the general effects of the remedy.

Case 4.—Richard B., aged 18. Seventh day. Pulse 96, of good volume and power. Tongue moist. Pupils $\frac{1}{8}$. *Injected $\frac{1}{96}$ grain sulphate atropia.* After $\frac{3}{4}$ hour, pulse 104, full and bounding. Tongue unchanged. *Felt comfortable.*

Eighth day.—Pulse 90, good and full. Tongue moist. Pupils $\frac{1}{8}''$. *Injected $\frac{1}{48}$ grain sulphate atropia.* After 10 minutes no change. After 30 minutes, pulse 100; no change in the tongue or pupils. *Inclined for sleep.* After $1\frac{1}{2}$ hour, pulse 96, regular; tongue not quite so moist. *Pupils $\frac{1}{9}$, sleepy*; when aroused $\frac{1}{8}$, unchanged. *Slept soundly during the last hour.* After 2 hours, pulse 94. Still sleeping, and on raising the eyelids the pupils were found to be *contracted as before to $\frac{1}{9}$.* Convalescent 10 days afterwards.

Case 5.—Elizabeth B., sister of the above, aged 14. Seventh day. Pulse 120. Pupils $\frac{1}{4}''$. Tongue moist. Bowels very loose. Injected $\frac{1}{96}$ *of a grain of sulph. atropia.* After 20 minutes, pulse 112; pupils *contracted* to $\frac{1}{8}$; tongue quite moist; had been sleeping tranquilly. After $\frac{3}{4}$ hour, pulse 108, fuller; was sleeping soundly; and on raising the eyelids the pupils were directed towards the inner canthus and *contracted to* $\frac{1}{11}''$. On waking they dilated to $\frac{1}{7}$. Tongue unchanged. After 1 hour, was still sleeping, with the pupils contracted to $\frac{1}{11}$. Pulse resumed its initial rate, 120. Convalescent 12 days after.

Case 6.—William A., aged 16. *Tenth day.* Pulse 84. Pupils $\frac{1}{8}''$. Tongue very red and glazed, with prominent papillæ, dry along the centre, moist at the sides. Injected $\frac{1}{48}$ *grain sulphate atropia.* After 12 minutes, sleeping; *pulse* 80, *fuller*; pupils $\frac{1}{10}$, when aroused $\frac{1}{8}$; tongue unchanged. At the 25 and 35 minutes, pulse 88; still *sleeping, pupils* $\frac{1}{10}$, when aroused $\frac{1}{7}$. *Tongue inclining to moisten.* After 1 hour, pulse 88; pupils $\frac{1}{6}$ when awake. *Tongue quite moist.* After $1\frac{1}{4}$ hour was still sleeping, and on waking him, the *tongue* was *still moist.* Convalescent 9 days afterwards.

Case 7.—Elizabeth H., aged 17. About the *fortieth day.* Excessive prostration and apathy; excretions voided in bed; severe tympanites. Pulse 120, small and feeble. Tongue dry, brown, and cracked, edges slightly moist. Injected $\frac{1}{48}$ *grain sulphate atropia.* After $\frac{1}{2}$ hour, pulse 120, a little fuller and more resistant. Tongue quite *moist* and covered with a little creamy fur. No sleep. She died 5 days afterwards. The fæces were solid and the intestinal lesions nearly healed.

III. **Typhus.**—The influence of belladonna in modifying some of the more prominent symptoms of the disease is seen in the following cases:—

Case 8.—Thomas H., aged 30. Admitted, apparently after an attack of *typhus,* in a state of extreme prostration, semi-conscious, and apathetic. Pulse 110, very small and feeble. Pupils $\frac{1}{10}$. Tongue brown, hard, wrinkled, and destitute of a trace of moisture. Ordered ℥ xv. of brandy in the 24 hours, with an ammonia mixture. He continued in this state several days without manifesting the least improvement, and the dryness of the mouth was such that a piece of bread was not moistened when masticated. *Injected* $\frac{1}{32}$ *of a grain of sulphate atropia.* After 10 minutes, pulse 112, soft; pupils $\frac{1}{7}''$; *tongue moistening.* After 20 minutes, pulse 112, pupils $\frac{1}{4}$, *tongue quite moist.* During the next hour, the pulse attained a maximum acceleration of 120 and became full and soft, the patient was aroused, manifested a little pleasing delirium and said his tongue felt more comfortable, and, putting his hand upon the cardiac region, 'that his heart was so puffed up.' In the course of the hour the *tongue* thrice showed *a tendency* to become *dry again.* After $1\frac{1}{4}$ hour, pulse 112, only slightly fuller, and scarcely of better power than before the injection. *Tongue quite moist.* The next day he had relapsed into his previous condition, and died two days afterwards.

Case 9.—Mary W., aged 41. *Twelfth day.* Pulse 110. Pupils $\frac{1}{10}''$. Tongue dry and cracked. *Injected $\frac{1}{60}$ grain sulphate atropia.* After 24 minutes, pulse 118, unchanged else; pupils $\frac{1}{9}$; tongue *moist* round the edges; was very comfortable.

After 50 minutes, pulse 112, pupils $\frac{1}{9}$, tongue *quite wet* everywhere; felt inclined for sleep. After $1\frac{1}{4}$ hour, pulse 112, pupils as before, *tongue drying* at the centre; continued to doze.

Two days afterwards, pulse 96, tongue inclining to moisten, the cracked epithelium separating; pupils $\frac{1}{8}$. *Injected $\frac{1}{48}$ of a grain.* After 10 minutes, pulse 100, fuller; no change in either the tongue or pupils. After $\frac{3}{4}$ of an hour, pulse 96, of good volume, pupils $\frac{1}{7}$; tongue *moister* around the edges; was comfortable, but had not slept. Convalescent 5 days afterwards.

Case 10.—George P., aged 35. *Ninth day*, pulse 110, soft, full, and intermittent; pupils $\frac{1}{8}''$. Tongue completely dry, brown, and wrinkled; hard palate glazed. Respiration 32. Injected $\frac{1}{48}$ *grain sulphate atropia.* During the next hour the *pulse gradually decreased* to 96, the volume, power, and frequency of the intermission remaining unchanged. Respiration 30. *Pupils unchanged.* At first a *little moisture* appeared on the back parts of the tongue and palate, but soon disappeared again.

Case 11.—James M., aged 21. *Tenth day.* Pulse 80, of fair power and volume. Pupils $\frac{1}{7}''$. Tongue red, dry, glazed, and cracked, edges moist. Hard palate dry and glazed. Injected $\frac{1}{48}$ *grain sulphate atropia.* After 17 minutes, pulse

82, fuller; pupils $\frac{1}{6}$; moist margin of tongue a little wider; hinder part of the hard palate quite moist. After $\frac{3}{4}$ hour, pulse 82; a few drops of clear moisture on the centre of the tongue; moisture advancing from the back and sides. Injected $\frac{1}{60}$ *grain* more. After $\frac{1}{4}$ hour, pulse 92, pupils $\frac{1}{4}$, tongue and palate as at $\frac{3}{4}$ hour. Respiration unchanged. Continued to sleep tranquilly.

IV. **Acute Nephritis.**—If belladonna really influences congested bloodvessels in the way above indicated, in any tissue of the body indifferently, it is to be expected that its power in relieving inflammatory conditions of the gland, which is solely instrumental in eliminating medicinal doses, will be even more rapid and decided.

Whether these expectations are realised or not in the following cases, I must leave the reader to determine. I wish that I could have given further illustration of this interesting question; but I can only offer such as the requisite leisure for the tedious enquiry has allowed me to accumulate.

In nephritis the quantity of albumen in a given sample of urine will depend on two conditions:—1, the ingestion of nitrogenised diet; and 2, the activity of the body. After a full meal and exercise the albumen will attain the maximum, and after a period of rest and fasting the minimum amount. If a dose of atropia be given when the exsudation of albumen is at the maximum, it will be observed to undergo little or no diminution so long as the belladonna symptoms continue fully developed; towards the close of its operation, however, a gradual diminution will be observable. On the other hand, if the dose be administered when the exsudation of albumen is at its minimum, the first effect will usually be a slight increase, unless the dose be a very small one; and if the dose be excessive the albumen will be increased to the maximum amount. If the dose be not excessive the albumen begins to decrease as the symptoms of the belladonna action decline, and after a few hours the beneficial influence of the drug becomes apparent. This is strictly in accordance with what has been observed of the effects of large and small doses on the heart and larger bloodvessels, and in using belladonna in renal disease, we must be careful to avoid large doses, remembering that the action of the drug is not simply confined in this case to the relief of congestion, but that it also excites the functional activity of the kidneys—the whole of a medicinal dose of belladonna being eliminated by these glands.

Atropia passes freely through the kidneys even when they are in the most advanced stage of fatty or albumenoid degeneration, and the diuretic effects of the drug are as marked in nephritis as in the healthy condition of the glands. The first result of the diminution of congestion in the acute form of the disease is the increase in the quantity of the water itself. This effect is well marked in the following cases.

As far as the remedial effect is concerned, it is wholly due to belladonna. I have refrained from prescribing a laxative, or even a draught of water, as Dr. Dickinson recommends, in any case, and have allowed the patients to go out as usual.

In estimating the quantity of the albumen in the urine, I have adopted the following process. Make three small filters from the same sheet of paper, and cut the other two down to the weight of the lightest, and mark the filters with a pencil A, B, and O. Take 1,000 grain measures of urine A, and having boiled it, pour it while hot upon filter A. Treat urine B in the same way. As soon as the urine has passed, the filters and albumen are thoroughly washed with warm distilled water from all traces of urine. After this has been effected, an ounce of water containing two drops of nitric acid is poured upon the albumen, and the filter and its contents are subsequently washed free from acid. The filter marked O is placed in the first instance between one of the other filters and the funnel, and is thus equally saturated with urine and acid, and equally washed from both. All three filters are dried together, and in determining the weight of the albumen, the empty filter is used as a counterpoise to the others.

It will be observed that the increase of phosphates is not so apparent in the analyses of the albuminous urine; this is chiefly due to the precipitation of the excess of the phosphates with the albumen; and thence the need for washing the latter with nitric acid. In these observations the quantity of phosphates must therefore be regarded as only approximatively correct.

Case 12.—*Acute nephritis, treatment commenced as the dropsy was appearing. Recovery.*

Joseph P., aged 15, a robust youth of apparently healthy constitution, and previously free from disease, came under my care *March* 24, 1868.

A few days before, on the 17th or 18th, he was chilled while working in the open air without his coat. *On the* 19*th* the urine was observed to be 'dark red as if mixed with blood.' *On the* 20*th*, there was nausea. He continued at his work, but the *next day*, being Sunday, the nausea continued and he vomited once, he therefore remained in bed, and had alternate chills and heats. He felt faint, and there was pain across the belly and in the loins. *On the* 24*th*, he came to me for advice. He had had no medicine, the bowels had acted regularly and naturally; there was no pyrexia, but some thirst. He complained of pain in the legs, and there was slight pitting over the tibia. Pulse 64, regular, of good volume and power. Tongue wet and a little furred; bowels acting as usual twice a day, and the motions solid. The urine had been and still continued scanty, less than 20 ounces in the 24 hours. It was now of the colour of very weak coffee, and very turbid from the presence of a chocolate-coloured flocculent matter which on settling occupied about $\frac{1}{4}$ the bulk of the urine. It was chiefly composed of light brown molecular matter, vast numbers of shrivelled and disinteg rated blood corpuscles, and fragments of fibrinous tubular casts. The chemical analysis of three samples of urine voided on the 24th, and the subsequent progress of the case as indicated by the condition of the urine, is subjoined.

I commenced treatment by injecting the $\frac{1}{48}$ *of a grain of sulphate of atropia* beneath the skin. It produced full effects; the pulse was sustained between 128 and 120 for an hour; the mouth was completely dried, and the voice rendered very husky. The action was attended by an increase in both the water and the solid constituents (see urine 3). The absolute increase could not be determined, for dysuria resulted, and the bladder could not be completely emptied.

The subsequent treatment consisted in the administration of ♏v.—♏xx. *succi belladonnæ*, morning and night, and the occasional injection of atropia beneath the skin. The doses were adapted to keep up moderate effects. At 11 A.M. the pulse ranged between 70 and 80, and was of good power, the skin was warm and naturally moist, the mouth dryish, and exhaling a slight belladonna fœtor, the tongue dryish and rough, the pupils dilated from $\frac{1}{8}$ to $\frac{1}{5}$, the vision obscure.

This treatment was discontinued at the *end of the third week*, when the patient, having regained his strength, returned to his work.

Owing to the reappearance of a trace of albumen a few days afterwards (April 24) the belladonna was resumed, and he was allowed to work half time only. Nine days afterwards the medicine was finally discontinued. On scanning the following account it will be observed that the quantity of the urine rapidly increased. This was an immediate result of the first injection; the diuretic influence of the belladonna is well seen in urines 6 and 7, ℥xiij. being secreted in 2¾ hours. The urine 9, March 30, contained only a faint cloud of flocculent matter. It was composed of a few altered blood discs, a little free renal epithelium, and a few large granular nucleated corpuscles, and here and there a fragment of a granular cast with adherent renal epithelium. In the course of a few days the organic deposit disappeared.

The albumen disappeared on the 10*th day of the treatment* (April 3). The urine was subsequently examined every second or third day; it continued quite free from albumen and abnormal deposit until April 24, when it reappeared as a slight trace in the evening urine after a long and unusually hard day's work. On resuming the medicine the albumen finally disappeared three days afterwards. The urine was examined at intervals for a month after the medicine was discontinued, and when he was in full work; and it constantly remained quite healthy, and the patient has continued well up to the present time.

CASE XII.

March 24. No food taken since dinner, at 12 noon.	1. Secreted between 3 and 5.30 P.M. ʒij. sp. gr. 1022, acid turbid, deposits a large quantity of brown flocculent matter, the supernatant fluid resembling very weak coffee. 1000 grain measures contained: Grs. Chlorine . 2·22 Urea . . 26·66 Sulphates and Phosphates 11·66 Albumen . 2·22	2. Secreted between 5.30 and 7.15 P.M. ℥j. ʒij. sp. gr. 1022, more acid than 1, a little paler: deposits a cloud of lithates upon a bulky deposit of brown organic matter; supernatant fluid smoky. 1000 grain measures contained:— Grs. Chlorine . 2·62 Urea . . 26·66 Sulphates and Phosphates 10·5 Albumen . 2·2	3. Secreted between 7.15 and 9.45 P.M., during the action of $\frac{1}{48}$ grain Sulph. Atropia (by the skin) ℥iv. sp. gr. 1020·3, acid, deposits a white cloud, less bulky than that of 1 or 2. Supernatant fluid bright, dark sherry-coloured. 1000 grain measures contained:— Grs. Chlorine . 3·23 Urea . . 25·33 Sulphates and Phosphates 9·08 Albumen . 2·5	
March 27. Bread and butter, and one pint of tea, at 8.30 A.M.; then walked two miles. Passed ℥xx. more of urine after 1.30 P.M. March 28. Passed 40 oz. of clear urine.	4. Passed on rising at 7.30 A.M. sp. gr. 1015·6, of a pale sherry colour, quite clear and bright, with only a faint deposit like the natural cloud of mucus in healthy urine. 1000 grain measures contained 0·85 grs. of albumen.	5. Passed at 10.45 A.M. sp. gr. 1018·4, acid, full sherry-coloured, slightly turbid, deposits a small white cloud, organic. 1000 grain measures contained:— Grs. Chlorine . 4·47 Urea . . 16·33 Sulphates and Phosphates 6·41 Albumen . 2·05	6. Secreted between 10.45 A.M. and 12.20 during the operation of $\frac{1}{48}$ grain Sulphate Atropia (by the skin), ℥v. sp. gr. 1012·8. neutral, paler, and with a smaller cloud than 5. 1000 grain measures contained 0·50 grain of albumen.	7. Secreted between 12.20 and 1.30 P.M., and towards the close of the operation of the atropia, ʒvij. sp. gr. 1010·8, neutral, otherwise as 6. 1000 grain measures contained:— Grs. Chlorine . 2·82 Urea . . . 12·0 Sulphates and Phosphates 4·33 Albumen . 0·90
March 29. Passed 44 oz. of urine.	8. That voided at 10.30 P.M., half an hour after supper, sp. gr. 1020·8, acid, a faint flocculent white cloud, clear and bright, contained an excess of uric acid. 1000 grain measures contained 0·6 grain of albumen.			
March 30. Bread and butter, and one pint of tea, at 8.45 A.M.	9. Passed at 8.30 A.M. on rising, ℥xx. sp. gr. 1012·8, acid, natural colour, faint flocculent cloud, uric acid in excess. 1000 grain measures contained 0·3 grain of albumen.	10. Secreted between 8.30 A.M. and 10.30 A.M. ℥vij. sp. gr. 1012·4, bright and clear, natural colour, no appreciable cloud; uric acid an excess. 1000 grain measures contained 0·15 grain albumen.	11. Secreted between 10.30 A.M. and 1.55 P.M. during the operation of $\frac{1}{60}$ grain of Sulphate Atropia. ʒv. passed with difficulty, a portion retained sp. gr. 1016. Alkaline. A slight cloud; supernatant urine bright, much triple phosphate after four hours. An excess of uric acid. 1000 grain measures contained 0·8 grain of albumen.	
March 31. Passed 50 oz. of pale urine.	12. Voided at 10.30 P.M. sp. gr. 1015·6, acid, clear, pale sherry-coloured, with a slight organic cloud. Uric acid in excess. 1000 grain measures contained 0·27 grain of albumen.			
April 1. Passed 57 oz.	13. Passed at 8.30 A.M., on rising, ℥xv. clear, but faintly opalescent, sp. gr. 1008. 1000 grain measures contained 0·15 grain of albumen.	14. Secreted between 8.30 A.M. and 11.45 A.M. ℥xii. sp. gr. 1014·4, contained a mere trace of albumen.		
April 2. Passed 50 oz.	15. Passed at 10.30 A.M. sp. gr. 1018, strongly acid, no deposit, bright pale sherry-coloured. 1000 grain measures contained 0·27 grain albumen.			

Case XII.—*continued.*

April 3. Passed 76 oz. of urine in 24 hours.	16. Passed between 10.30 P.M. and 7.30 A.M. ℥xxiij. sp. gr. 1011·6, pale and clear, a very small faint cloud; uric acid a large excess; no albumen. Heat and nitric acid cause a fluorescence, not amounting to opalescence, and no trace of flocculi.	17. Secreted between 7.30 and 11.45 A.M. ℥xvjss. sp. gr. 1007·2, acid, pale whey-coloured, bright; no trace of albumen. No fluorescence with heat and nitric acid. 1000 grain measures contained:— Grs. Chlorine . 2·20; Urea . . . 6·66; Sulphates and Phosphates 2·66; Uric acid, an excess.	18. Secreted between 11.45 A.M. and 2.45 P.M., during the action of $\frac{1}{75}$ grain of Atropia Sulphate (by skin), ℥vijss. sp. gr. 1013·2, faintly alkaline, sherry-coloured, clear, but on close inspection a fine scattered flocculent matter can be seen. No trace of albumen. 1000 grain measures contained:— Grs. Chlorine . 3·54; Urea . . 12·66; Sulphates and Phosphates . 7·0; Uric acid, an excess.

April 22.—At 6 A.M. ℥viijss. sp. gr. 1020, acid, clear, no albumen. At 11.30 A.M. sp. gr. 1014, faintly alkaline, slightly turbid from spermatic products.

" 24.—At 10 P.M. sp. gr. 1020, strongly acid, contained *a distinct trace of albumen*; separable by heat or nitric acid as minute flocculi.

" 25.—At 8 A.M., on rising, ℥xx. sp. gr. 1020, freely acid, clear and bright, no cloud nor deposit, *no trace of albumen.*	No. 23.—At 7.25 P.M. ℥xj. pale clear urine, sp. gr. 1009·2, neutral, contained a *faint trace of albumen.*	No. 24.—Secreted during the action (two hours) of the $\frac{1}{75}$ gr. Atrop. Sulph. by skin. Quantity? dysuria, sp. gr. 1024. Alkaline, deposited much triple phosphate, phosphatic opalescence on heating; cleared by nitric acid, leaving a *few fine flocculi of albumen.*

" 26.—Passed 54 ounces of urine; that at bedtime faintly opalescent, but free from cloud or deposit; sp. gr. 1023·6, acid, *a faint trace of albumen.*

" 27.—Secreted 52 ounces; that at 8 A.M. bright, strongly acid, sp. gr. 1026·4; after heat and nitric acid, the faintest opalescence and separation of a few scum like filmy *flocculi of albumen.*

Secreted between 6 and 8.15 P.M. ℥xjss. sp. gr. 1010, bright, faintly alkaline, *free from any trace of albumen.*

" 28.—Secreted 60 ounces, that between 8 and 11 P.M. ℥x. sp. gr. 1018, acid, slightly fluorcescent, but *entirely free from albumen.*

" 29.—Secreted 66 ounces, on rising ℥xviij. sp. gr. 1013·2, acid, no cloud or opalescence, *no trace of albumen.* Between 8 and 11 A.M. ℥xviij. sp. gr. 1006, acid, whey-coloured, bright, *free from albumen.*

" 30.—Secreted 58 ounces: that at 9 A.M. sp. gr. 1016, *free from albumen.*

May 1.—Passed ℥xx. on rising, sp. gr. 1016, acid. Between 8·30 and 11·30 A.M., secreted and passed in my presence ℥xxvj. sp. gr. 1006, bright, both *free from albumen.*

" 2-14.—Secreted from 40 to 70 ounces in the 24 hours; sp. gr. from 1013 to 1018; *no trace of albumen.*

" 15.—At 11 P.M. on 14th took ♏xxx. Succi Belladonnæ. Passed on rising ℥xv. sp. gr. 1020, freely acid, *no trace of albumen.* 1000 grain measures contained:—

Chlorine	4·21 grains.
Urea	22·33 "
Phosphates and Sulphates . .	12·82 "

Case 13.—*Desquamative Nephritis following an attack of Scarlatina two months previously. Treatment by Belladonna. Recovery.*

Thomas H., aged 17, applied to me on *April* 25, 1868. He was in a state of extreme anæmic cachexia, and brought a bottle containing dark-red turbid urine, to illustrate the disease from which he was suffering. From his account of an ulcerated throat and feverish cold, it appeared to me that he was attacked with scarlatina, on *January* 16, 1868. After nursing himself for a fortnight he went out, and four days afterwards was very sick, and the face, throat, and legs became much swollen. Two days subsequently pain was experienced across the kidneys, and the water became coffee-coloured and scanty. During the next six weeks he was under medical treatment at home; the dropsy gradually passed away, the urine became natural in colour and copious, and sometimes it was necessary to rise in the night to void it.

On *March* 2 he returned to work (bookbinding); but after a week he had a relapse, the water became thick and coffee-coloured again; and there was pain, after exertion or walking, across the loins. At the end of a fortnight these symptoms had disappeared, and he again returned to work, but found that the urine became dark-brown and turbid if he walked far, or if the bowels were confined. He, however, continued at his work for three weeks, at the end of which time (*April* 10), he was again seized with pain in the loins, and vomiting. The urine became coffee-coloured, as before, and there were indications of pleurisy. He recovered from the more acute symptoms, but the urine remained in the same condition, and at this stage of his disorder he came under my care. The lips were almost bloodless, the tongue coated, and the breath foul; he was weak and emaciated. There was no trace of œdema of either the legs or the face. The appetite was defective, the bowels sluggish. The urine was entirely free from crystalline deposits. On standing, a light chocolate-coloured organic matter was deposited from a pale smoky-coloured fluid. This sediment was chiefly composed of brownish molecular casts of the uriniferous tubules, and of these casts filled with granular nucleated corpuscles. Numbers of the latter were free and actively proliferating; they measured $\frac{1}{300}$ to $\frac{1}{800}$ of an inch. There was, in addition, a large quantity of loose brown molecular matter. In a state of rest the quantity of albumen varied, during the day, from 0·5 to 3 grains in 1,000 grain measures.

On *April* 27, I injected $\frac{1}{48}$ of a grain of sulphate of atropia beneath the skin, and prescribed ♏xx. to ♏xxx. succi belladonnæ every night. The beneficial effects of the medicine were soon apparent. The composition of the urine, before and after the first dose of belladonna, is seen in the following analyses:—

April 27, 1868. Milk and coffee a pint and half, bread and butter, at 8.30 A.M. $\frac{1}{48}$ grain Atrop. Sulph. beneath skin at 11.15 A.M.	A. ℥vj. secreted between 8.30 and 11.15 A.M. sp. gr. 1014, acid, very turbid; after standing three hours a light chocolate organic matter subsided. Its bulk was one-third that of the whole fluid. 1000 grain measures contained:— Grs. Chlorine . . 2·02 Urea . . . 17·33 Sulphates and Phosphates . . . 5·83 Uric acid, an excess. Albumen . . 0·52	B. ℥ij. secreted between 11.15 and 1 P.M., turbid, but less so than A, and the deposit was of a drab colour. Sp. gr. 1014, acid. Contained excess of uric acid, was voided with difficulty.	C. ℥iij. voided with difficulty at 2 P.M. Only slightly turbid, with a light drab-coloured matter, which on settling occupied about ¼ of the volume of the whole fluid. Sp. gr. 1016, acid. 1000 grain measures contained:— Grs. Chlorine . . . 1·62 Urea . . . 18·33 Sulphates and Phosphates . . . 12·50 Uric acid, more than A. Albumen . . 1·10

Afterwards, during successive periods of 24 hours, he passed 70, 60, 65, and 60 ounces of urine respectively. That of the last period corresponded to the *4th day* of treatment. The urine passed at bedtime on this day had sp. gr. 1013·2, and 1000 grain measures contained 0·26 *grain albumen*; the urine on rising, sp. gr. 1016, *albumen* 0·10 *grain*; and that secreted after breakfast, between 8 A.M. and 11.30 A.M., sp. gr. 1012·4, and 1000 grain measures contained 0·65 *grain albumen*. The urine secreted under the influence of a full dose (♏xl.) of belladonna juice, during the next 2 hours, was pale, and contained only a trace of colourless organic deposit, and 0·16 *of a grain of albumen* in 1000 grain measures.

The same quantity of urine continued to be voided; the sp. gr. varied from 1012 to 1016; the deposit continued colourless, and gradually decreased; and on 15*th day* of treatment it did not exceed, in bulk or opacity, that of the mucous cloud in healthy urine. 1000 grain measures of the evening, the night, and midday urine, contained 0·20, 0·12, and 0·17 *grain of albumen* respectively. The faint cloud was composed of the same elements as at first, but casts were very rare, the deposit being almost wholly composed of granular nucleated corpuscles. *On the* 17*th day* the acid evening urine presented only the *faintest opalescence*, but no perceptible particles on boiling; and the addition of nitric acid nearly cleared it, by the separation of the organic matter, into a few minute scum-like films. The reaction with excess of nitric acid in the cold was the same. The *night* urine was absolutely *free from a trace of albumen*; that secreted at *midday* had sp. gr. 1020, and was equally *free from albumen.* The light mucous cloud contained only a few free granular corpuscles.

The patient was now much improved in health and strength, the lips had recovered a healthy colour, and the anæmic appearance had given way to a pinker hue. The appetite was good, the skin active. Pain had left the side and loins. He had taken ♏*xxx. of the succus belladonnæ every night.* Thirteen hours after the dose the pulse was 76, regular, of good volume, and fair power; the tongue clean and moist, the pupils at the light ⅛″. He stated that the medicine made the mouth very dry, and caused dimness of the vision during the morning and early part of the afternoon. He continued to take belladonna in the above-mentioned dose every night for the *next* 2 *months*, when he resumed work. During the *next month*, the medicine was taken every other night. The urine was constantly examined during these 3 months. It remained in the condition last described—that excreted at the end of the day's work presenting the faintest opalescence without the separation of visible particles; and that excreted during the night and first half of the day, remaining bright under the use of reagents for the detection of albumen. The quantity became a little less, the daily average being 50 ounces. The patient had recovered his usual health at the end of the sixth week, but I deemed it prudent to keep him from work, and continue the use of the belladonna, for six weeks longer. At the end of the third week one of the samples of urine which he brought to me contained the faintest trace of albumen, in the form of flocculi as separated by heat and nitric acid. It was the evening urine, and had been voided after a prolonged game of quoits. A similar trace of albumen was detected after a walk of 12 miles a fortnight later. The return to work had no bad effect, and he has continued well, and able to follow his usual employment, up to the present time.

V. Chronic Nephritis.—In this condition belladonna can scarcely be expected to effect either rapid improvement or recovery. Still there is no remedy that can be brought to bear so directly and powerfully on the diseased bloodvessels as atropia; and the results in the following cases are sufficiently encouraging to induce us to continue its use so long as the albumen decreases. Up to a certain point the remedial influence of belladonna is more rapid and decided than could have been anticipated; and the prolonged use of the drug may probably be attended with most salutary changes in the morbid structure, provided that the degenerative changes are not too far advanced. In *Case* 15 there was evidence of fatty degeneration, and yet the beneficial action of the medicine was unmistakable. The result of the operation of belladonna in these cases must be accepted as the best proof of the condition of the bloodvessels generally during that operation. It is quite clear that there is no impediment from contraction of the arteries on the one hand, or from dilatation of the capillaries on the other, to the flow of blood through the kidney. On the contrary, it appears that the vessels of the gland are aroused by the action of the drug into a healthy state of excitement—a condition highly favourable for the nutrition of the organ, and the removal of chronic disease.

The following case, in which recovery appears to have been completed, properly stands midway between the acute and chronic varieties:—

Case 14. *Nephritis of three months' duration.*—Charles E., aged 35, had been under my care for three months for an acute attack of nephritis, commencing with excessive œdema of the legs, and exsudation of albumen. At the end of this time (*January* 10, 1868) he had so far improved, under the influence of astringent chalybeates and hydragogue purgatives, that there remained but slight pitting of the integuments over the tibia; and the urine, when boiled and treated with nitric acid, gave only a small precipitate of albumen—enough, however, to render the fluid completely opaque, from the presence of small flocculi of albumen. At 8.30 P.M. on this day, I injected the $\frac{1}{48}$ *of a grain of sulphate of atropia* beneath the skin; and he passed at that time urine A. The atropia produced full effects; and at 10.30 P.M., when these had passed off, he voided urine B with some difficulty, and in small driblets. *Urine* A had a specific gravity of 1022·4, and contained exactly a grain of albumen in 1000 grain measures. *Urine* B was of specific gravity 1024·4, and contained only half the quantity of albumen present in urine A. Four days afterwards, the patient presented himself at the hospital, and reported himself quite well. The œdema of the legs was entirely gone. He passed urine in the prescribing-room; and repeated examination, by my clerks and myself, showed that the albumen had quite disappeared. The patient did not attend again, from which I infer that he has continued well. He had presented himself regularly at the hospital the previous three months, and the urine was examined on each occasion. The albumen was observed to be slowly diminishing in quantity, but it had never been absent from the secretion. It appeared, in this case, that the single dose of atropia had given the kidneys a sudden impulse to healthy action.

Case 15. *Chronic Nephritis of two years' duration. Belladonna treatment at the end of this time, and progressive diminution of the Albumen.*—John B., aged 25, a stalwart navvy, engaged on the Thames Embankment, took a chill, and came under my care nine days afterwards for acute nephritis, with great œdema of the legs and scrotum. He continued under treatment for more than two years, working at first half-time, but was able during the last year to follow his employment as usual. The quantity of the albumen and œdema gradually decreased, and there remained only a little puffiness of the ankles, and slight pitting over the tibia. He was treated with astringent chalybeates and hydragogue purgatives during the whole of this time. At the end of two years (*January* 1, 1868) the following was his condition:—Appetite very good, muscular strength unimpaired, pulse good, face rather pale, no trace of œdema. Felt quite strong and well. The urine was copious, and he usually rose two or three times in the night to void it. It deposited a small cloud, composed of large fatty casts, smooth waxy casts, a few large dark granular corpuscles, and a little granular renal epithelium. The albumen had not undergone any diminution for the last six months, and the quantity varied during the day from 5 to 8 grains in 1,000 grain measures. At this time I commenced the belladonna treatment, and gave him ♏ xxv. *succi belladonnæ* every morning, and an occasional dose of atropia by the subcutaneous tissue. The albumen now began to diminish rapidly. The subsequent progress of the case, up to the time when he left London for occupation elsewhere, is seen below :—

Jan. 1, 1868. 8 A.M. Breakfast, bread and butter, and tea (one pint). 9.15 A.M. last voided urine. 10.30 A.M. $\frac{1}{75}$ grain Sulph. Atrop. (by skin).	1. Urine at 10.30, time of injection, ℥iiij*s*. sp. gr. 1020, strongly acid, pale sherry colour, faint turbidity after standing. 1000 grains contained :— Grs. Chlorine . . . 3·84 Urea . . . 16 Sulphates and Phosphates . . . 3·7 Albumen . . . 8.06 Uric acid, no excess.	2. Urine 3¾ hours after the dose, and after 10 minutes' persevering trial, ℥v. sp. gr. 1020, strongly acid, and in all other respects resembling 1 ; remains bright on standing. Dropped into the eye, it caused dilatation of the pupil within one hour and 20 minutes. 1000 grain measures contained :— Grs. Chlorine . . . 5·05 Urea . . . . . 14·66 Sulphates and Phosphates 5·4 Albumen . . . 6.02 Uric acid, no excess.

Feb. 17. 8 A.M. Breakfast as above. Afterwards smoked ʒj. tobacco 10.10 A.M. injected $\frac{1}{35}$ grain Atrop. Sulph. beneath skin. Not passed urine since breakfast.	3. Urine at 10.10, time of injection, ʒix. sp. gr. 1017·4, bright pale sherry-coloured, freely acid, a slight flocculent precipitate on standing. 1000 grain measures contained : Grs. Chlorine . . 2 82 Urea . . 20· Sulphates and Phosphates . 6·83 Albumen . 2·93 Uric acid, no excess.	4. Urine 2¾ hrs. after the injection. ℥vj*s* acid. sp. gr. 1020, otherwise resembling 3. 1000 grain measures contained :— Grs. Chlorine . . 3·27 Urea . . . 23·66 Phosphates and Sulphates . 7 Albumen . 4·12 Uric acid, no excess.	After a quarter of an hour's persevering trial, he succeeded in passing ℥vij of urine ¾ of an hr. after the injection. Its sp. gr. was 1020, agreeing with 4 in its physical characters, but depositing crystals of uric acid after 48 hours. Urine 4 was removed by a catheter, as he failed to pass a single drop after long and repeated trials.

March 18, evening	urine, sp. gr.	1016.	Albumen in 1000 grain measures,	1·84 grain.
„ 19, night	„ „	1016·8	„ „ „	1·25 „
„ 27, evening	„ „	——	„ „ „	2·40 „
„ 28, night	„ „	——	„ „ „	0·90 „
April 3, evening	„ „	1015·2	„ „ „	0·84 „
„ 4, night	„ „	1014·8	„ „ „	0·37 „

The belladonna always produced decided effects ; and the pupils were dilated, and the mouth dryish, during the early morning. The general health continued good.

Case 16. *Chronic Albuminuria of unknown duration. At the end of sixteen months at least, Belladonna treatment and progressive decrease of the Albumen.*

Thomas F. R., aged 27. Had scarlatina at 8 years of age, followed by dropsy ; and kept bed about seven weeks. Ten years ago had a distinct attack of pain in the back, with dark-coloured urine, but no dropsy. *September* 23, 1866, got wet at sea, and sat several hours in his wet clothes. *Four days* afterwards noticed swelling of the legs and face. There was no pain in the loins, nor pyrexia—no apparent disorder of the renal function, nor change in the appearance of the urine.

He came under my care on *September* 29, when there was considerable soft œdema of the legs, and puffiness, with great pallor of the face. The urine was of the natural tint, pale, bright, and free from any deposit; but it contained a very large quantity of albumen. A chalybeate tonic, and an occasional dose of compound jalap powder, were prescribed. The œdema entirely disappeared during *the next three weeks*, but the pallor remained. The general health and strength were gradually restored, and there was no subsequent tendency to dropsy. The urine continued normal in amount, and so free from deposit that no microscopical particles could be detected. During the *first six months* the quantity of albumen slowly decreased, but throughout *the following year* there was no appreciable diminution; and the daily variation was from 0·5 to 1·5 grain in 1,000 grain measures of the urine secreted during the night (*night urine*), to 3 or 4 grains in the same quantity of the urine secreted towards the close of the day (*evening urine*). For the last year he had been taking 2 grains of dried sulphate of iron twice a day, in the form of pill. On *February* 7, 1868, I began the use of belladonna, and, having ascertained the effect by subcutaneous injection, prescribed ♏ xx. to ♏ xxx. *succi belladonnæ* every evening, at the end of the day's work. The patient's occupation included both reading and writing, and at first the hypermetropic effect was a source of so much difficulty that it was necessary to intermit the use of the medicine. This occurred during the early part of March, and it is to be observed that the albumen was more abundant at this time than during the previous month. By the use of strong spectacles, and taking the medicine a little earlier, and in diminished doses, he was able to resume the belladonna, and continue it excepting when he was under great pressure of work. During the vacation (from *August* 10 to *September* 1) he was able to take the belladonna more regularly, and in full doses;

and when he returned to town, the albumen had dwindled down to a mere trace. On *September* 17, he was suffering from catarrh, to which the increase of albumen at that time was probably due. According to my experience, the months of February and March are the most unfavourable to recovery from, as they are most favourable to the development of nephritis; and the good influence of belladonna in this case is manifest, both in the very decided improvement which immediately followed its use at this season of the year, and also in the continuance of the improvement during the early frosts of October.

During the months of November and December the patient was unable to continue the belladonna regularly, and to this cause, in some measure, and to the damp cold weather chiefly, the considerable increase of the albumen is due. Still the improvement is substantial.

In estimating the therapeutical value of belladonna, we must not lose sight of the influence of season. It is true that there was no improvement observable during the summer of 1867, but the summer of 1868 was a very unusual one for this island, and the great and protracted heat was most favourable for the relief of chronic renal disease.

The duration of the albuminuria is uncertain in this case. It could hardly have dated from the attack of scarlatinal dropsy. But with such predisposition it may have existed since the subsequent lumbar attack.

VI. Suppression of Urine, and Uræmia.—The direct and simultaneous action of belladonna on the circulation and the renal function declares this drug to be the appropriate means of arousing the former and re-establishing the latter in these morbid conditions. For this purpose, the dose should be moderate, and not exceed the $\frac{1}{48}$ of a grain of sulph. atropia by the skin. Three or four hours should elapse between the doses.

VII. In Rheumatism and Gout, acute and chronic, belladonna is serviceable, both as a means of promoting oxidation and as an anodyne. I have employed it in rheumatic fever with marked success. I inject the $\frac{1}{60}$ or $\frac{1}{40}$ of a grain of

CASE XVI.

Date		U. 1	U. 2	U. 3	U. 4	U. 5
Feb. 7, 1868.	$\frac{1}{24}$ gr. Atrop. Sulph. (skin), at 7.30 P.M., 4½ hours after dinner: all three urines were remarkably bright, clear, and free from microscopical particles. After standing, 3, 4, and 5 deposited crystals of uric acid, 5 most of all.	U. 1, at 8 A.M., on rising. From 12 midnight to 8 A.M. ℥x. sp. gr. 1015·6, freely acid. 1000 grain measures contained:— Grs. Chlorine . . 2·62 Urea . . . 13·33 Phosphates and Sulphates . . 7·2 Albumen . . 0·50	U. 2, at 6 P.M. same day. ℥vij. sp. gr. 1022·8, freely acid. 1000 grain measures contained:— Grs. Chlorine . . 4·92 Urea . . 20 Sulphates and Phosphates . 9·71 Albumen . 1·30	U. 3, at 7.30 P.M., time of injection, ℥iii½; sp. gr. 1023·2, freely acid. 1000 grain measures contained:— Grs. Chlorine . . 4·64 Urea . . 22·33 Sulphates and Phosphates . 10·0 Albumen . . 3·34	U. 4, at 9.45 P.M., 2¼ hours after injection. ℥iijss. sp. gr. 1022·8, freely acid. 1000 gr. measures contained:— Grs. Chlorine . . 4·50 Urea . . . 23·33 Sulphates and Phosphates . . 11·06 Albumen . . 3·29	U. 5, between 9.45 P.M. and 9 A.M. next day, and after a light supper, ℥xivss. sp. gr. 1023·6, freely acid. 1000 grain measures contained: Grs. Chlorine . . 3·33 Urea . . . 28·33 Sulphates and Phosphates . . 13·95 Albumen . . 1·44

Feb. 13. Night urine contained, in 1000 grain measures, 0·92 grains of albumen; and the evening urine, 2·6 grains.

Date		U. 6	U. 7	U. 8	U. 9
Feb. 17.	$\frac{1}{24}$ gr. Atrop. Sulph. (skin) at 6.45 P.M., 4½ hrs. after dinner (cold meat, bread and water). Was very busily engaged in mental and bodily occupation all day.	U. 6, at 5.30 P.M. to-day, sp. gr. 1026·0, freely acid, bright and clear: after 18 hrs. a whitewash film of lithates on surface of glass. 1000 grains contained:— Grs. Albumen . . 2·42	U. 7, at 6.45 P.M., time of injection, sp. gr. 1024·8, acid. 1000 grain measures contained:— Grs. Chlorine . . 4·16 Urea . . . 23·33 Sulphates and Phosphates . . 11·45 Albumen . . 2·47	U. 8, at 8.30 P.M., 1¾ hour after injection, sp. gr. 1024·2, bright and clear as usual. 1000 grain measures contained:— Grs. Chlorine . . 4·79 Urea . . . 24·66 Phosphates and Sulphates . . 10·66 Albumen . . 2·56	U. 9, between 8.30 P.M. and 9 A.M., supper, a pint of soup, bread, and half-pint of bitter beer. ℥xiv. same physical character as 6, 7, and 8, sp. gr. 1024·8, freely acid. 1000 grains contained:— Gr. Albumen . . 1·35

Date	Urine	sp. gr.	Albumen in 1000 gr. measures
Feb. 24.	Night urine, sp. gr. 1022, contained in 1000 gr. measures		0·55 gr. Albumen.
	Afternoon ,, ,,	—	1·8 ,,
28.	Night ,, ,,	1020·2	0·43 ,,
	Evening ,, ,,	1024·0	1·4 ,,
March 4.	Afternoon ,, ,,	1024·4	2·25 ,,
16.	Night ,, ,,	1016·8	0·85 ,,
	Afternoon ,, ,,	1027·2	3·80 ,,
	Evening ,, ,,	1022·4	2·0 ,,
25.	Night ,, ,,	1014·0	0·4 ,,
	Afternoon ,, ,,	1023·2	1·9 ,,
30.	Night ,, ,,	1013·6	0·45 ,,
	Afternoon ,, ,,	1024·0	2·5 ,,
April 16.	Night ,, ,,	1014·4	0·48 ,,
	Afternoon ,, ,,	1025·2	1·8 ,,
	Evening before B. ,, ,,	1015·6	1·47 ,,
	Evening after B. ,, ,,	1017·8	0·86 ,,
28.	Night ,, ,,	1013·2	0·64 ,,
	Afternoon ,, ,,	1024·0	1·6 ,,
May 8.	Night ,, ,,	1015·2	0·75 ,,
	Afternoon ,, ,,	1024·8	1·55 ,,

Date	Urine	sp. gr.	Albumen in 1000 gr. measures
June 15.	Early aft. urine, sp. gr. 1021·6, contained in 1000 gr. measures		1·0 gr. Albumen.
	Late afternoon ,, ,,	1021·0	0·6 ,,
July 13.	Night ,, ,,	1016·8	0·62 ,,
	Afternoon ,, ,,	1022·8	1·20 ,,
21.	Night ,, ,,	1021·2	1·4 ,,
	Afternoon ,, ,,	1021·8	0·42 ,,
Aug. 10.	Night ,, ,,	1020·8	0·55 ,,
	Afternoon ,, ,,	1020·0	0.85 ,,
Sept. 4.	Night ,, ,,	1017·2	a trace ,,
	Afternoon ,, ,,	1022·3	0·18 ,,
7.	Night ,, ,,	1013·0	0·6 ,,
	Afternoon ,, ,,	1024·2	1·5 ,,
Oct. 5.	Night ,, ,,	1012·0	0·37 ,,
	Afternoon ,, ,,	1024·0	0·36 ,,
19.	Night ,, ,,	1012·8	0·28 ,,
	Afternoon ,, ,,	1025·6	0·87 ,,
Nov. 17.	Night ,, ,,	1020·0	0·6 ,,
	Afternoon ,, ,,	1028·0	2·22 ,,
Dec. 8.	Night ,, ,,	1012·0	0·3 ,,
	Afternoon ,, ,,	1027·2	2·2 ,,

the atropia salt into the integument over the affected joint, as soon as the first indication of inflammatory action arises, and repeat the injections, if need be, at intervals of 12 hours. Not only is the pain relieved, but the inflammation is subdued, and, if the injection be timely performed, it may be prevented. The effects of belladonna on the pulse, tongue, and pupils, when the fever is high, are usually very slight. As far as I have been able to ascertain, it has no particular influence either in checking or increasing the perspiration.

VIII. **Neuralgia.**—In painful affections of the nerves, whether arising from functional disorder or inflammatory action, atropia used subcutaneously is the most valuable remedy that we possess. In my hands it has never failed to bring relief, and in all cases but one has finally removed the affection. Its action, as I have said before, is not merely anodyne; it is remedial, and its beneficial effects are often marvellously speedy. It rarely happens that the remedy can be injected into the affected part. When it can, greater benefit may be expected to result than when the agent comes in contact with the seat of irritation after being diluted with the blood. Although the pain from distension of the integument over the painful spot was for several minutes very intense, the subject of *Obs.* 25 preferred to have the injection placed in the finger, because she found it more effectual than when lodged in the arm.

XI. **In Spasmodic Asthma,** originating in peripheral or centric nervous irritation, the subcutaneous use of atropia is often followed by speedy and long-continued relief. And similar benefit may be expected in every variety of spasmodic action arising from undue irritation of the spinal centres. In cases of **Epilepsy** which are traceable to emotional excitement, I have found belladonna very serviceable.

In removing pain, and relaxing the contraction of the circular fibres of the ducts or intestines, belladonna is a most beneficent agent in facilitating the passage of **Calculi.** And, by virtue of a similar action, it is invaluable in **Enuresis** resulting from irritability of the muscular tissue of the bladder.

CHAPTER VII.

COMBINED ACTION OF OPIUM AND BELLADONNA.—The simultaneous operation of these two drugs, on the same individuals in whom the separate action of each has been already ascertained, will now be considered in detail. And in order to avoid many possible sources of doubt or error, the same drugs have been used as were employed in the previous observations.

The enquiry will consist in the simultaneous or successive injection of atropia, and each of the active constituents of opium, or of laudanum itself, into the subcutaneous tissue.

I. MORPHIA AND ATROPIA.

On the Horse.—*Obs.* 85.—Effects of $\frac{1}{2}$ *grain morphiæ acet. and* $\frac{1}{12}$ *grain sulph. atropiæ*, introduced by separate punctures, on the subjects of *Obs.* 62, p. 194. (From $\frac{1}{2}$ to 1 grain of the morphia salt, alone, produced no appreciable effect on either animal.) *After* 35 *minutes* the pulse had attained the maximum acceleration in all, and amounted in the grey horse to 38, in the colt to 12, and the chestnut pony to 32 beats. The tongue and lips were quite dry; the pupils dilating—those of the grey horse oscillating, contracting one minute, and opening widely the next. *After* 1 *hour* maximum dilatation of the pupil in all: in the grey horse permanent and complete dilatation, in the other two nearly complete dilatation. *After* 2 *hrs.*, excepting a fall of 4 beats in the grey horse, the maximum acceleration of the pulse continued in all. *After* 3 *hrs.* the acceleration in the grey amounted to 18, in the pony 14; the max. accel. continued in the colt. The dryness of the mouth and dilatation of the pupils continued as at 1 *hr.* *After* 5 *hrs.* accel., 6, 8, and 2 beats, in the horse, the colt, and pony respectively. This was the only symptom that remained. Each was eating food, and the mouth was wet and healthy in all.

The general effect on all was tranquilising. An hour after the dose, each animal showed *a tendency to sleep.* The pulse was increased in volume and power; the kidneys acted freely. In the grey horse the urine was decidedly increased; in the other two there was sneezing between the first and second hours. Referring now to *Obs.* 62 (p. 194), it will appear that ½ grain of acetate of morphia, which *alone* produced no effect, *increased all the atropia symptoms, and prolonged* them for at least an hour. 2. That the combination produced a soporific effect, which did not follow the atropia when given alone.

Obs. 86.—Effects of 4 *grains of morphiæ acet.*, and $\frac{1}{12}$ *grain of atropiæ sulph.*, introduced by separate punctures. *After* 35 *minutes,* cardiac acceleration in both animals (the grey horse and the colt), 24; pupils dilating; a little restlessness. *After* 40 *min.*, cardiac acceleration in both, 36; mouth quite dry; pupils dilated. *After* 1 *hr.*, pulse attained the maximum acceleration of 38 beats in both; mouth quite dry, pupils completely dilated; restlessness increased. The animals were not seen again until 4¼ *hrs. after the injection,* when both were deliriously restless, and impatient of restraint; the max. accel. of the pulse continued in both; skin hot; conjunctivæ injected; mouth wet and slobbering. Both had staled copiously. *At the 9th hr.* the delirium and restlessness had nearly passed off. Pulse of good volume and power in both: in the grey horse accelerated 12 beats, in the colt 18. Pupils returned to their initial dimensions. Mouth clean and wet. *After* 22 *hrs.* both animals had dunged copiously. They were in their usual condition. Pulses natural at the initial 36. They were quiet and comfortable.

On comparing this with *Obs.* 1 (p. 100), it will appear that, in the last experiment, a result—*plus* dilatation of the pupils and dryness of the mouth—was obtained in 35 minutes, equal to that produced by the morphia alone at the end of an hour. *After* 40 *minutes* the effects on the brain and the pulse were much greater than those which occurred at any time during the action of the opium alone. There was no effect of the action of either morphia or atropia that was not considerably increased and prolonged by the combination.

Obs. 87.—The effects of 2 *grains atropiæ sulph.* and 4 *grains morphiæ acet.*, introduced by separate punctures, were exactly the same in the two animals. (*a*) In the grey horse, *after* 10 *min.*, an acceleration of the pulse 24 beats; pupils dilating. *After* 12 *min.* acceleration 32 beats, the pulse being now exactly doubled; pupils completely dilated; mouth getting dry. *After* 20 *min.* pulse increased two beats more; skin hot, and beginning to perspire; general muscular tremor, twitching of the head, confusion of vision, slight delirium. *After* 40 *min.* acceleration of the pulse 44 beats; mouth completely dry; delirium, and great restlessness, stepping up and down with the forelegs and a forward impetus—the halter being constantly on the stretch, and the head pushed against the wall. *After* 1 *hr.* restlessness and delirium very severe; conjunctiva and sclerotic much injected, and very red; pupils fixed; mouth parched; the back and sides damp with sweat; the pulse unchanged, full, and strong. *After* 1½ *hr.* rigidity of the whole of the muscular system; the body was sweating profusely. *After* 2½ *hrs.* the pulse attained the maximum acceleration of 60 beats—now numbering 92. It was regular, and moderately full and strong. *For the next* 4 *hrs.* the symptoms continued unabated, and the animal was in a state of *extreme delirium* and *excitement*, sweating profusely. Shortly after the *first* 1½ *hr.*, all the extremities were affected with occasional *twitchings. Alkaline urine was freely voided every* 10, 20, *or* 30 *minutes*, and on these occasions only the horse became quiet for a miuute, drooped the eyelids, and showed a strong tendency for sleep, and once or twice he *actually slept.* At the *end of the* 4*th hr.* the intervals of quietude and somnolency became more frequent and prolonged, and the delirium, restlessness, sweating, and frequency of micturition began to abate. The pulse also was decreased 4 beats, and the conjunctiva less injected. *After* 5 *hrs.* the previous delirium and distress had given place to exhaustion; the pulse was 80, having decreased 12 beats from the maximum, quite regular, but very weak. The mouth was parched and dry, and the tongue dry, brown, and wrinkled, as in extreme cases of typhus. The conjunctiva still much injected, and the head hot. The iris reduced to a mere line, and the pupils fixed. The skin of the prominent parts of the face was bruised and abraded from rubbing against the wall. The coat was matted together by the previous perspiration. The animal was generally dull, sleepy, and exhausted, and he now remained quite still, and slept for intervals of 10 or 15 minutes. *During the next* 10 *hrs.* the restlessness gradually decreased, and the intervals of quietude became longer. *After* 18 *hrs.* the pulse was weak, but regular, and accelerated 12 beats; the mouth and tongue pale, clean, and moist; the pupils still completely dilated, and continued so for the next 24 hours.

(*b*) In the colt the effects were even more severe, for, as there was less tendency to somnolency, the delirium and distress continued undiminished while the grey horse had intervals of repose. After 10 *hrs.* the effects began to decline. At the end of 14 *hrs.* they had not wholly subsided, and the animal was in a state of extreme exhaustion.

On referring to p. 197, it will be found that both the *general* and the *particular* effects of atropia were more severe in this last observation than when the atropia was given alone. The delirium was much more intense, the acceleration of the pulse almost double, the dryness of the mouth even more complete. Nor was there any sensible perspiration after either the atropia or the morphia when given alone; but after the combined dose, profuse perspiration followed, as a consequence of the extreme bodily and mental excitement, and the exhaustion attending it. On turning to p. 100, *Obs.* 1. we recognise slight symptoms—restlessness, a tendency to sweat, dilatation of the pupil, and acceleration of the pulse, which under the influence of the atropia rose to that degree of intensity, and some of them even beyond, which was manifested during the action of three times the dose of morphia. (See *Obs.* 3, p. 102.) Further, the atropia developed hypnotic effects, which were only observed after much larger doses of morphia. (See pp. 102, 103.)

When given by the stomach, the two drugs increase each other's effects, thus :—

Obs. 88.—Gave the grey horse 2 *grains atropiæ sulph.*, and 4 *grains morphiæ acet.*, together, in a pint of water. *After* 20 *min.*, pulse accelerated 20 beats. *After* 25 *min.*, attained a maximum acceleration of 28 beats; pupils almost fully dilated; mouth dryish. *After* 1¼ *hr.*, pulse as at 25 min.; pupils fully dilated; mouth generally dry, but nowhere completely free from moisture. Dunged, reaction acid. *After* 2¼ *hrs.*, pulse accelerated twelve beats; mouth moist. *After* 3 *hrs.*, pulse still accelerated 12 beats; pupils dilated as at 25 min.; urine alkaline. *After* 18 *hrs.*, lively and well; pupils contracted; pulse natural, but soft. *The general effect* was tranquilising; the animal remained dull and quiet during the whole of the time. Compare the effects of 4 grains of atropia, and 12 grains of morphia, given alone by the stomach (see pp. 197, 104).

Conclusion.—In the horse, there is no effect produced by either opium or belladonna alone (see pp. 105, 198), which, when they are given together, is not both intensified and prolonged.

On the Dog.—The action of morphia and atropia simultaneously introduced on the bitch (see p. 106), is illustrated in the following :—

Obs. 89.—½ *grain morphiæ acet.* and $\frac{1}{48}$ *grain atropiæ sulph.* were given by one puncture; pulse 120; pupils contracting to ⅛″. At the 3*rd min.*, the pulse rose to 168. At the 4*th min.*, gulped a little food without previous effort, and then lay down. *After* 8 *min.*, pulse 210; pupils, in bright sunlight, nearly completely dilated; nose, lips, gums, and tongue quite dry. *After* 20 *min.*, pulse 240; resp. 30. *After* 1 *hr.*, pulse 216; resp. 20; the free hind-leg jerked inwards at each inspiration. *After* 1½ *hr.*, pulse 176; resp. 16. *After* 2 *hrs.*, pulse 120; resp. 16. *After* 3 *hrs.*—Up to this time the pulse was of good volume and power, and perfectly regular; and the resp. regular, and of good capacity. Now the former began to manifest the respiratory irregularity; the cardiac systoles numbered 114, 108, 96, and 103 during successive minutes, the contractions falling towards the end of *expiration* to 90, and becoming accelerated with each *inspiration* to 120; the breathing was now more frequently interrupted by long-drawn sighs. *After* 4 *hrs.*, cardiac systoles during the short insps. 82, during the prolonged exps. 60; resp., one min., 15 regular, another 12, with a long-drawn inspiration. Up to this time the dryness of the mouth continued; the parts now became moist; the pupils, which throughout had been dilated to $\frac{1}{3}''$, and insensible, were now $\frac{1}{4}$, and contracted to $\frac{1}{8}$ on the approach of a taper. The effects of the belladonna had therefore almost wholly passed off. *After* 5 *hrs.*, pulse 104, and now regular, but accelerated by the occasional sighs; resp. 15. *After* 11 *hrs.*, pulse 120, regular; resp. 22; pupils at a gas-lamp contract to $\frac{1}{7}'$. Recovered from the action of the opium, and now experienced some uncomfortable sensations, expressed by restlessness and whining, from the after-effects of the drug.

General Effects.—From the 15*th min.* after the injection to the end of the 4*th hr.*, she lay narcotised, flaccid, and motionless in any position in which she was placed. On separating the eyelids, and touching the cornea, there was no reflex movement, unless the pencil came in contact with the margins of the lids. At the end of the 4*th hr.*, when the belladonna symptoms were almost passed off, she opened the eyes and raised the head when loudly called, but instantly laid it down again, and relapsed into her former state, and 15 min. afterwards seemed as deeply narcotised as ever. She made no movement when called, or when the pads and nose were pricked or pinched. She continued in this state until the end of 5½ *hrs.*, when, on being disturbed, she awoke, crawled a few paces away, and then lay down and slept soundly, and without movement, for ¾ of an hour more. After this she began to wake up at intervals of 20 or 40 minutes, and change her place; and when disturbed was nervous and timorous. Drowsiness and stupor continued up to the 10*th hour.*

On turning to pp. 106 and 198, and comparing the separate effects of ½ grain of morphia and $\frac{1}{48}$ grain atropia on the animal, with those recorded in this last observation, we shall find

—first, with respect to the atropia, that its action on the pulse was neither quite so strong nor so prolonged when given with morphia in the above proportions as when given alone—the dilatation of the pupils, and the dryness of the mouth, however, were both greater and more prolonged; 2ndly, that the deranging effects on the vagus, the vomiting, and respiratory irregularity of the pulse, were almost wholly counteracted so long as the action of the atropia continued. The vomiting, indeed, still occurred, but it was a mere instantaneous gulp of a small quantity of matter, and not a violent evacuation of the whole contents of the stomach, preceded by prolonged retching, as occurred when morphia was given alone. As soon as the belladonna action ceased, the pulse resumed the respiratory character. The other effects of opium were decidedly increased: during the first 3 or 4 hours the narcotism was as complete in the one case as in the other; but if the condition of the animal in the two experiments between the 4th and 6th hours be compared, it will be seen that there was complete narcotism after the combined dose, at a time when, after morphia alone, the animal was easily disturbed.

The three following observations illustrate the action of small and large doses of atropia, in animals previously subjected to the influence of opium:—

Obs. 90.—At the end of 3 hours, and when the action of $\frac{1}{2}$ *grain morphiæ acet.* was declining (see p. 108), $\frac{1}{96}$ *grain atropiæ sulph.* was injected beneath the skin, the pulse being 66, respiration 15, and the pupils $\frac{1}{7}$. *After* 8 *minutes*, pulse accelerated 42 beats; respiratory character very marked, 4 or 5 quick beats following each inspiration; respiration 19, regular. *After* 15 *min.*, pulse accel. 150 beats, quite regular; pupils dilating; mouth become wet with a copious alkaline secretion. *After* $\frac{1}{2}$ *hr.*, pulse accel. 174 beats, of exactly the same character as if under the influence of atropia alone; resp. 20, regular; pupils fully dilated; mouth still wet. *After* 1$\frac{1}{4}$ *hr.*, pulse attained a maximum accel. of 234, now numbering 300. *After* 1$\frac{1}{2}$ *hr.*, pulse varying from 300 to 180. *After* 1$\frac{3}{4}$ *hr.*, pulse 148, regular; resp. 14; mouth quite dry, secretion acid; belladonna fœtor. *After* 4 *hrs.*, the atropia effects had passed off; pulse 90, regular, of moderate volume and power; resp. 22; mouth quite moist, secretion alkaline; pupils contracted to $\frac{1}{8}$.

General Effects.—For 3 hours after the injection of the atropia, the animal was in a state of narcotism. She lay motionless on the table during the whole of the time, and made no resistance to my repeated examinations, made no movement when I raised an eyelid, and approached a light to within 3 inches of the eye. In fact, she took no notice of any disturbance whatever. At the end of the 7th hour from the injection of the morphia she was still very heavy and stupid; lax, feeble, and cold; and lay dozing, with the tip of the now moist tongue protruded.

The influence of a minute dose of atropia on the dog, when under the influence of a full dose of morphia, is seen in the following experiment:—

Obs. 91.—1¼ hour after the injection of 1 *grain morphiæ acet.* beneath the skin of the bitch (see *Obs.* 7, p. 109), pulse, respiratory, 43; respiration irregular, 18 to 14; pupils $\frac{1}{8}''$. Injected $\frac{1}{240}$ *grain atropiæ sulph. After* 2 *minutes*, pulse 56. *During the* 3rd *min.*, suddenly increased to 84. *After* 6 *min.*, to 100, with a distinct and regular pause at the end of each expiration; resp. 16, regular; pupils dilated to $\frac{1}{4}''$. *After* 13 *min.*, pulse 200, and quite regular in the femoral artery. *After* 18 *min.*, pulse 240 = a maximum acceleration of 197; resp. 16, regular; pupils, on the nearest approach of a taper as the animal lay narcotised, $\frac{1}{4}''$. *After* 1 *hr.*, pulse 150; resp. 16. *After* 2¼ *hrs.*, pulse 90, regular, of fair volume and power; resp. 14, regular, or 11, and one long-drawn sigh; mouth quite moist. At this time the belladonna effects had nearly passed off.

On comparing the foregoing with the effects of $\frac{1}{96}$ grain of atropia, in the natural state (see p. 198), it will appear that the action of this drug loses nothing by association with opium. The dilatation of the pupils after this $\frac{1}{240}$ of a grain was even greater than after the $\frac{1}{96}$ alone. ['Resp., 8 to 14,' in *Obs.* 7, p. 109, should be 'resp., 18 to 14.']

The effects of a large dose of atropia, at a time when the influence of opium was waning, is seen in the following:—

Obs. 92.—¼ *grain atropiæ sulph.* was injected beneath the skin of the bitch 2 hours after the injection of ♏xl. *tincture of opium* (see *Obs.* 8, p. 110); pulse respiratory, about 70; resp., 17 to 24. *After* 5 *min.*, was completely narcotised; pulse 360, regular; the heart's action strong; resp. 16; pupils completely dilated, and fixed. *After* 1¾ *hr.*, the maximum acceleration having been sustained for 1½ hour, the pulse at this time had fallen to 220; resp. 16, regular; sclerotic a little injected; mouth quite dry, but not parched. *After* 3 *hrs.*, pulse 200; resp. 15, regular and full; mouth destitute of moisture, excepting under the tongue, where there was just enough moisture to damp a little bibulous paper, and show an alkaline reaction. At the *end of the 4th hr.* from the injection of the atropia, and 6 *hrs. from* the injection of the 1*st dose of laudanum*, she raised her head, and opened her eyes for a second, and then sunk into deep sleep again. Pulse 212; resp. 15; mouth very dry.

General Effects.—During the whole of these 4 hours, the animal remained motionless, and insensible to sound, light, or touch; she evinced no feeling when I pricked or pinched the nose, ears, or pads; and made no resistance when I poked my finger down the throat as far as I could push it. At the end of the 4th hour of the atropia, I shook her, and rolled her about to arouse her, and she then walked the length of her body, dragging the hind-legs after her, and threw herself along, and continued to sleep motionless, and without taking the slightest notice of anything going on around her, for the *next* 4 *hours*. She continued drowsy for some hours longer, occasionally giving way to a little restlessness and whining. 14 *hours* after the first injection, she was dull and quiet, but very nervous, and whined occasionally. Pulse 92, regular; pupils dilated, and fixed. Mouth not quite so moist as usual; nose parched; refused food.

The two observations next following illustrate the action of a full dose of opium, and a still larger dose of morphia, respectively, when the effects of atropia alone were fully developed:—

Obs. 93.—Injected ¼ *grain atropiæ sulph.*, and after 15 minutes, when the belladonna symptoms were fully developed (see p. 201), ♏xx. of the same sample of *tincture of opium* as was used in *Obs.* 8, 9, and 12. 30 *min. after* the last injection, had slowly lapsed into sleep, and was now sleeping in the position in which I placed her ¼ hr. previously; resp. 22. *After* 50 *min.*, resp. 16. *After* 1½ *hr.*, resp. 14, regular; pulse 200, regular and strong. *After* 5 *hrs.*, pulse 160; resp. 14, shallow, and apparently diaphragmatic. *After* 6 *hrs.*, mouth and nose moistened. *After* 7 *hrs.*, pulse 120, regular; resp. 14; mouth moist and clean. *After* 9 *hrs.*, pulse 98, regular. *After* 10 *hrs.* from the atropia injection, pulse 96; pupils as throughout fully dilated and fixed. Voided urine twice during the last ½ hr.

General Effects.—During the *first* 4 *hrs.* she lay motionless, and such was the depth of the narcotism, that pricking the nose, or lips, or pushing a needle through a fold of skin, failed to elicit the slightest movement. Nor could she be aroused by rough handling, or by the introduction of a finger down to the larynx. *At the* 5*th hr.* she made the first movement by abruptly raising the head for a second, but immediately relapsed into a state of complete apathy; and from this time to the 9th hr., continued to sleep soundly and without motion, except at the 6th and 7th hrs., when, upon some disturbance, she got up once on each occasion, and, having crawled a few paces, settled down to sleep again. *At the* 10*th hr.*, she appeared to be nearly recovered; she was free from nervousness or restlessness, and was simply dull, quiet, and drowsy.

Remarks.—During the 4 hrs. following the injection of the laudanum, the narcotism was as great as that produced by ½ grain morphia alone, or in combination with atropia. That it was determined in this case by the atropia, appears conclusively from a consideration of *Obs.* 8 and 9 (pp. 109, 110), in which it will be seen that the same dose, and even double the quantity of laudanum alone, failed to produce sound sleep in this animal. It is further to be observed that the atropia completely prevented nausea, vomiting, and the respiratory pulse in this case. On the other hand, none of the atropia symptoms were diminished.

The effects of morphia on the terrier (see p. 111), when under the influence of atropia, and subsequently the converse, are seen in the following:—

Obs. 94.—An hour after dinner, injected $\frac{1}{6}$ *grain atropiæ sulph.*, and *after* $\frac{1}{2}$ *hr.*, when the effects were fully developed (p. 200), $\frac{2}{3}$ *grain morphiæ acet.* So far from producing any quieting effect, the atropia made the animal excitable and irascible, so that it was no easy matter to effect the injection of the morphia. But *within* 3 *min.* of its introduction beneath the skin, and *without the least tendency to nausea previously*, the animal was completely narcotised and flaccid; the pulse 200, unchanged, full and strong; the resp. 22. *During the next* 2 *hrs.* he remained in this condition, completely unconscious of any disturbance, and the only reflex movement that could be elicited was a feeble contraction of the orbicularis on touching the cilia, together with a slight advance of the forepaws; pupils completely dilated and fixed; pulse gradually decreasing from 200 to 180, the heart making a rapid to-and-fro dub-a-dub-a-dub-a sound, like a winnowing-machine in full action. At this time he was carried a distance of a quarter of a mile, and shortly afterwards the narcotism gave way, and he then gradually passed into the miserable condition described at p. 111; the pulse fell to 150, and became *respiratory*. *After* $2\frac{1}{2}$ *hrs.*, urine began to dribble, and during the next 3 hrs. his condition was most distressing; he continued to cry piteously the whole of the time, and to drag himself about on the wet belly, and, being apparently quite blind, he ran against every object in the way; his nervousness was extreme, and the slightest noise caused him to start away. *After* $6\frac{1}{2}$ *hrs.*, he still continued uncontrollably noisy, restless, and distressed; pulse 72; pupils completely dilated, and the eye apparently insensible to light; mouth quite clean and moist. At this time I injected $\frac{1}{48}$ *grain atropiæ sulph.* In a few minutes he fell off to sleep in my lap, and *after* 10 *min.* actually snored. The pulse rose to 120 during the first 5 min., and in 10 min. it was 200, and the heart thumped regularly against the chest-walls as at the first; the respirations 12, full and regular. *After* 20 *min.*, the nose and mouth were dry. Having remained sleeping thus heavily for about $1\frac{1}{2}$ hr., he became moderately restless again; but this gradually declined, and after a few hours more he appeared to be quite recovered. The next day he was well and frolicsome. There was neither vomiting nor defecation throughout; the distress which came on, as the effects of the atropia passed off, was due to faintness and delirium.

Conclusions.—I. In the dog, belladonna, when administered simultaneously with opium, more or less completely prevents nausea and vomiting; and when given previously, entirely prevents these effects. II. Whether given previously, simultaneously, or subsequently, atropia completely counteracts the respiratory restraint on the free action of the heart, which is so prominent an effect of the operation of opium. We can wish for no more perfect an illustration of the beneficial influence of a medicine under suitable conditions than that afforded by the simple and direct action of atropia in relieving the impending syncope, which often persists for many hours after a dose of opium. At first the cardiac systoles are doubled, but a regular expiratory pause remains. During the next 60 or 90 seconds the systoles become stronger, and each one distinctly perceptible to the finger; and at the end of this time the inspiratory intermissions cease, and no trace of their presence remains (see *Obs.* 91, &c.). The effect is even more marked under the influence of the combined action of morphia and thebaia. III. While the spinal effects of opium on the muscles of organic life are thus counteracted by the stimulant action of atropia on the sympathetic, the cerebral and anæsthetic effects are intensified and prolonged by belladonna, and hypnosis is converted into narcosis. IV. On the other hand, all the effects of atropia, excepting, perhaps, the influence on the heart, are increased and prolonged by opium, and the cerebral effect in particular, the insomnia which results from excessive doses, is converted into narcotism, or a mixture of narcotism and delirium. V. The simultaneous action of opium retards the evacuation of urine, but in no degree interferes with the elimination of atropia by the kidneys.

On Man.—I shall confine my observations on the combined action of belladonna and opium on man, to those individuals in whom the effects of opium have been already determined (see p. 124 *et seq.*). I do so for the following reasons:—1.

They were all ablebodied, and in good general health; 2. In reference to opium, they represent every variety of constitutional peculiarity; and 3. The neuralgic affections from which they suffered required such frequent use of morphia and atropia, that I have been able to note the effects of these drugs, alone, and in every variety of combination, several times in all, and many times in some. The effects, moreover, which were observed in these individuals, are such as were common to all who have come under my notice. In reference, then, to the action of opium and belladonna, alone or in combination, these cases may be regarded as fully illustrative of the general effects of the drugs.

The first subject, Samuel M. (see pp. 124, 185, &c.), may be taken as the type of those individuals who are easily influenced by the hypnotic effects of opium, and who are therefore readily narcotised.

The following observations—95 (*a*), (*b*), (*c*), (*d*)—show the effect of the *simultaneous introduction* of morphia and atropia on this patient. Pulse 74; pupils $\frac{1}{9}''$; respiration 19–20:—

Obs. 95.—(*a*). Injected $\frac{1}{4}$ *grain morphiæ acet.* and $\frac{1}{100}$ *grain atropiæ sulph.*, by one puncture. *After* 25 *minutes*, pulse accelerated 10 beats; pupils $\frac{1}{12}''$; great somnolency. *After* 1 *hr.*, pulse accel. 20 beats; resp. 17; somnolency continued; conjunctivæ injected. *After* 2 *hrs.*, pulse and resp. as at 1 hr.; pupils still contracted, as at 25 min.; slight dryness of the mouth and throat; soporific effect continued.

(*b*). $\frac{1}{4}$ *grain morphiæ acet.* and $\frac{1}{96}$ *grain atropiæ sulph.*, by one puncture. *After* 1 *hr.*, pulse accelerated 28 beats; pupils $\frac{1}{16}$; tongue dry and brown; palates dry and glazed; throat dry; voice husky; conjunctivæ injected. *After* 2 *hrs.*, pulse accel. 24 beats; pupils $\frac{1}{10}''$; throat still dry; mouth moist and clammy; soporific effect = (*a*).

(*c*). $\frac{1}{4}$ *grain morphiæ acet. and* $\frac{1}{40}$ *grain atropiæ sulph.* Somnolency came on within 4 min. *After* 15 *min.*, pulse accel. 46 beats; pupils unchanged. *After* 1 *hr.*, pulse still 120; pupils unchanged; conjunctivæ injected, and wet; mouth and throat felt dry, but there had been no persistent dryness yet; resp. 16. *After* 2 *hrs.*, pulse accelerated 38 beats; resp. 16; pupils $\frac{1}{7}$, sideways just over $\frac{1}{6}''$; mouth and throat had been very dry and parched; the latter had now become moist and clammy, and exhaled a strong belladonna fœtor. An equal dose on another occasion was followed by the same effects; but at the end of an hour the pulse had risen to 160, or more than double the initial rate. At the end of 2 hours, the acceleration was only 38. On both occasions the soporific effect was very powerful and persistent.

(*d*). ½ *grain morphiæ acet.* and $\frac{1}{40}$ *grain atropiæ sulph.*, by separate punctures. Somnolency came on at the *third min.* *After* 5 *min.*, pulse accel. 16 beats; resp. 17. *After* 30 *min.*, pulse accel. 56 beats, of undiminished volume, but strong and resistant, one or two intermissions in the minute; resp. 16, regular; tongue dry, and brown down the centre; both palates dry and glazed; throat very dry, voice husky; eyes suffused; face and scalp flushed. *After* ¾ *hr.*, resp. 14; pulse still 130. *After* 1½ *hr.*, pulse accel. 54 beats; resp., as he slept, 13, regular. *After* 2 *hrs.*, pulse accel. 46 beats, contracted, strong, and regular; resp. 12 some min., 14 others. *After* 3 *hrs.*, pulse still 120, as at 2 hrs., regular, and of good power; respiration during the last hour, from 10 to 9, full and regular; tongue, excepting a narrow margin, and the palates, as at 30 min. The pupils remained unchanged throughout. The suffusion of the head and conjunctivæ still continued. Shortly after this, at 11.30, he retired to bed, and slept without awaking until 8 a.m. The soporific influence remained during the following two days. From the 3rd min. to the end of 3½ hours, the general effect was that of irresistible sleep. The sleep was deep, tranquil, accompanied by snoring, and not far removed from narcotism.

The following observations illustrate the action of the drugs when *successively administered.* In 96 (*a*), (*b*), and (*c*), the atropia was given *before;* in (*d*), *after* the morphia. Subject, Samuel M.:—

Obs. 96. (*a*). $\frac{1}{48}$ *grain atropiæ sulph.*, and after 1¼ hour, ¼ *grain morphiæ acet.* 1 hr. 10 min. after the atropia, pulse accelerated 26 beats, regular, of natural volume; pupils ⅙; somnolency and giddiness considerable; mouth and throat parched and dry. At this time the morphia was injected.

After 15 *min.*, pulse accel. 36 beats, otherwise unchanged; tongue and palates were completely parched; pupils unchanged. *After* 30 *min.*, pupils, the right $\frac{1}{8}''$, the left $\frac{1}{6}$. *After* 1 *hr.*, pulse has continued as at 15 min., but it was now decidedly contracted; moisture returning to the mouth. *After* $1\frac{1}{2}$ *hr.*, pulse accel. 26 beats, and a trifle fuller; pupils, the right $\frac{1}{8}$, the left $\frac{1}{6}$. *After* $1\frac{3}{4}$ *hr.*, and when thoroughly aroused, the right pupil measured $\frac{1}{6}''$ at the light, the left a trifle more; mouth moist and clammy, the throat parched, the voice husky; pulse accel. 14 beats. He fell asleep 15 min. after the injection of the morphia, and continued sleeping by my side, and in front of a gas-lamp, breathing tranquilly, from 14 to 12 a minute, during the whole of the time, and then went home, and retired to bed at 11 p.m., and slept without awaking until 7 a.m. The soporific effect was equal to that of *Obs.* 95 (*c*). The observations were taken without arousing him; and the contraction of the pupil was partly, if not wholly, due to the effect of sleep (see *Case* 4, p. 250). Inequality of the pupils is a phenomenon which I have frequently observed under the influence of atropia in combination with opium. A paroxysm of tic coming on in this patient during the action of the drugs, usually determined contraction of the pupil of the same side.

(*b*). $\frac{1}{2}$ *grain morphiæ acet.* failed to produce any effect on the pupils, which had dilated to $\frac{1}{6}''$; $\frac{3}{4}$ of an hour after $\frac{1}{40}$ *grain atropiæ sulph.* The soporific effect was irresistible, and continued for many hours.

(*c*). $\frac{1}{30}$ *grain atropia sulph.*, and after 1 hour $\frac{1}{2}$ *grain morphiæ acetatis.* At the end of 20 minutes the atropia produced a maximum acceleration of 36 beats, as in *Obs.* 76, p. 206. At the end of 1 hr. the acceleration amounted to only 22 beats, and the pulse was small and weak; resp. 17; the other symptoms as at p. 206. The morphia was now injected. He went off to sleep in the course of 5 minutes. *After* 25 *minutes* pulse as before, 96, a trifle stronger; tongue and palates, excepting a little dry patch on the latter, not only moist but wet. *After* 1 *hr.* pulse 98, still a little stronger; mouth and tongue everywhere quite moist, as before the injection of the atropia; conjunctivæ injected.

Two hrs. after the injection of the morphia, pulse as before; tongue dry and brown, from tip to base; both palates completely dry and glazed. Pupils, as he slept, $\frac{1}{6}$; awake, they had remained unchanged from the time of the injection of the morphia—viz., $\frac{1}{8}$" at the light, $\frac{1}{4}$ sideways. Dropped off to sleep again directly after he was aroused.

The respiration and general effects did not differ outwardly from those of *Obs.* 95 (*d*); but he stated that the soporific effects were greater than any he had experienced from this or any other injection. He slept very heavily throughout the night, and drowsiness persisted for many hours afterwards.

(*d*). $\frac{1}{4}$ *grain morphiæ acet.*, and after 1 hr. $\frac{1}{48}$ *grain atropiæ sulph.* An hour after the injection of the morphia, pulse accelerated 14 beats, regular, of good volume and power; resp. 15, regular; pupils at light $\frac{1}{12}$", sideways $\frac{1}{10}$; tongue moist, as at first. The atropia was now injected. *After* 12 *minutes*, pulse 122, being accelerated 48 beats, slightly increased in force; resp. 14, regular; pupils at light, $\frac{1}{10}$. *After* 25 *min.*, pulse accelerated 54 beats; resp. 14; pupils $\frac{1}{10}$, sideways $\frac{1}{7}$; no dryness of mouth. *After* 30 *min.*, pulse accelerated 46; resp. 14; pupils $\frac{1}{8}$, sideways $\frac{1}{6}$; sclerotic injected; face and scalp a little flushed; anterior half of the tongue quite dry; both palates completely dry and glazed. *After* 1 *hr.*, pulse and pupils as at 30 min.; tongue now completely dry; respiration 12 as he sleeps. *After* 1½ *hr.*, pulse still accelerated 46 beats; respiration, one minute 12, the next 14, full and easy; Now (11 p.m.) retired to bed, and did not awake until 7 a.m.

The drowsiness, which was great before the injection of the atropia, increased, 15 min. afterwards, to inability to keep awake; and he continued to sleep soundly by my side, and opposite a gas-lamp, breathing tranquilly and regularly, about 13 times a minute, during the whole of the time. He stated that the soporific effect was as great as when the same doses were given simultaneously, and, as far as I could judge, such appeared to be the case.

In the following observation, the simultaneous injection

of full doses of morphia and atropia caused increased soporific effects, with dilatation of the pupils.

Obs. 96 *bis.*—Subject, John L—— (see p. 124); pulse 74; pupils $\frac{1}{8}''$. $\frac{1}{4}$ *grain morph. acet.* with $\frac{1}{40}$ *grain atrop. sulph.*, by one injection, caused a maximum acceleration of 26 beats, lasting for an hour; dryness of the mouth and throat, with almost an inability to articulate, and a progressive dilatation of the pupils, until at the end of $2\frac{3}{4}$ hrs. they measured rather more than $\frac{1}{6}''$. He slept soundly for three hrs. after the dose, and then went home. The giddiness, drowsiness, and dryness of the throat continued for several hours.

That atropia increases the hypnotic effect of opium is obvious from the following observation. $\frac{1}{4}$ grain of morphia alone failed to produce somnolency (see *Obs.* 21, p. 127) in this patient :—

Obs. 97.—Charles V.—— Pulse 76; pupil $\frac{1}{10}''$, sideways $\frac{1}{7}$. The effects of $\frac{1}{48}$ *grain atropiæ sulph.* on this patient were as follows :—*After* 1 *hr.* a maximum acceleration of the pulse, 24 beats; tongue and roof of mouth only dryish; pupils at light $\frac{1}{8}$, sideways $\frac{1}{6}$. The medicine 'made him feel gapy, but not sleepy.' Cheeks and forehead cool. *After* 2 *hrs.*, pulse accelerated only 4 beats; pupils just over $\frac{1}{8}$, sideways $\frac{1}{6}$; mouth moist, as before the injection. No heat nor flushing of face, nor any sleepiness throughout. The composition of the urine secreted before and during the action of this dose is given at p. 214, *Case* 1.

The effects of $\frac{1}{4}$ *grain morphiæ acet.*, and after 1 hr. $\frac{1}{48}$ *grain atropiæ sulph.*, were as follows :—55 *min. after the injection of the morphia* the pulse had fallen 4 beats, regular; pupils contracted to $\frac{1}{12}''$; a few vessels of the sclerotic slightly injected; no heat or flushing of face; no heaviness nor inclination for sleep. 15 *min. after the atropia injection*, pulse increased 18 beats, unchanged else. *After* 20 *min.* increased 38 beats; tongue dry at the tip. *After* 25 *min.* pulse increased 52 beats; pupils dilated to $\frac{1}{10}$; no somnolency or giddiness yet. *After* $\frac{3}{4}$ *hr.* pulse accelerated 48 beats; tongue and mouth dryish; moderate somnolency. *After* $1\frac{1}{2}$ *hr.* pulse increased 44 beats; mouth and throat everywhere dry, but not parched. *After* $2\frac{1}{2}$ *hrs.* pulse increased 36 beats; mouth clammy and dry. *After* 3 *hrs.* pulse accelerated 24 beats; pupils $\frac{1}{8}''$, sideways $\frac{1}{8}$; mouth still clammy and dryish; injection of the conjunctiva gone. 4 *hrs.* after the injection of the atropia, pulse accelerated 8 beats, of good volume and power; pupils at the light nearly $\frac{1}{7}''$; mouth regained its moisture, but the throat still felt dry; somnolency passed off. Thirty minutes after the atropia injection, and for the following 3 hrs., there was moderate somnolency, sufficient to have ensured sound sleep, if he had remained undisturbed. The following analyses of the urine voided before and during the action of the above doses, show the effect of morphia alone, and in combination with atropia, on the renal function:—

	1. Urine secreted between 8.30 and 11.20 A.M. ℥xviij. sp. gr. 1008, acid, bright on boiling. 1000 grain measures contained:	2. Urine secreted during the action of the morphia, one hour, ℥vij. sp. gr. 1007, faintly acid, bright on boiling. 1000 grain measures contained:	3. Urine secreted during the four hours following the injection of the atropia. ℥xviij. sp. gr. 1011·2, bright on boiling. 1000 grain measures contained:
8.30 A.M. voided urine, and took one pint of tea, with bread and butter.	Grs.	Grs.	Grs.
11.20 A.M. $\frac{1}{16}$ morphiæ acet.	Chlorine . 2·42	Chlorine . . 1·82	Chlorine . 3·64
12.20 A.M. $\frac{1}{30}$ grain atropiæ sulph.	Urea . . 6·20	Urea . . 5·66	Urea . . 8·76
	Sulphates and Phosphates 2·33	Sulphates and Phosphates 1·46	Sulphates and Phosphates 3·80

Urine 1 is diluted by the fluid taken at breakfast, and, together with urine 2, must be taken to represent the chylous urine. Urine 3 presents an increase of all the constituents, as is observed after the action of atropia alone. 12 drops of this urine—that is, assuming it to contain the whole of the atropia injected, 12 drops of a solution of sulphate of atropia in 414,720 parts of water—dropped into my right eye during an interval of two hours, caused dilatation of the pupil from $\frac{1}{8}''$ to very nearly $\frac{1}{6}''$, and maintained it so for 5 hours at least.

The remaining observations show the influence of atropia in preventing the distressing effects which result from derangement of the pneumogastric nerve:—

Obs. 98.—Subject, Michael Egan (see p. 124). Pulse 78; pupils $\frac{1}{10}''$. $\frac{1}{96}$ *grain atropiæ sulph.* caused, *in* 38 *minutes*, an acceleration of the pulse 20 beats; pupils $\frac{1}{6}$; dryness of the tongue, and hard palate, but they were not absolutely devoid of moisture. *After* 1 *hr.* 40 *min.*, pulse accel. 10 beats;

pupils $\frac{1}{8}$; mouth as moist as before the dose. The combined effects of $\frac{1}{4}$ *grain morphiæ acet.* and $\frac{1}{96}$ *grain atropiæ sulph.*, administered by one injection, were as follows:—*After* 15 *min.*, pulse accel. 20 beats; felt the face and head becoming hot; giddiness. *After* $\frac{3}{4}$ *hr.*, pulse attained max. accel. of 32 beats; pupils unchanged; tongue dry, brown, and hard, except the margins; the whole of the hard and soft palates completely dry and glazed. *After* 1$\frac{1}{2}$ *hr.*, pulse accel. 14 beats; pupils unchanged; away from the light they dilated as at the first to $\frac{1}{6}''$; mouth as at $\frac{3}{4}$ hr., but the dryness had decreased the width of the moist margins of the tongue. *After* 1$\frac{3}{4}$ *hr.*, tongue just becoming moist, both palates still as at $\frac{3}{4}$ hr; pulse accel. 10 beats. *After* 2 *hrs.*, mouth quite moist and clammy; pupils unchanged. *After* 2$\frac{1}{4}$ *hrs.*, pulse returned to 78, of undiminished volume and power; tongue and palates had become a little dry again; pupils dilated to $\frac{1}{8}''$.

General Effects.—Great drowsiness, heaviness, and giddiness accompanied the action of the medicine. Dryness of the mouth continued 5 hours longer, and was accompanied by intense sleepiness. He went to bed at 3.30 p.m., and slept soundly until 5, when he was aroused, took a cup of tea, and after 1$\frac{1}{2}$ hr. went to sleep again, and continued to sleep soundly throughout the night. He did not experience nausea, depression, or faintness, as on the previous occasion. There were no after-effects.

The effects of $\frac{1}{48}$ *grain atropiæ sulph.* alone on this patient were *less* than those of $\frac{1}{96}$ in the combined dose; the acceleration of the pulse was 2 beats less; the dryness of the mouth was neither so complete nor prolonged. At the end of 2$\frac{1}{2}$ *hrs.*, the pupils were $\frac{1}{5}''$. It produced much giddiness and heaviness, and induced a dreamy, dozy condition. The analysis of the urine voided before and during the operation of this quantity of atropia is given at p. 214, *Case* 3.

Obs. 99.—Mary B. (see *Obs.* 23, p. 128). Pulse 80; pupils $\frac{1}{7}''$. *The effects of* $\frac{1}{96}$ *and* $\frac{1}{48}$ *grain atropiæ sulph.* agreed with those recorded in *Obs.* 71 and 73 (p. 203), the maximum acceleration after the former dose being 30, after the latter 50; either dose dried the mouth completely, and at the end of

2 *hrs.* the pupils measured $\frac{1}{8}''$. The smaller dose did not produce any cerebral effects; the latter only giddiness and heaviness, not amounting to somnolency.

(*a*). $\frac{1}{4}$ *grain morphiæ acet.* and $\frac{1}{96}$ *grain atropiæ sulph.*, by one puncture, produced the following effects:—A max. accel. of the pulse 40 beats; extreme dryness of the mouth, giddiness, and hypnosis. She slept soundly after the dose for 2 hrs., then rode home, and went to bed, and slept continuously for the next 8 hrs. There was no change of the pupils during the action of the medicine, neither any tendency to nausea or faintness. The effect was charming, and she was free from any unpleasant effect the next day.

(*b*). $\frac{1}{6}$ *grain morphiæ acet.* and $\frac{1}{48}$ *grain atropiæ sulph.*, by one puncture, produced, on another occasion, equally satisfactory effects. The pulse attained a max. accel. of 70 beats, being almost doubled; the volume and power were good. The soporific effects were immediate, and rather more powerful than the previous dose. She was quite comfortable the next day. At the *end of* 2 *hrs.* the pupils were dilated to $\frac{1}{6}''$.

(*c*). $\frac{1}{4}$ *grain morphiæ acet.* and $\frac{1}{96}$ *grain atropiæ sulph.*, introduced by one injection, caused the following effects:—*After* 40 *min.*, pulse accel. 48 beats, full and regular; tongue and palates completely dry; great somnolency. *After* 2 *hrs.*, pulse accel. only 8 beats, strong, full, and regular; tongue completely dry; palates dry and glazed; pupils slightly contracted—just under $\frac{1}{8}''$. The soporific effect was greater, she said, than that of the last dose. There was no tendency to nausea, or any other unpleasant sensation. At the end of $2\frac{1}{2}$ hrs., she went home, and slept soundly and continuously for 8 hrs. On rising next morning, she was sick, and experienced nausea during the whole of the morning. She considered it to be a common bilious attack; but I have no doubt that it was due to the dose, the proportion of atropia being insufficient. If the nausea were really due to the after-effects of the morphia, the injection of a little atropia before rising would have entirely prevented it.

Obs. 100.—Mrs. E. W. (see *Obs.* 25, p. 129). Pulse 80; pupils $\frac{1}{10}''$, sideways $\frac{1}{6}''$. (*a*). Effects of $\frac{1}{96}$ *grain atropiæ sulph.* and $\frac{1}{12}$ *grain morphiæ acet.*, introduced by one puncture:—

After 20 *minutes*, pulse accelerated 22 beats, regular; tongue dry at the tip, palates quite dry; pupils unchanged; felt giddy and a little faint a minute ago, but it had passed off; very sleepy. *After* 1 *hr.*, pulse accel. 20 beats, no change in the power or volume; pupils $\frac{1}{6}$; had been sleeping very comfortably. *After* $2\frac{1}{2}$ *hrs.*, pulse accel. only 10 beats; pupils $\frac{1}{6}''$, dilating to $\frac{1}{4}$ at dark side of room; mouth moist and clammy. *After* 2 *hrs.*, pulse returned to the initial rate, of good power; pupils $\frac{1}{8}$; mouth still dryish; slept comfortably since 20 min. after the dose. The drowsiness and dryness of the mouth continued for 5 hrs. more, and she was quite comfortable, and free from any tendency to faintness, nausea, or illusion. This combined dose was given thrice with the same satisfactory results.

(*b*). The *same dose of morphia* given *with* $\frac{1}{60}$ *grain atropiæ sulph.*, produced the same effects. The pulse attained a max. accel. of 32 beats.

(*c*). $\frac{1}{12}$ *grain morphiæ acet.*, and *after* 1 *hr.*, $\frac{1}{60}$ *grain atropiæ sulph.* The morphia injection was followed by the symptoms described under *Obs.* 25 (*c*), p. 130, and at the end of the hr. she had vomited, and still continued sick and faint; the face was pale and anxious, and the pulse very weak; pupils unchanged. 30 *min. after the atropia* injection, the nausea was slightly relieved, and she had not vomited again; pulse accel. 32 beats; pupils $\frac{1}{8}$; anterior part of the tongue and the hard palate dry; felt sleepy, and as she dozed, 'saw strange figures pass before her.' *After* 1 *hr.*, nausea less; pulse increased 14 beats, regular, and of much better power; a little colour and warmth had returned to the cheeks; had been sleeping tranquilly. *After* $1\frac{1}{2}$ *hr.*, nausea quite gone; had continued to sleep quite comfortably, and without dreaming, and now felt quite comfortable; the pulse was accel. 10 beats, and was regular, and of good volume and power; mouth quite moist; pupils, at the light $\frac{1}{8}$, towards the dark side of the room $\frac{1}{4}$. She continued dozy and comfortable during the next 6 hrs., slept comfortably during the night, and felt no after-effects.

(*d*). $\frac{1}{48}$ *grain atropiæ sulph.*, and *after* $\frac{3}{4}$ *hr.*, $\frac{1}{12}$ *grain morphiæ acet.* $\frac{3}{4}$ *hr.* after the atropia injection, pulse increased 32 beats; tongue dry at the tip; the hard and anterior part of the soft palate dry and glazed; pupils $\frac{1}{8}''$. 15 *min. after the morphia*, pulse a little fuller; pupils $\frac{1}{7}''$; felt more giddy, and a little faint; mouth dryer. *After* $\frac{3}{4}$ *hr.*, the feeling of faintness went off in a few minutes; pulse accel. 24 beats; pupils contracted to $\frac{1}{8}''$ again; mouth as dry as before; great somnolency. *After* $1\frac{1}{2}$ *hr.*, pulse accel. 14 beats; pupils again dilated to $\frac{1}{7}''$; the soft palate had become clammy and moist. The drowsiness and dryness continued as on former occasions, and no unpleasant symptom followed.

(*e*). $\frac{1}{48}$ *grain atropiæ sulph.*, and *after* $1\frac{1}{4}$ *hr.*, $\frac{1}{7}$ *grain morphiæ acet.* $1\frac{1}{4}$ *hr. after the atropia* injection, the pulse was accel. 34 beats, the pupils just over $\frac{1}{8}''$, and the tip of the tongue and anterior part of the hard palate dry. 20 *min. after the morphia* injection, felt a little giddy and faint; the pulse was accel. only 26 beats, regular, and unchanged in volume or power; tongue dry anteriorly; both palates dry and glazed; pupils further dilated to $\frac{1}{7}''$, sideways $\frac{1}{8}$. *After* 1 *hr.*, pulse accel. 25 beats, full and regular; all but the tip of the tongue and the hard palate moist again; faintness and giddiness passed off, and she continued to sleep comfortably. *After* $1\frac{1}{2}$ *hr.*, pupils nearly $\frac{1}{6}''$; mouth nearly moist again. Soon afterwards she walked home, and experienced no unpleasant effects whatever.

Conclusions.—It appears, from a comparative examination of the phenomena detailed in the preceding observations, and those attending the operation of morphia and atropia on man when given alone—1. That in medicinal doses, the essential effect of morphia (hypnosis) is both increased and prolonged by the action of atropia, whether induced previously, or at any time during the operation of the former.—2. That atropia relieves, and, if given simultaneously or previously, prevents the nausea, vomiting, syncope, and insomnia, which frequently result from the action of opium.[1]—3. That in a sufficient proportion (for most individuals $\frac{1}{48}$ grain sulph. atropia to $\frac{1}{4}$ grain acetate morphia), atropia neutral-

[1] It is remarkable that this conclusion is exactly the opposite of the statements made by M. Erlenmeyer, 'Archiv. gén. de Méd.' (1866); and Drs. Mitchell, Keen, and Moorehouse, 'Amer. Jour. Med. Sc.' vol. l. p. 74 (1865).

ises the contractile effect of opium on the pupils, but in larger doses dilatation takes place as if no morphia had been given. It is also to be observed, that if the quantivalent doses are *successively* introduced, the drug last administered exhibits for a short time a counteracting effect: compare *Obs.* 94 (*a* and *d*), and *Obs.* 117 (*a* and *b*).—4. That all the other effects of atropia are intensified and prolonged by the action of morphia, induced previously, or at any time during the operation of the former. If, however, the dose of atropia be small, and the morphia produce considerable deranging effects on the vagus, the rapidity of the pulse is not greater than when the atropia is administered alone.

2. NARCEINE AND ATROPIA.

Obs. 101.—Subject, Samuel M. (see *Obs.* 33, p. 149). Pulse 72; pupils $\frac{1}{9}''$; resp. 20. Injected ♏xx. sol. 2=1 *grain narceine, together with* $\frac{1}{40}$ *grain atropiæ sulph.*, by one puncture. *After* 30 *min.* pulse accelerated 26 beats, fuller; great somnolency; dryness of the throat. *After* 1½ *hr.* pulse accelerated 28 beats; pupils dilated to $\frac{1}{7}''$; throat very dry, and the voice husky; tongue, excepting wide margins, dry, hard, and brown; palates dry and glazed. Had slept comfortably. *After* 2 *hrs.* pulse increased 12 beats, unchanged in volume and power; pupils $\frac{1}{6}''$; mouth moist, throat very dry still; somnolency decreasing. The respiration was regular throughout, and gradually fell to 14. Urine passed at the third hour dilated the pupil. The general effect differed in no respect from that of morphia and atropia. The hypnosis was for two hours as great as that produced by $\frac{1}{8}$ grain morphia acet. and $\frac{1}{40}$ grain atropiæ sulph. The atropia effects were as decided as those produced by the injection of $\frac{1}{40}$ grain alone, although it is apparent that the absorption must have been slower on account of the local irritation. The effect on the pulse was also less, from the same cause. The soporific influence was as persistent as after morphia, if not more so. There was no dysuria. Five days afterwards the integument around the seat of the injection was red, swollen, and prominent; and the central part was soft, as if matter had formed. The inflammation, however, subsided, and there was no such issue.

3. MECONINE AND ATROPIA.

Obs. 102.—Subject, Samuel M. (see *Obs.* 38, p. 154). ♏xxx. of the solution prescribed at p. 154 (=1 *grain of meconine*) and $\frac{1}{40}$ *grain atropiæ sulph.* were injected by two punctures. The effects on the pulse, breathing, pupils, and mouth corresponded exactly with those in the previous observation; nor was there any appreciable difference in the soporific effect.

Remarks. — Compared with the action of narceine and meconine alone, the combined doses had a more powerful soporific effect. This is indicated by the respiration, which was slower under the influence of sleep than when there was only a tendency to it.

4. CRYPTOPIA AND ATROPIA.

Observations of the combined effects of these two drugs have been made on the dog and on man:—

On the Dog. — *Obs.* 103. Injected an acetous solution of *cryptopia* (=1½ *grain*) and $\frac{1}{10}$ *grain atropiæ sulph.*, by separate punctures, beneath the skin of the beagle (see *Obs.* 40, p. 159, &c.). At this time the bladder was emptied of a large quantity of urine. *After* 7 *min.*, pupils completely dilated; remained quiet, and apparently quite comfortable. *At the* 10*th min.* he began to look searchingly from side to side, and two min. afterwards dashed forwards, tail erect, panting loudly, and with restrained whining, and scoured the room round and round, and from side to side, in a succession of wild movements, which he seemed to be impelled to perform when he was using all his strength to restrain them. At first he came up to me when called, and wagged the tail cheerfully; but the moment I ceased to caress him, the body was jerked backwards and forwards as the legs were firmly set on the ground, the head was rapidly rotated with eager glances from side to side, and in a few seconds the excitement was as intense as ever. He soon became perfectly wild and uncontrollable, rushing stiffly and jerkedly in every direction, and knocking everything over in his way; and although he showed no desire to bite, he kept up a subdued growl. He continued in this state for 10 min., and then became quiet and lay on the belly, panting loudly from 300 to 400 times a min.; the tongue protruded, rough, dry, and fiery red; the pupils fully dilated; the heart's action very rapid and strong. *After* 22 *min.* he came up to me when called. *During the next hour* he lay on the abdomen, occasionally raising himself on the legs, and seemed wholly occupied with the rapid panting. *After* 1½ *hr.*, the panting ceased, and he now sat on the haunches, breathing quietly, but quickly and irregularly; pulse 182, the whole body affected with choreic movements; the head was turned from side to side, now suddenly advanced, and now jerked backwards; the angles of the mouth and the eyebrows were frequently twitched. He came readily and with a wag of the tail when called, but walked carefully, and as he advanced the head was occasionally ducked or twisted aside. Now and then the breathing rose into a short pant. He was quite lively, and appeared comfortable, and in his natural temper. Pulse 182; pupils fully dilated and fixed; mouth completely dry. *After* 2 *hrs.* there was a strong tendency to sleep; the nodding head slowly sank on the carpet, and he continued dozing for the next ¾ hr. Resp. 26, regular. During this time 4 or 5 slight twitches a min. affected some part of the muscular system; the sleepy head was frequently raised, and twisted or slightly jerked in every direction. *After* 2¾ *hrs.* he got up, shook himself, and passed a large quantity of yellowish-green alkaline urine.

During the next 2 *hrs.*, the dog was tired, gapish, and sleepy, and lay down and dozed nearly the whole of the time. *After* 3½ *hrs.*, the pulse was 182, and the atropia symptoms continued. *After* 4½ *hrs.*, the pupils were still fully dilated, and the mouth was dryish; but he seemed quite recovered, and ate a plate of meat voraciously. *At the* 6*th hr.*, he was quite lively and well, and followed me downstairs nimbly.

The urine voided at 2¾ hrs. had a sp. gr. of 1034·4, became solid with excess of nitric acid from precipitation of urea, and deposited a large cloud of phosphates when heated,

Three drops placed in my eye dilated the pupil from $\frac{1}{8}''$ to $\frac{1}{6}$, and maintained it so for 5 hrs.

The excitant effect in this *Obs.* differed in no respect from that produced by 2 grains of cryptopia alone (see p. 160). The intensity was, perhaps, a little greater in this last experiment, but the duration was a few minutes shorter. The after-effects—the somnolency and chorea—were only observed in this last experiment.

On Man.—*Obs.* 104. *One grain cryptopia* dissolved by the aid of HCl. in spirit and water, and $\frac{1}{48}$ *grain atropiæ sulph.*, injected simultaneously by separated punctures, produced the following effects in the subject of *Obs.* 20, p. 126:—Pulse 78; pupil $\frac{1}{8}''$; resp. 17. *After* 20 *min.*, pulse increased 10 beats; anterior half of tongue dry; drowsy. *After* 1 *hr.*, pulse increased 22 beats fuller; pupil $\frac{1}{5}$; resp. 16; tongue from apex to base, and the margins also, completely dry, rough, and brown; both palates completely dry and glazed; had slept soundly; the face was flushed and warm. *After* $2\frac{1}{4}$ *hrs.*, the dryness rapidly passed off $\frac{1}{4}$ hr. ago; pulse only increased 4 beats, regular and contracted; pupil just over $\frac{1}{4}$; could not see the lightning-conductor; resp. 15 one minute, 16 the next.

The general effect was *sound tranquil sleep, without dreaming or starting.* All the symptoms exceeded those produced by the simultaneous injection of $\frac{1}{4}$ grain of acetate of morphia and $\frac{1}{48}$ grain sulphate of atropia, and equalled those of $\frac{1}{2}$ grain morphiæ acet., and after $1\frac{1}{2}$ hr. $\frac{1}{30}$ grain atropia sulph.; but the dilating effect on the pupil and the dryness of the mouth were even greater.

On turning to my notes of the effect of atropia upon this patient, I find that I have twice injected the $\frac{1}{40}$ grain alone. At the end of $2\frac{1}{4}$ hrs. the pupil was $\frac{1}{5}''$ on one occasion, and just over $\frac{1}{5}$ on the other. I thus unexpectedly discovered fresh evidence of the dilating influence of cryptopia on the pupil.

5. CODEIA AND ATROPIA.

On the Dog.—*Obs.* 105. Injected 1 *grain codeia*, acidified with HCl, and *after* 1 *hr.* $\frac{1}{6}$ *grain of atropiæ sulph.*, beneath the skin of the bitch (see *Obs.* 49, p. 170). 1 *hr. after* the injection of the codeia, pulse 60, regular; resp. 22; mouth wet; no appreciable change in the pupils; she was dull and quiet. 25 *min. after* the atropia injection, pulse 240, regular, and of good power; pupils fully dilated and fixed; resp. 20; nose dry; mouth clean and wet; very dozy and lethargic. *During the next* 5 *hrs.* she continued to lie along on the side, with the head on the carpet apparently dozing, with the eyes nearly closed. On approaching her she raised the head, and when called pricked the ears and looked up. She got up four times to void urine. 1 *hr. after* the injection of the atropia, the pulse was between 300 and 400; the resp. 14, and subsequently continued at this rate. The pulse maintained the max. accel. for 1 hr., and then slowly declined. The nose and lips were dry, but the mouth and tongue continued quite clean and *wet.* 5¼ *hrs. after* the atropia injection, the doziness had passed off, and she sat still and dull; the pupils fully dilated and fixed, the tongue quite wet. At this time she had nausea, and thrice gulped up a little clear fluid. *At the 7th hr.* of the codeia and the *6th hr.* of the atropia, the pulse was 180, regular, and of good power; resp. 16; pupils still dilated and fixed; the nose dry; the tongue clean and wet, with abundance of glairy fluid. The animal showed no inclination for sleep, was quite intelligent, but dull and still. She remained in this state for 3 hrs. more. All the urines contained atropia.

Remarks.—On comparing these effects with *Obs.* 49 and 69 (pp. 170, 199), it will appear that the codeia did not interfere with the general action of the atropia. The tongue and mouth were prevented from drying by the constant outpouring of saliva, due in the first instance to fright, and subsequently to reflex irritation of the salivary glands associated with a tendency to nausea (see *Obs.* 8 and 9, p. 109). The slobbering excited in the horse after morphia (see *Obs.* 3), is due to the same cause. On the other hand, the atropia counteracted in a measure the nauseating effects of the codeia, and increased its soporific effect.

On Man.—The effects of the simultaneous introduction of

codeia and atropia into the subcutaneous tissue are illustrated in the following:—

Obs. 106.—Subject, Samuel M. (see *Obs.* 53, p. 176). Pulse 74; pupils $\frac{1}{9}''$; resp. 20.

(*a*). Injected $\frac{1}{2}$ *grain codeia* and $\frac{1}{48}$ *grain sulphate atropia*, one into each arm. *After* 30 *min.*, pulse increased 36 beats, otherwise unchanged; began to feel a little heavy. *After* 1 *hr.*, pulse increased 34 beats; pupils dilated to $\frac{1}{6}''$; dorsum of tongue and both palates completely dry; was very drowsy. *After* $1\frac{1}{2}$ *hr.*, pulse increased 26 beats; greater volume and power. *After* $2\frac{1}{4}$ *hrs.*, pulse increased 6 beats—80; pupils $\frac{1}{6}$; tongue, excepting a broad edge, brown, dry, and hard; both palates dry and glazed.

(*b*). *One grain codeia and* $\frac{1}{48}$ *grain sulphate atropia*, one drug in each arm. *After* 30 *min.*, pulse accelerated 46 beats; tongue and palate partially dry; somnolency came on after 10 minutes, and continued. *After* $1\frac{3}{4}$ *hr.*, acceleration 26 beats; regular, of usual volume and power; pupils $\frac{1}{6}$; throat and mouth very dry; moisture beginning to return.

(*c*). *Two grains codeia and* $\frac{1}{48}$ *grain sulphate atropia* injected by one puncture three months after the last dose. *After* 10 *min.*, pulse increased 26 beats; somnolency. *After* $\frac{3}{4}$ *hr.*, pulse accelerated 56 beats, contracted, but regular and strong; pupils unchanged; tongue and palate partially dry and brown. *After* $1\frac{1}{2}$ *hr.*, pulse accelerated 58 beats, regular, and a little fuller; pupils at light $\frac{1}{7}''$, sideways $\frac{1}{5}''$; tongue dry, brown, and hard from back to front; palates entirely dry and glazed; face and scalp hot and flushed, and the bloodvessels full; the temporal artery and its ramifications dilated; the skin generally hot and dryish; the scalp perspiring; conjunctiva slightly injected. 'Very giddy, and never felt more sleepy.' *After* 2 *hrs.*, pulse accelerated 62 beats, regular, and now full and strong; otherwise as at $1\frac{1}{2}$ hour. The pulse now began to decrease, and, excepting the dryness, the symptoms quickly subsided. He went to bed at the third hour (11 p.m.), and slept soundly until 4 a.m.

Remarks.—The hypnotic effect on the above occasions was much greater than when codeia was given alone, and it was associated with much giddiness. For the first two hours, the soporific influence was as great as that produced by $\frac{1}{4}$ the quantity of acetate of morphia combined with the same quantities of atropia; but the effect was comparatively evanescent. The stimulant effect of codeia on the pulse increased that of the atropia: compare *Obs.* 93 (*e*). Excepting the effect on the pupil, all the atropia symptoms are increased and prolonged by codeia. The contractile effect of codeia on the pupil is almost as strong as that of morphia, and yet we see (in *b*) that the effect of 1 grain is completely counteracted, at the end of $1\frac{3}{4}$ hour, by the $\frac{1}{48}$ grain of sulphate of atropia. On each occasion the atropia completely counteracted any nauseating or otherwise unpleasant effect of the codeia. Hypnosis and giddiness were the only cerebral effects of the combined action.

6. THEBAIA AND ATROPIA.

On the Dog.—The combined action of these drugs on the beagle (see pp. 159 and 181) is illustrated in the following:—

Obs. 107.—Injected an acetous solution of *thebaia* used in *Obs.* 56, 57 (=$\frac{3}{4}$ *grain*) and $\frac{1}{10}$ *grain sulphate atropia*, by a single puncture. *After* 13 *minutes*, dryness of the mouth; was uncomfortable and restless, and panted at intervals; tongue dry, and covered with a white fur; pupils completely dilated; heart's action rapid and strong. *After* 22 *min.*, vomited about ℥iij. of clear water; continued restless, walking stiffly, and panting rapidly. The muscles of the face and trunk now began to be twitched; and if he stood still for a second, a momentary spasm of one or other of the already stiffened legs caused him to start. During the next hour, the spasms increased in frequency and intensity; the whole muscular system was rigid, and the twitchings of the legs were so severe that they nearly threw him down. Yet the dog seemed reluctant to lie down; he did so once or twice, but did not retain the position for many seconds. *After* 1 *hr.* from the injection, the spasms attained their maximum; the dog walked but little, and very stiffly and awkwardly. The muscles were very rigid, and when he stood still, the body seemed to get more rigid, and was inclined to one or other side as if he was about to fall over; then a jerk of the leg caused him to totter, and he threw himself down, but instantly regained the legs; wagged the tail when called, but seemed fearful of being touched; and on approaching a hand, he blinked the eyes as the head and shoulders were twitched aside. *After* 2¼ *hrs.* he began to lie down, and retained this posture most of the time; pulse 216, regular and strong. panting ceased; resp. good; decreasing successively from 100 to 70; continued to doze with the eyes closed until disturbed by a twitch, when he got up and changed posture. At these times the breathing was reduced to 60, and became more regular. *After* 2½ *hrs.*, and during the next 40 minutes, he slept soundly, and, excepting a slight occasional twitch of the legs, without movement; pulse 160; resp. 66. *After* 3½ *hrs.*, resp. 29, regular; pupils fully dilated; tongue moist. At this time he got up, and twice voided a few drachms of urine. He now seemed recovered, and walked about as usual. *After* 3¾ *hrs.* he again lay down and slept soundly, awaking up with a start or twitch at intervals of ½ an hour, *until the end of the sixth hour.* The resp. meantime, which had increased to 80, fell to 24; pupils fully dilated; pulse 120. At the end of this time he awoke quite well, and ate a good dinner, but would not take water. A drop of the fluid obtained from the urine, in the manner described at p. 202, dilated my pupil in the course of an hour from $\frac{1}{10}''$ to $\frac{1}{4}$; and after 18 hrs. it was still $\frac{1}{9}$. On comparing this *Obs.* with that recorded at p. 181, it appears that the convulsive effects were about equal in intensity and duration in the two experiments. The soporific effects of the two drugs, on the other hand, were increased each by the other.

General Considerations and Conclusions.—In the dog we have seen that nausea, retching, and vomiting are the earliest and most constant effects of opium. Continued peristaltic action of the intestines and complete evacuation of their contents frequently follow. The respiratory movements are soon diminished, and the breathing becomes shallow and irregular. The action of the heart is simultaneously depressed, and the pulse assumes the respiratory character. Such also are some of the results of the action of opium, usually in a less degree, on many individuals of human kind. These effects, together with contraction of the pupil, are purely local, and the result of cramping or spasmodic excitations conveyed by the third pair, the pneumogastric and spinal nerves, to the circular contractile fibres of the pupils, the stomach, the lungs, and the intestines respectively. The effect on the heart is chiefly, if not entirely, due to contraction and impending collapse of the lungs, and to the restrained action of the diaphragm and respiratory muscles generally. The cardiac branches of the vagus and the trunk of the sympathetic may possibly convey similar excitations from the cranio-spinal axis to the heart itself; but we know too little of the *direct* influence of these branches on the heart to speak positively on this point; and since the causes above indicated are of themselves sufficient to account for the phenomena, this possible influence has been hitherto disregarded (pp. 134, 191).

Granting, however, that the heart is both directly and indirectly affected by the action of opium, the derangements which result therefrom, as well as those which affect the stomach, the lungs, and intestines, are wholly removed by the action of a sufficient dose of belladonna. On the same parts over which opium exercises a cramping influence, belladonna, in sufficient doses, has a dilating action. In some individuals the pupil may be taken as an index of the degree of contraction of the smaller bronchial tubes and pylorus; but in the majority it would appear that the contraction of the pupil is altogether out of proportion to that of the other circular fibres, and this is readily accounted for by the theory which I have advanced at p. 137.

A consideration of the action of opium on the horse serves to strengthen these conclusions. The excitant effects on the cerebro-spinal and sympathetic nervous system are so nearly balanced in this animal, that none of the above-mentioned effects, resulting from a preponderating action on the spinal system, are observable. The horse, indeed, never vomits;

but the stimulant effect on the heart and the pupils, and the absence of influence on the respiratory movements, all prove that the stimulant action on the sympathetic is even greater than that on the cranio-spinal axis.

From these general considerations, and a survey of the combined action of opium, as illustrated in the foregoing observations, I deduce the following conclusions:—

A.—*As to the influence of Atropia on the action of Opium.*

1. Atropia increases the cerebral and anæsthetic effects of opium. The increase of the deliriant action is well seen in the horse; the increase of the hypnotic in the dog and in man. Combinations of atropia with narceine, codeia, cryptopia, and even thebaia, have a marked soporific effect on individuals in whom these substances alone have either no hypnotic effect at all, or produce only a slight tendency to sleep. The increase of the anæsthetic effect is common to all.

2. Excepting in those parts of the body where we recognise two sets of involuntary contractile fibres—the one occluding and under the influence of the spinal nerves, the other dilating and under the influence of the sympathetic system—atropia has no influence in diminishing the cramping and convulsant effects of opium, but, on the contrary, slightly increases them.

3. By virtue of a more powerful stimulant action on the sympathetic nervous system than on the spinal, atropia in sufficient doses is able to counteract and overcome the cramping influence of opium on the occluding contractile fibres.

4. The influence of belladonna in removing the respiratory difficulty is slight and ineffectual, since it extends only to the release of the bronchial tubes, without affecting the diaphragm or external respiratory muscles.

5. That by removing the restraint due to partial collapse of the lungs, atropia thus indirectly relieves the distended heart; while the direct and powerful stimulant action of the drug on the heart itself greatly facilitates and completes this result.

B.—*As to the influence of Opium on the action of Belladonna.*

6. Excepting where the spinal and sympathetic nerves meet in muscular antagonism, the actions of opium and belladonna are concurrent, each intensifying the other. Just as belladonna will often convert the restlessness and delirium caused by opium into tranquil sleep, or even narcotism, so does opium reciprocally convert the insomnia resulting from excessive doses of belladonna into a similar effect.

7. The antagonism, therefore, which exists between opium and belladonna is purely local, and dependent, not on a depressing influence on one nervous system, and an excitation of the other, but on a stimulant action common to both, and which, in the case of each drug, affects them unequally. The dilatation and contraction of the pupil under the influence of belladonna and opium, respectively, merely indicates a *difference in the degree* of the stimulation of the two nervous systems by each drug; and the condition of the pupil, under the influence of combined doses, further shows how far the action is balanced.

8. In man, a full medicinal dose of belladonna is required to neutralise the spinal effects of a full medicinal dose of opium on the pupil, the lungs, and the stomach.

9. It is impossible to neutralise the *local effects* of the action of belladonna, or opium, above specified, without increasing the *general action* on the rest of the cerebro-spinal and sympathetic nervous systems twofold.

Two other subjects remain to be considered—the medicinal use of opium and belladonna in combination, and the antidotal or antagonistic action between opium and belladonna:—

The Medicinal Use of Opium and Belladonna, in combination.—The foregoing conclusions illustrate the use of belladonna as an *antispasmodic,* and indicate, at the same time, those conditions in which opium should be avoided. If, however, the use of the latter cannot be dispensed with, it should be combined with belladonna, in quantities sufficient both to counteract the cramping effects of the opium, and to relieve the spasm for which the medicines are prescribed. The belladonna should always be in sufficient excess to cause dilatation of the pupil. It has been shown that belladonna increases the activity of the kidneys, and that its action on the liver is also stimulant. To these properties it adds a dilating action on the longitudinal layer of the muscular fibres of the ducts and intestines. It is therefore admirably adapted to prevent some of the objectionable effects of opium. Given in combination, I have never found constipation result. In the treatment of *acute disease*, opium may be most beneficially combined with belladonna; and the appropriate mode of administration consists in the immediate introduction of these remedies beneath the skin, as soon as the inflammatory action has been detected. The relief to the constitutional irritation, and probably pain, and the retractile and stimulant action on the circulation, which immediately follow this use of these drugs, will often arrest with wonderful rapidity an inflammatory action (*e. g.*, in the lungs), which, from its severity and extent, threatens soon to become dangerous to life. My own experience induces me to believe that, by means of this treatment, inflammation in its earliest stages lies completely within our control. I have usually introduced a moderate dose of the drugs (acetate of morphia from $\frac{1}{8}$ to $\frac{1}{4}$, and sulphate of atropia from $\frac{1}{96}$ to $\frac{1}{40}$ of a grain) by a single puncture, at intervals of eight or twelve hours. If the soporific influence of the first dose remain, or if pain and insomnia be absent, the occasional use only of the opium is needed.

In the treatment of *neuralgia* and *insomnia*, the best effects are obtained by a combination of these drugs; and by judicious combinations, persons who cannot otherwise endure a dose of opium may be brought under its beneficial influence. Morphia, indeed, should never as a rule be injected alone, unless we have reason to know that the patient will experience no ill effects. To counteract those distressing and sometimes dangerous effects which follow the subcutaneous use of morphia, combination with $\frac{1}{96}$ of a grain of sulphate of atropia will usually be sufficient; and, in some cases, it may be necessary to first induce the atropia action by such a dose, and then a quarter of an hour afterwards introduce the combined dose.

Antidotal or Antagonistic Action between Opium and Belladonna.—We turn now to examine those cases of poisoning by either of these drugs in which the other has been used as an antidote, in order to find evidence of a general antagonistic action, which has been proved to be absent in the horse, the dog, and in man, when given in the doses mentioned in the preceding observations. The idea of antagonism dates as far back as the year 1570, but the cases which are advanced to prove it are neither very numerous nor very satisfactory. The conclusions, however, which have been drawn from them by some medical writers, are sufficiently decided, as the following statements will show:—'The great therapeutical fact that belladonna acts as a direct antidote to morphia, becomes every day more and more fully verified.'[1] 'The mass of evidence in favour of the belief in the antagonism of atropia and morphia is now considerable; we assume, therefore, that there is such a peculiarity of power in these two alkaloids as to enable them in man to neutralise one another physiologically, as acid and alkali may do chemically.'[2] After a review of many of the cases contained in the following tables, Dr. Norris concludes thus:—'The foregoing cases conclusively show that in opium-poisoning, belladonna, in doses which in a state of health would certainly poison, may be administered with impunity, and be followed by a rapid subsidence of the symptoms produced by the former drug; and, *vice versâ*, that opium rapidly and safely counteracts the poisonous influence of belladonna.'[3] In an excellent *résumé* of the subject, in the '*Bulletin gén. de Thérap.*' (vol. lxx. 1866), the author concludes, from a review of the *physiological* action of the two drugs, that it is impossible to regard them as altogether antagonistic (p. 530). But, after quoting some of the cases contained in the accompanying tables, he remarks: 'Now, from all these facts, it appears to result that the antagonism of opium and belladonna is sufficiently proved *clinically*. In fact, when to a subject poisoned by opium we give bella-

[1] 'Medico-Chirurg. Rev.' July 1867, p. 265.

[2] Drs. Mitchell, Keen, and Moorehouse, 'Amer. Jour. Med. Sc.' vol. l. p. 70 (1865).

[3] 'Amer. Jour. Med. Scien.' vol. xliv. p. 405 (1862).

donna, we observe at first that the phenomena proper to belladonna do not appear, in spite of the enormous quantity which we may give. Besides, the phenomena produced by opium are not aggravated, which would not fail to happen if the poisons did not counteract each other. Lastly, patients have, on the contrary, been cured very promptly, in spite of the enormous proportions of opium taken in a very short time. The same proposition is true, inversely, when a case of poisoning by belladonna is treated by opium' (p. 539).

Now, with regard to a statement which is repeated in these extracts, it is to be observed—First, that when a large dose of opium is taken on an empty, and also on a full stomach, only a portion is usually absorbed. Evidence to this effect is very common, and the poison is often rejected several hours after it was ingested. In *Case* (*d*), mentioned below, this occurred after nine hours. (See also *Cases* 4, 6, and 18, Table I.) It appears, then, pretty conclusively that the stomach gradually loses the power of absorbing the poison until this function is altogether arrested. Secondly, that emesis has usually preceded the administration of the belladonna. The toleration, therefore, of the large quantities of belladonna given in the cases alluded to, may be fairly attributed to the extreme slowness of absorption—due partly to the paralysing influence of the opium, and partly to the sickening effect of the emetic.

Whenever the poison has been introduced by the stomach, the skin is the appropriate medium through which the antidote should be conveyed to the blood. In the single case (3) in which this was done, we are enabled to appreciate its action.

The question of the antagonism of opium and belladonna has become a very important one, and it is desirable that the evidence should be thoroughly sifted. With this view I have brought together, in the accompanying tables, the cases which have been advanced to show an antagonism, and, as far as the reports allow me, have given the symptoms at the time the treatment commenced, and the subsequent progress of the cases as to time. I have not included those cases in which *toxic* phenomena have been assumed to be present, from the subcutaneous use of $\frac{1}{30}$ grain of sulphate of atropia. The effect of such doses in combination with opium has been already fully considered. As an introduction to the examination of these tables, I will simply quote a few examples of opium-poisoning, which I happen to have found in the very volumes from which the tables themselves were compiled. Thus associated, these examples will serve very appropriately to steady our gaze while we are looking at the complex phenomena which attend the combined action :—

(*a*). An infant, 7 months old, took between 2 and 3 grains of opium as laudanum, and retained it. Stupor rapidly followed, and the child continued in complete coma all day, and afterwards recovered.[1]

(*b*). A child, aged 2¼ years, took 1 grain of acetate of morphia in an ounce of oxymel of squills. The medicine remained undisturbed in the system for 2½ hours, at the end of which time an emetic produced free vomiting, and under the use of ordinary remedies the child recovered.[2]

(*c*). A child, nearly 6 years old, took by mistake 7½ grains of opium in a powder. The patient was seen 14 hours afterwards, when the narcotism was profound. It gradually wore off, and at the end of 3 days had entirely disappeared.[3]

(*d*). A man, aged 72, healthy, and unaccustomed to opium, took ʒxij. tincture of opium with suicidal intent. He passed a sleepless night; the laudanum was spontaneously rejected after 9 hours, and the patient subsequently recovered.[4]

(*e*). A female adult took by mistake ℥ij. tincture of opium in an enema, and retained it. She remained in a state of coma for 24 hours, and was in too feeble a condition to allow of any treatment. She recovered.[5]

(*f*). A man, aged 80, drank a wineglassful of tincture of opium, by mistake for tincture of aloes. After 1½ hour he lay down and went to sleep. He was discovered asleep 5 hours after the dose, and could not be aroused. At the 7th

[1] Dr. O'Rorke, 'Gaz. des Hôp.' November 21, 1867.
[2] Mr. Winterbottom, 'Lancet,' 1863.
[3] Dr. Hays, 'Amer. Jour. Med. Scien.' April 1859, vol. xxxvii. p. 367.
[4] Dr. G. D. Gibb, 'Lancet,' 1857, vol. ii. p. 80.
[5] Dr. Morell Mackenzie, 'Med. Times and Gaz.' 1863, vol. i. p. 278.

hour he was profoundly comatose, the pulse slow and feeble, the breathing stertorous. Shortly afterwards the breathing had nearly ceased, the radial pulse could not be felt, and the surface was cold and clammy. At the 16th hour he was sufficiently recovered to converse. Recovery.[1]

Of the twenty-one cases of opium-poisoning treated by belladonna, contained in Table I., only ten were treated by belladonna alone, the remainder being subjected to a more or less complex treatment; and in some the influence of belladonna must have been comparatively slight. Of the whole number, *three* (Nos. 7, 9, and 20) *died*, at the 27th, 21st, and 13th hrs. respectively. In CASE 7 the remedy was administered between the 9th and 12th hrs. The *somnolency passed into narcotism* at the 10th hr. At the 13th hr. there was evidence of *full belladonna action*, with *an increase of the poisonous symptoms*. At the 17th hr. the belladonna had relieved the cardiac depression, but there was no relief of the cerebral symptoms. In CASE 9, Thayer's active preparation of belladonna was given in full doses between the 6th and 7th hrs.; *the stupor increased soon after*, and thenceforward the symptoms increased. CASE 20 cannot be fairly used either to prove or disprove the antagonism in question, as death was chiefly caused by the disease.

In CASE 11 the quantity of opium was unknown, and the symptoms were not more severe than those which often follow the subcutaneous use of ¼ grain of morphia (compare with *Obs*. 23, p. 128). Although the belladonna treatment began at the 2nd hr., *the opiate effects did not disappear until the* 15*th hr*. In CASE 16 the quantity of opium was too small to prove fatal if it had been retained. After the free evacuation of the poison, *the recovery was not more rapid than might have been expected.* In CASE 18 the quantity of opium was not known, otherwise the same remarks apply as in *Case* 16. In CASE 21 the symptoms may more reasonably be referred to the 'hepatic colic,' for which the *two poppy-heads* were taken, than to this minute quantity of opiate. The relaxation of the bile-duct by the belladonna equally accounts for the calm. In CASE 17 the dose was small;

[1] Dr. A. B. Shipman, 'Amer. Jour. Med. Sc.' 1840, vol. xxvi. p. 508.

emesis was induced within ½ an hour; the symptoms never amounted to narcotism, and the belladonna may be fairly regarded as the least effective of the means used to recover the patient. In CASE 10 the quantity of opium is uncertain; the symptoms, it appears, never amounted to narcotism; *the belladonna treatment was not commenced till* 14½ *hrs.*, nor does it appear to have relieved the somnolency, for *the patient was still drowsy at the* 34*th hr.*

As far then as any beneficial influence of belladonna is concerned, we may leave the *foregoing nine cases* altogether out of consideration. Of the remaining twelve, *five only* (Nos. 2, 5, 8, 13, and 14) were treated by belladonna alone.

In CASE 2 the coma was unrelieved, *and continued from* 10 *to* 14 *hours after the use of the belladonna.* In CASE 5 belladonna was not given *until the* 14*th hr.*, and *the coma*, it appears, *did not pass off until the* 17*th hr.* at the earliest. In CASE 8 free vomiting was induced by strong coffee at the 3rd hr. The patient *regained consciousness at the* 11*th hr.*, *six hrs. after* the administration of *the belladonna*, and at a time when the issue of the case is commonly decided one way or the other. In CASE 13 the patient had *recovered from the poisonous effects* of opium. The antagonistic influence to such after-effects has been fully recognised. In CASE 14 the necessary details as to time are omitted. Dr. Lee considers that the antidote would have proved fatal in a state of health. We have often given a larger dose to a child in a state of health without the production of any unpleasant symptoms. In CASE 33 *the symptoms are quite exceptional*, and more resemble those of prussic acid than belladonna.

In the remaining seven cases (Nos. 1, 3, 4, 6, 12, 15, and 19), belladonna was not the only remedy used. In CASE 1 the patient himself took an antidote, probably not very long after swallowing the dose. Emesis also preceded the *belladonna, which was given only between the 3rd and 5th hrs.*, without diminution of the somnolency. *At* 5½ *hrs.*, and when 'fully' under the influence of the belladonna, he fell into *a state of torpor, in which he continued, with a quick pulse and slow and shallow stertorous breathing, up to the* 13*th hr.*

CASE 3 is just such as was needed to complete the history

of the combined action of opium and belladonna in man given in the foregoing pages. *At 4¼ hrs., the man answered questions* correctly when aroused; during the next 1½ hr., he swallowed ʒvj. *tincture belladonna, and the stupor increased. Between the 8th and 11th hrs.*, he received the enormous quantity of ⅓ *grain of atropia* under the skin (=1⅓ grain by the stomach), in order to combat a quantity of morphia(=1¼ grain) introduced by the skin. *The coma continued to deepen*, and from ¾ *hr. after the first injection of atropia up to the 19th hr., the patient could not be aroused by the strongest electrical currents; and during the whole of this time*, if not before, it is clear that *the patient was fully under the influence of atropia, yet the respiration fell, and was only sustained by the free use of the battery.* The only effect, apparently, of opium that remained at this time was twitching; the other effects of the drug had also been prolonged and intensified, and thenceforward they were *superseded by those of belladonna, from which the patient began to recover at the 47th hr.*—i. e., 42 *hrs. after the first dose of belladonna, and* 36 *hrs. after the last dose of atropia.* A comparison of the symptoms during the time the belladonna action continued, with those following a larger dose of atropia (see p. 207), will show how far opium can be considered an *antidote* to belladonna. In CASE 6, opium was removed, and the stomach washed out, at 2¾ hrs., and strong coffee administered for the next 5 hrs. The belladonna was given between the 8th and 11th hrs.; but what of the influence of the electro-magnetism during the same period? In CASE 4, much of the opium appears to have been rejected at 1½ hr. *The belladonna was given between 2½ and the 4th hr. At 1½ hr. she could be roused. At the 5th hr., and up to the 17th hr. at the earliest, she was insensible.* In what does the antidotal action of belladonna consist in this case? ℥ij. of laudanum have been *retained*, and recovery effected in the absence of any remedy, in 24 hrs.; whereas in this case a considerable portion of the poison was rejected, and yet recovery was postponed to the 20th hr. CASE 12 is an ordinary one, in which the usual means were adopted to prevent narcotism. The effect of the belladonna is not indicated. In CASE 15 we are not informed when the opiate effects disappeared. In CASE 19 the evacuation of the contents of the stomach and the application of electricity conduced something towards the recovery, which under the circumstances is not remarkable for the rapidity with which it occurred.

Having now called attention to the chief points in these cases, and emphasised by italics those which bear directly on the question of antagonism, I shall confine further comments to the following conclusions:—

1. That the evidence of antagonism in any given case is inconclusive.

2. Taken individually or collectively, the cases show that belladonna has no influence whatever in accelerating the recovery from the poisonous effects of opium.

3. That somnolency, stupor, narcotism, and coma—the essential effects of the action of opium—are both intensified and prolonged by the concurrent action of belladonna.

4. That belladonna is powerless to obviate the chief danger in opium-poisoning—viz., the depression of the respiratory function. This is well illustrated in *Cases* 1, 3, 7.

5. That the results of the combined action of opium and belladonna are the same, whether given in medicinal or toxical doses (see p. 300). While, therefore, belladonna cannot in any sense be regarded as an antidote against opium, but in large doses the exact reverse, it may, under certain conditions, mentioned below, and always in very small doses, be used in conjunction with other remedies as a means of aiding the recovery.

Treatment of Opium-poisoning.—Death is due to depression of the respiratory function, and the stomach becomes sooner or later paralysed. Hence we must try to arouse the spinal cord, and expect nothing in advanced cases from the introduction of antidotes by the stomach:—1. Complete evacuation of the stomach by mustard and hot water, or by the stomach-pump, and the occasional introduction of hot fluids into the stomach and bowels, with a view of arousing the gastro-pulmonary and the cardiac plexuses, and in the lower portion of the alimentary canal the spinal nerves. Sinapisms and heat to the epigastrium.—2. Mild and continuous currents of electricity, from the back of the neck downwards, around the margins of the chest to the epigastrium.—3. When the heart shows indications of failing power, the subcutaneous injection of $\frac{1}{96}$ of a grain of sulphate of atropia at intervals of two hours. It has been repeatedly shown that the full stimulant effects of belladonna may be induced by this or a smaller dose, when the individual is in a complete state of narcotism. If larger doses be given, or if small doses be too often repeated, the beneficial effects of belladonna will be converted into a depressent and narcotising influence. In cases where narcotism is absent, or has been relieved, and where nausea and gastro-pulmonary distress prevail, belladonna in the above mentioned-doses is the appropriate remedy.

After the careful review which has been given of Table I., little need be said of the cases in the succeeding Tables. In a critical examination of them, attention should be given to the cases quoted at pp. 207–8, in which the poisonous doses are as large as in any of the tabulated cases. Secondly, in reference to *Cases* 22 and 23, that the symptoms immediately follow the subcutaneous injection of the drug, and that they may generally be expected to be on the decline at 5 and 2½ hrs. respectively, when the morphia was given. Thirdly, in reference to *Case* 33 (in which the quantity of opium given is considered to be enough to cause a fatal result in the absence of the belladonna), that 120 drops of laudanum dropped from a thick-lipped ℥jss. bottle, by the side of the partly-withdrawn stopper, measure only 52 minims, and only 60 when dropped direct after the stopper is removed. Fourthly, in respect of *Cases* 40 and 41, in which the quantity of opium taken is large, it must be remembered that it was given on a sick stomach, and at intervals during a considerable period, under which circumstances very large quantities of opium may be given. With reference to the combined action of the drugs, the results in the main corroborate the conclusions already formed. Thus, in *Cases* 22, 23, 24, 36, and the remainder, sleep, narcotism, or actual coma followed the administration of the opiate.

All the cases in Table IV. (except 51) also prove that belladonna increases the narcotic effect of opium. In *Case* 51 it is assumed that ¼ grain of morphia and ʒij. of tincture of henbane by the stomach are equal to ⅛ grain of morphia by the skin. ʒij. of tincture of henbane is useless as a hypnotic, and 1 grain of morphia by the stomach is the equivalent of ¼ grain by the skin.

The influence of opium in converting the insomnia of belladonna into sleep, and the influence of belladonna in determining not only sleep, but narcotism, in individuals under the influence of opium, are illustrated in *Obs.* 92-3-7 (p. 276 *et seq.*). Some of the cases in Table II. serve to give greater force to these observations, and teach us that we must be careful how we employ opium as a means of converting the restlessness and insomnia following excessive doses of belladonna into quiet sleep.

In the treatment of Belladonna Poisoning, our efforts must be directed to sustain the breathing. Opium must be used, not as an antidote, but as a means of calming the nervous agitation when it is excessive; and we must not forget the fact, that the patient is much safer in a state of insomnia and restlessness, than he would be in a state of deep sleep. In the former condition the respiration is excited through the brain; in the latter it is debased. Narcotism is more to be dreaded in poisoning by belladonna than in poisoning by opium (see p. 243).

TABLE I.—SYNOPSIS OF CASES OF OPIUM-POISONING TREATED BY BELLADONNA.

No.	Sex.	Age.	Quantity of the opiate.	Symptoms when first seen.	Treatment.	Subsequent progress and result.	Reference to Author.
1	M.	19	Sulphate of morphia, 75 grains (?) in solution (*suicidal*).	After nearly 3 hrs.* great somnolency, unsteadiness in walking, slowness of speech. Pupils contracted. P. 80. Thirst.	At 2nd hr.* swallowed ℥ij. tannin as an antidote. Soon after sulphate zinc, and free vomiting. Between 3rd and 4th hr. coffee, and 20 grs. ext. belladonna in divided doses. During the 5th hr. 30 grs. more, and constant walking between two assistants. During 6th hr. cold shower-bath, friction, galvanism. Brandy injection; sinapisms. Between 6th and 14th hrs. injections of brandy and quinine, and galvanism.	At 3¾ hrs.* P. 100. 5th hr. pupils began to dilate, no diminution of somnolency. At 5½ hrs. pupils widely dilated, 'showing the full action of the belladonna.' P. 120. R. laboured and very slow. At this time he fell into a state of torpor, and was laid on the bed. From 7th to 13th hr. these symptoms persisted almost without variation. P. 112 to 114. R. 11 to 12, feeble, shallow, stertorous. Pupils dilated. At 15½ hrs. vomited. Pupils dilated and fixed. P. 114. R. 12. At 21st hr. lively and reasonable. R. 16, easy and natural. Bilious vomiting, and subsequent recovery.	Dr. W. F. Norris, 'American Jour. of Med. Science,' 1862, vol. xliv. p. 395; 'Archives générales de Médecine,' 1864, p. 584.
2	M.	adult	During 36 hrs. about 9 grains muriate of morphia in solution.	After 36 hrs. from first dose, profound coma. R. 4 to 5, stertorous. P. slow, feeble. Pupils contracted.	Between the 36th and 40½ hrs. ℥vj. of tinct. of belladonna, given at intervals, in ℥j. doses.	After the 3rd dose, at the 38th hr. pupils began to dilate. At the 50th hr. coma entirely gone. R. 20 to 25. P. nearly 120, increased in force. Pupils dilated. Skin flushed and warmer. On the morrow convalescent.	'L'Union médicale,' 1859, tome i. p. 313; Dr. T. Anderson, 'Edin. Month. Jour.' 1854, p. 377.
4	F.	38	℥ij. tincture of opium (*suicidal*).	After 1½ hr. profound torpor, almost insensible to external impressions, and after being roused sank back into a comatose sleep; P. 70, full and strong. R. stertorous. Face flushed. Pupils contracted.	At 1½ hr. sulphate of zinc and ipecacuanha, and 3 grs. tartarised antimony. Vomiting of dark fluid smelling strongly of laudanum. Between 2½ and 4 hrs. ℥j. tinct. belladonna in divided doses. After the 5th, and up to the 19th hr. 20 grs. of extr. belladonna in divided doses, and in solution, per rectum.	At 5th hr. insensible to external impressions; pale, cannot swallow; skin clammy. P. almost imperceptible. Pupils mere points. 7th to 9th hr. no perceptible change. 11th hr. skin dry, extremities warm. 14th hr. R. easier. P. 100, a little stronger. 17th hr. still insensible: skin warm. 19th hr. P. 96, full and strong. Pupils act slightly. 20th hour pupils slightly dilated and active. R. almost natural. 20½ hr. started up in bed and asked for water. Pupils widely dilated. P. 96, full and strong. Was nearly rational. 27th hr. dimness of vision. Pupils large. Tongue dry. Fauces red. Recovered.	Dr. W. S. Duncan, 'Amer. Jour. Med. Sc.' 1862, vol. xliv. p. 277.

3	M.	59	Sulphate of morphia 5 grs.	After 2 hrs. lethargic, but could be roused. After 4 hrs. quite insensible; face flushed, extremities cold, convulsive twitching of the muscles. Pupils contracted.	At 4th hr. cold to the head, sinapisms to extremities. $4\frac{1}{2}$ hr. ʒj. tinct. belladonna per rectum. 5th hr. stomach emptied and washed out. At $5\frac{1}{2}$ hr. ʒiij. tinct. belladonna, and ʒij. more at $5\frac{3}{4}$ hr. At 8th hr. subcutaneous injection of [illegible] grain atropia, repeated at $9\frac{1}{2}$, $10\frac{1}{4}$, and $10\frac{3}{4}$ hrs. At 26th hr. 20 drops tinct. belladonna. Between the 6th and 10th hrs. free use of galvanic battery. Between the 27th and 37th hrs. free use of alcohol.	At $4\frac{1}{2}$ hrs. sensible enough to answer correctly all the questions put to him. P. 170, very feeble; left quiet, lapses into stupor. 7th hr. lost power of swallowing and walking. 8th hr. more sluggish. P. 150. R. 40. $8\frac{3}{4}$ hr. worse, intense sleep, insensible to the most violent current of electricity applied to the nose, lips, and nipples. 9th hr. more comatose. R. 10. P. 140. 10th hr. R. raised —by continual application of galvanism —from 16 to 20. P. 140. Pupils slightly enlarged. A little colour in cheeks. At 11th hr. scarlet from head to foot. Tongue dry. Pupils about middle size. P. 150. R. 20, stertorous and puffing. Slight twitching of muscles. $19\frac{1}{2}$ hrs. Stupor lessened and breathing easier, after last date. Now seems conscious and swallows. 20th hr. able to speak—voice thick and whispering. P. 134. R. 20. Slight delirium: constant tremor of muscles, like a man in *delirium tremens*. Paralysis of the bladder. 25th hr. P. 124. R. 20. Twitchings still: dozes. Pupils midway. 26th hr. tendency to sleep deepening. 27th hr. busy delirium with twitching. Tongue dry. Pupils unaltered. P. 160. R. 22. 30th hr. P. 130. 35th hr. P. 124. R. 24. Paralysis of bladder. 47th hr. P. 104. R. 19. Had ceased to twitch. Paralysis of bladder. Recovery.	Dr. S. W. Mitchell, New York 'Med. Jour.' vol. iv. p. 116.
5			℥jss. tinct. of opium.	After 12 hrs. deeply comatose. Pupils contracted.	Between 14th and 17th hours ʒxjss. of tinct. belladonna.	At 17th hr. effectual dilatation of the pupil, and subsequent rapid recovery from both poison and remedy.	Dr. Motherwell, Austral. 'Med. Jour.' Oct. 1861; 'Med. Times and Gaz.' 1862, vol. i. p. 18.
6	F		ʒiss. tinct. of opium (*accidental*).	After $2\frac{1}{2}$ hrs. very drowsy. R. heavy and oppressed. P. 96. Pupils contracted. Skin cold. Face livid. Could be roused, and understood what was said to her.	$2\frac{3}{4}$ hrs. ʒij. zinci sulph. as an emetic, vomiting of fluid containing faint traces of laudanum. Stomach washed out at intervals, and strong coffee administered in the interim. 8th to 11th hr. electro-magnetism and gr. xvj. extr. belladonna. Enemata of brandy and ammonia.	6th hr. comatose. R. stertorous, 6. P. 100, small, feeble. Pupils much contracted. Could not be roused; subsultus of muscles of the arms. 7th hr. no improvement, but was roused by the electricity so as to swallow the first dose of belladonna. 9th hr. arose and got off the couch to walk, by the aid of two persons. Took a cup of coffee in her hand and drank it. P. 104. R. 6. Less stertor when asleep. Pupils less contracted. 10th hr. intelligence returned. P. 136. R. 8. No stertor, face much flushed. From 12th to 17th hr. slept lightly. At 17th hr. P. 140 to 150. R. 11. Pupils moderately dilated. Next morning pupils much dilated. Recovery.	Dr. P. Lucas, 'Med. Times and Gaz.' 1865, vol. i. p. 195.

* From the time the poison was taken. This applies in every case, and in every column, unless otherwise stated.

TABLE I.—*continued.*

No.	Sex.	Age.	Quantity of the opiate.	Symptoms when first seen.	Treatment.	Subsequent progress and result.	Reference to Author.
7	M.	55	℥j. tincture of opium (*suicidal*).	After 8 hrs., in a deep sleep. P. very feeble. R. stertorous. Pupils contracted.	At 8th hr. sulphate zinc ℨss. and ipecacuanha wine ℥j., and slight vomiting in ¾ of an hr. Afterwards extr. belladonna gr. xvijss. given in 4 doses at the 9th, 10th, 11th, and 12th hrs. Between 14th and 18th hrs. galvanism, sinapisms, and artificial respiration; alcohol, and lastly alcoholic injections.	10th hr. the attendants could not keep him awake. 12th hr. P. very feeble and rapid. R. stertorous. Pupils contracted. 13th hr. pupils dilated, showing the full action of the belladonna, but R. slower, 7. P. at wrist imperceptible. Heart beats 120. 17th hr. P. 120, now felt at wrist. R. 16, stertorous: continued thus till the 27th hr., when he sank and died.	Dr. W. F. Norris, 'Amer. Jour. Med. Science,' 1862, vol. xliv. p. 397; 'Archiv. gén. de Méd.' 1864, p. 354.
8	F.	24	℥j. tincture of opium (*suicidal*).	After 3 hrs. excessive somnolency. At 5th hr. comatose, stertor. P. 50. Complete contraction of pupils. Surface cold.	At 3rd hr. strong coffee, which produced vomiting. At 5th hr. gr. vij. extr. belladonna dissolved in water, but it was swallowed imperfectly. At 6th hr. ℥j. tinct. belladonna was injected into the stomach.	At 7th hr. pupils dilated to thrice their previous diameter. P. R. and temp. improved. At 8th hr. skin warm. P. 100. R. easy. Appeared to be in a quiet sleep. 11th hr. awoke, and complained that she could not see or stand. At 13th hr. no belladonna symptoms.	Dr. W. H. Mussey, 'Cincinnati Med. Observer,' 1856; 'Boston Medical and Surg. Journ.' 1856, vol. liv. p. 56; 'Amer. Jour. Med. Science,' 1862, vol. xliii. p. 58.
11	F.	23	ℨvijss. (it was assumed) tincture of opium (*suicidal*).	After 2 hrs. restlessness, nausea, and vomiting. Vertigo. Pupils contracted. R. normal. P. 120, hard. Skin hot and sweating.	At 2nd hr. ♏xv. tinct. belladonna repeated at intervals until the 15th hr. In all ℨiijss. were given.	At 11th hr. dryness of the throat. At 15th hr. dilatation of the pupils, and disappearance of the opium symptoms.	Dr. C. Paul, 'Bulletin général de Thérap.' lxxii. 320.
12	M.	60	ℨvij. tincture of opium.	After 2 hrs. coma, insensibility, pallor, contraction of the pupils. R. slow. P. 72, feeble. General muscular relaxation.	At 2nd hr. sulphate zinc; no vomiting. Removal of a little fluid by stomach-pump. Between 3½ hr. and 9th hr. ℨiijss. tinct. belladonna in divided doses. Forced exercise meanwhile.	From 2nd to 5th hr. no improvement. 7th hr. understood questions, and showed the tongue. 8th hr. told what quantity of poison he took, and thenceforward improved.	Mr. Adamson, 'Brit. Med. Jour.' 1866; 'Bull. de Thérap.' lxx. 138.
13	F.	adult	About ℨv. of tincture of opium.	After 5 hrs. narcotism, nausea, retching, and vomiting, continuing up to the 24th hr., when the P. was 54. R. normal. Pupils contracted.	At 3rd hr. vomiting induced, and coffee and brandy subsequently given. 24th hr. subcutaneous injection of sulphate atropia.	After the injection (24th hr.), instantaneous suppression of the last symptoms of opium-poisoning—viz., the constant sickness.	T. Dodeuil, 'Bull. de Thérap.' lxix. 277.

19	F.	50	ʒij. and after 1½ hr. ʒiij. more tincture of opium.	After 4½ hrs. could not be roused. R. 10, stertorous. P. feeble. Pupils mere points. Extremities rather cold.	At 4th hr. evacuation of the contents of the stomach, and application of electricity. 5¼ hrs. ʒj. tinct. belladonna. 5¾ hrs. ʒij. more = ʒx. in all.	At 7th hr. alteration in size of pupils. R. 12. P. stronger. Continued to improve till 10th hr., when all indications of opium-poisoning had disappeared. Pupils large and fixed. P. 100, of good power.	'Edin. Monthly Jour.' 1855 (?); Cazin, 'Monog. de la Belladone,' p. 58.
14	—	2	An unknown quantity of tincture of opium (*homicidal?*).	After — hrs. profound coma, skin cold and clammy. R. slow and laborious. P. 40. Pupils contracted.	60 drops of tinct. belladonna in doses of 15 drops at intervals of 20 minutes.	After second dose, slight elevation in temperature of skin; after third, blush over neck and face: pupils active. R. 25. P. 86. Intelligence returning. After 4th dose, scarlatinous tint of skin; dilatation of pupils. One hr. after, unpleasant symptoms gone.	Dr. C. C. Lee, 'Amer. Jour. Med. S.' 1862, vol. xliii. p. 57.
16	F.	adult	ʒiss. tincture of opium (*accidental*).	After 1½ hr. pallor, skin cold. P. slow, small, intermittent. Pupils contracted. Constant nausea and somnolency.	Immediately took coffee and vomited. At 1½ hr. took 10 drops tinct. belladonna, but it did not remain on the stomach. During the next 4 hrs. took 50 drops more = in all ʒj.	Belladonna caused dryness of mouth, increased the volume and frequency of the pulse, and the temperature of the skin. At 5th hr. was nearly recovered, but subsequently had a sleepless night.	Dr. Léon Blondeau, 'Archives gén. de Méd.' 1865, p. 203; 'Bull. de Thérap.' lxviii. 135; 'Gaz. Hebdomad.' 1865, vol. ii. p. 211.
15	F.	adult	Unknown quantity of tincture of opium (*suicidal*).	After 4 hrs. drowsiness. On trying to walk her about she fell down like an inert mass. Skin cold. P. slow and feeble. Pupils contracted. R. shallow.	Between 4th and 5th hrs. 70 drops = ʒj. tinct. belladonna, in doses of 10 drops; afterwards striking of the chest with the moistened end of a towel.	After last dose of belladonna (5th hr.), dilatation of pupils, and increase in the force and frequency of the P., and of the depth of the inspirations. The use of the towel soon caused contraction of the muscles of the face.	Dr. Blondeau. (See previous case.)
20	M.	4	A teaspoonful = ʒj. of tincture of opium. (*accidental*). Pneumonia and bronchitis after measles.	After less than 1 hr. could not be roused. R. very heavy. P. 100. Pupils contracted to points; becoming black in the face. The patient afterwards was in a state of alternate stupor and consciousness.	10 grains ipecacuanha immediately, but no vomiting. At 45 min. 3 drops Thayer's ext. belladonna per rectum; and between the 2nd and 6th hrs. 13 drops more—partly by mouth and partly by rectum = in all 16 drops. Afterwards stimulant enemata.	At 1½ hr. quite sensible and wide awake, asking for drink. Pupils contracted. At 2½ hrs. stupor came on again. R. 8. Lips dark. At 5½ hrs. consciousness returned under the influence of belladonna. Collection of mucus interfered with R. At 8th hr. P. 100. Pupils natural. R. 24. 13th hr. died asphyxiated.	Dr. James Blake, 'Pacific Med. and Surg. Jour.' April 1862; 'Amer. Jour. Med. Science,' 1862, vol. xliv. p. 280; 'Archives gén. de Méd.' 1864, p. 588.
18	M.	40	Tincture of opium. Quantity? (*suicidal*).	After — hrs. somnolency, nausea, and vomiting when disturbed.	Some of the poison was rejected by spontaneous vomiting. After he came under treatment, $\frac{1}{12}$ of a gr. of ext. belladonna.	After the belladonna, sleep more tranquil, and cessation of the vomiting. Recovery on the morrow.	Dr. Béhier, 'L'Union méd.' 1859, p. 19; 'Bull. de Thérap.' lvii. 41.

TABLE I.—*continued.*

No.	Sex.	Age.	Quantity of the opiate	Symptoms when first seen.	Treatment.	Subsequent progress and result.	Reference to Author.
21	F.	54	A decoction of two poppyheads (per rectum).	Partial syncope and incessant vomiting. 4th hr. pale. P. small, hard, and frequent. Skin cold. Pupils contracted. Intelligence intact.	One pill containing not more than ½ grain of ext. belladonna.	Produced a complete calm.	M. Béhier. (See previous case.)
9	M.	30	ʒj. tincture of opium (*suicidal*).	Somnolency, but no approach to coma until 12 hrs. after treatment. Walked with the aid of assistance.	At 2½ hrs. sulph. zinc gr. x., ipecacuanha powder ʒj.; tickling fauces; strong coffee, followed by copious vomiting. From 2½ to 6th hr. exercised in walking between assistants. At 6th hr. ʒj. Thayer's fluid extr. belladonna, and at 6½ hr. ʒj. more. Stimulants and galvanism.	At the 6th hr., after the copious vomiting and continuous exercise, he was able to walk and resist the assistants; but the stupor returning soon after, the belladonna was given. No marked change followed the first dose, but dilatation of the pupils occurred after the second. A few hours after he began to sink, and died at the 21st hr.	Dr. John G. Blake, 'Boston Med. and Surg. Jour.' 1864, vol. lxx. p. 29.
17	F.	39	About ʒiij. tincture of opium.	After ½ hr. very sleepy, said she could not see. Pupils contracted and fixed. P. small, slightly accelerated. R. shallow. Skin very cold.	After 20 min. zinc emetic. Afterwards strong coffee and exercise. At 4th hr. enema of turpentine and castor-oil. Sinapism to scrobiculus cordis, and ♏x. tinct. belladonna every ½ hr. Continue strong coffee.	After 8 hrs. pupils moderately dilated; was frequently aroused by the nurse during the night, and there appears to have been no further opiate effect of any consequence. Recovered.	Dr. Radcliffe, 'Lancet,' Sept. 5, 1868, p. 312.
10	F.	60	Tinct. opii, nearly two tablespoonsful? (*suicidal*).	After 12 hrs. could be aroused by shaking. P. feeble. R. shallow. Pupils completely contracted. Skin very cold.	After 12 hrs. mustard emetic, enema of strong coffee; frequent shaking and flapping with cold cloths. At 14½ hrs. enema of turpentine and castor-oil. Sinapism to scrob. cordis. ♏x. tinct. belladonna every ½ hr.	After 34½ hrs. still very drowsy, but answered when spoken to. P. 66, weak; R. less shallow. Pupils slightly dilated (after about ʒiij. tinct. belladonna). Tongue dry and brown. Skin warm. Recovered.	"

TABLE II.—SYNOPSIS OF CASES OF BELLADONNA-POISONING TREATED BY OPIUM.

No.	Sex.	Age.	Quantity of the poison.	Symptoms when first seen.	Treatment.	Subsequent progress and result.	Reference to Author.
22	M.	adult	$\frac{1}{2}$ grain sulphate atropia used subcutaneously.	After 4½ hours, delirium, but tried to answer questions; agitation and constant moving of the hands, as if at his usual work. P. small and frequent; skin hot and bathed in sweat. R. hurried, face flushed.	At the 5th hour, injection of about $\frac{1}{3}$ grain of hydrochlorate of morphia, beneath the skin.	Almost immediately after the morphia, became calmer. From the 6th to the 7½ hr. continued to sleep, in one posture, and on awaking the flushing and congestion of the head were gone. P. fuller and less frequent. He continued to sleep composedly for the next 9 hrs., then awoke rather confused; again fell asleep, and awoke 13 hrs. after the morphia injection, free from mental disturbance.	Mr. Benjamin Bell, 'Edin. Med. Jour.' 1859, p. 1; 'L'Union médicale,' 1859, p. 378.
23	F.	adult	$\frac{1}{12}$ grain sulphate atropia used subcutaneously.	After 2½ hours, restless, moaning, feeling of want of power in the legs, dryness of throat, frequent pulse.	At 2½ hrs. about $\frac{1}{4}$ grain hydrochlorate of morphia was injected beneath the skin.	Was more comfortable after the morphia; at 4½ hrs. pulse still frequent, but had become full and soft; on the following day she was comparatively well.	B. Bell, 'Edin. Med. Jour.' 1859, p. 7.
26	F.	16	$\frac{1}{3}$ grain sulphate atropia used during 15 minutes as a collyrium.	Almost immediate dilatation of pupils; in 10 minutes dryness of throat, and inability to swallow; in 35 minutes, staggering, restless delirium, and great agitation, increase of P. and R.	At about 30 min. and during the next—$\frac{1}{8}$ grain of morphia was four times injected beneath the skin = altogether $\frac{1}{2}$ of a grain.	After the injections she slept for 1½ hr.: the P. and R. became slower. On awaking she was able to take drinks and walk round the garden.	Dr. Nieberg, 'Bull. de Thérap.' lxxii. p. 91.
27	M.	75	$\frac{1}{4}$ grain sulphate atropia in about ℥iij. of water (*accidental*).	After one hour, muscular weakness and inability to walk, dilatation of the pupils, dryness of the mouth.	At 1 hr. 6 drops of Rousseau's laudanum. Between 3rd and 4th hrs. 50 drops of Sydenham's laudanum; 5th hr. 10 drops more; 12th hr. 10 drops more = 70 drops altogether.	At 3rd hr. pupils moderately dilated; P. 108, full and hard. Skin hot; lies still, articulation indistinct, understands imperfectly. 4th hr., agitation of limbs, delirium. P. 120, skin warm and moist. 5th hr. symptoms much abated; 16th hr. vomiting, pupils almost normal. Rapid improvement.	Dr. Béhier, 'L'Union méd.' 1863, vol. xix. p. 100; 'Bull. de Thérap.' lxv. p. 135; 'Archiv. gén. de Méd.' 1864, p. 589.
33	M.	6	ʒj. *of Succus Belladonnæ* (accidental).	Soon after became scarlet in the face, tottered, and fell down insensible; flush deepening to violet hue; eyes fixed, pupils dilated to utmost; tongue dry; P. slow and bounding; almost comatose.	20 drops of laudanum, and the same quantity by the rectum; the dose repeated at intervals of $\frac{1}{2}$ an hr. until 120 drops altogether were taken.	At about 3rd hr. (after the 3rd dose) the pupils began to strongly contract, and the purple flush to fade, and a little later on the child was well and running about.	Dr. C. C. Lee, 'Amer. Jour. Med. S.' 1862, vol. xliii. p. 57; 'Bull. de Thérap.' lxiii. 233 and lxx. 532; 'Archiv. gén. de Méd.' 1864, p. 583; 'Dub. Med. Press,' 1862.

TABLE II.—*continued.*

No.	Sex.	Age.	Quantity of the poison.	Symptoms when first seen.	Treatment.	Subsequent progress and result.	Reference to Author.
29	F.	43	1½ grain sulphate atropia in solution (*accidental*).	After ¾ hour, restlessness, delirium, croupy voice, inability to articulate; pupils widely dilated; dryness of mouth, nose, and eyes.	80 drops tincturæ opii at once; 120 drops more during the next 14 hours.	At 3rd hr. was aroused with difficulty. P. 130. At 8th hr. tongue moistening. 9th hr. less stupor. 22nd hr. answered when spoken to; recovery from this time.	Dr. D. H. Agnew, 'Pennsylvania Hos. Rep.' vol. i. 1868, p. 356.
30	M.	2	Extract of belladonna, quantity?	After 5 hours delirious. P. strong; pupils dilated.	5th hr. an emetic, which caused vomiting; then an enema of castor-oil and turpentine. Between 6th and 15th hrs. 25 drops of laudanum.	After 15th hr. sleep and recovery.	Dr. Macnamara, 'Dublin Quart. Jour.' 1863; 'Amer. Jour. Med. S.' 1863, p. 521; 'L'Union méd.' 1863, vol. xviii. p. 286; 'Bull. de Thérap.' lxiii. p. 275.
31	F.	21	Extract belladonna (?) quantity unknown.	After 1 hour face and neck scarlet and hot, cold extremities; dilatation of the pupils; complete tendency to syncope; no delirium.	Between 1st and 2nd hr. a zinc emetic and vomiting; then ♏xx. tinct. opii, and repetition of the dose at about the 7th hr. No mention of further treatment.	After the second dose of opium, contraction of the pupils. At 24th hr. they had nearly regained their natural size. At 30 hrs. dilatation and recurrence of the other symptoms. The dilatation of the pupil remained for four days.	Dr. Fraser, 'Med. Times and Gaz.' 1864, vol. ii. p. 385.
32	F.	adult	Extract of belladonna, per rectum; about ⅓ of a grain daily for 6 days.	After 6 days, confusion of ideas, giddiness, dryness of the tongue, amblyopia.	After 6 days took ℨiv. 'sirop diacode,' and repeated the dose after a few hours.	The first dose of the syrup caused within ½ hr. subsidence of all the poisonous symptoms.	Dr. Béhier, 'Annuaire de Thérap.' Bouchard: 1860, p. 22.
34	F.	adult	A cupful of infusion of belladonna leaves.	Caused symptoms analogous to those of *delirium tremens*.	Cold affusion to the head, and nearly ½ grain of the gummy extract of opium every hour.	Recovery.	Cazin, *op. cit.* p. 12.
35	M.	23	10 ripe berries of atropa belladonna.	After 1½ hr. full effects of belladonna, followed by incoherency of idea and speech, restlessness and wakefulness.	4th hr. an emetic, followed by copious vomiting. Between 7th and 18th hrs. took ♏xliv. tincturæ opii.	7th hr. symptoms continued. 10th hr. slept a short time, but awoke still delirious. Between 19th and 21st hrs. slept, awoke quite collected, and soon recovered.	Mr. Seaton, 'Med. Times and Gaz.' 1859, vol. xix. p. 551.
36	M.	25	8 berries.	Symptoms equalled those of previous case. Insomnia and more or less delirium for 22 hrs.	4th hr. an emetic which caused free vomiting. Between 22nd and 26th hrs. took in repeated doses of ♏x. ʒiss. tincturæ opii.	After third dose of laudanum (26th hr.) fell asleep, and awoke next morning recovered.	The same.
37	F.	46	12 ,,	Belladonna symptoms appeared after an hour, followed by insomnia and delirium.	10th hr. *oleum ricini* was given, and vomiting ensued. Between 11th and 23rd hrs. took in small and repeated doses ʒiij½ tinct. opii.	After 23rd hr. sleep came on, and this was followed by recovery.	,,
38	M.	14	2 ,,	As the preceding case.	From 11th to 23rd hr. took in small and repeated doses ʒij½ tinct. opii.	After 23rd hr. fell asleep, and awoke at the 33rd hr. recovered.	,,

39	F.	14	Number unknown.	As the preceding case.	8th hr. vomited spontaneously. Between 12th and 21st hr. took in small and repeated doses 64 minims = about ʒj. tinct. opii.	From 12th to 21st hr. delirious, with intervals of complete unconsciousness. 21st hr. could not be roused, and remained thus till 26th hr., when she died.	The same.
40	M.	12	2 berries.	" "	Between 10th and 48th hrs. took in small and repeated doses nearly ʒiv. tinct. opii.	After the 48th hr. slept, and awoke recovered.	"
41	M.	8	5 "	" "	Between 12th and 28th hrs. took in small and repeated doses ♏ cxliv. = about ʒijss. tinct. opii.	After 28th hr. fell asleep, and awoke after some hours, recovered.	"
42	M.	7	6 "	" "	6½ hrs. spontaneous vomiting. Between 16th and 28th hrs. took in small and repeated doses ʒj. tinct. opii.	After 30th hr. fell asleep, awoke at 35th hr. recovered.	"
43		9	4 "	After 2½ hrs. P. 110, pupils dilated, skin hot; at 5th hour violent delirium.	15th hr. ⅙ grain hydrochlorate of morphia with ʒj. of brandy. Between — ½ grain of morphia, in divided doses.	Fell asleep after the first dose of morphia, and slept all night; awoke at — hr. delirious. Did not sleep again till about the — hr., and recovered.	J. Todd. 'Brit. Med. Jour.' 1861, p. 305.
29		child	35 grains extract belladonna by mistake for liquorice.	Symptoms of poisoning soon appeared.	Vomiting induced, ʒij. of olive oil given by mouth, and 12 drops of laudanum. Sinapisms and cold affusion: afterwards gr. xx. calomel and galvanism, strong coffee. This treatment continued for 14 hours.	Complete coma occurred after the opiate, and appears to have continued for many hours. At the 22nd hr. the pupils began to contract. Entire recovery followed.	Dr. Willey, American 'Med. Times,' No. 3; 'Med. Times and Gazette,' 1862, vol. i. p. 224.
28		3	⅛ grain sulphate of atropia in solution (*accidental*).	Exalted condition of the nervous system ensued and continued.	At 12th hr. came under treatment; ⅛ grain of opium was administered at intervals, until ⅝ths of a grain was taken.	After the 5th dose, tranquil sleep, ending in complete recovery.	Dr. Behn, 'Journal für Kinderkrank.' 1864; 'Dublin Med. Press,' 1864, vol. lii. p. 259; 'Med. Times & Gaz.' 1864, vol. ii. p. 238.
25	M.	2½	⅛ grain sulphate of atropia in solution (*accidental*).	After ½ hr. restlessness and irritability. Grave symptoms of belladonna poisoning followed; limbs powerless, cold; P. feeble; body covered with an eruption resembling scarlatina.	After — hrs. an emetic of zinc and ipecacuanha, which caused speedy vomiting; assafœtida injections ½ hr. after; ½ gr. of opium repeated every ½ hr. until 4 grs. were taken.	After the 5th dose of opium, the pupils began to contract, the eruption and other bad symptoms to disappear. After the 8th dose the pupils contracted to their normal size; the child went off to sleep, and awoke recovered.	'Philad. Med. Rep.'; 'Dublin Med. Press,' 1864, vol. lii. p. 509; 'Bull. de Thérap.' 1865, tome lxviii. p. 181.

TABLE III.—SYNOPSIS OF CASES OF DATURA-POISONING TREATED BY OPIUM.

No.	Sex.	Age.	Quantity of the poison	Symptoms when first seen.	Treatment.	Subsequent progress and result.	Reference to Author.
44	M.	adult	Took an intoxicating preparation of datura.	After some hours wakeful muttering delirium, and unconsciousness; eyes injected; pupils dilated and fixed. P. quick and small. Illusions and clutchings. Digestion good.	After another hr. 1 gr. morphia given, and repeated every hr. until 15 grs. were taken in 18 hrs.	After the 8th dose, the muttering could be arrested by loud talking and shaking. Insomnia. 16th hr. of treatment, delirium gone; tremulousness much relieved; pupils almost normal. 19th hr., and after 15 grs. morphia, sleep, which continued for several hrs. He awoke perfectly well, and was fit for duty two days afterwards.	Dr. Thomas Anderson, 'Edin. Med. Jour.' 1860, p. 1100; 'L'Union médicale,' 1859, tome i. p. 378; 'Bull. de Thérap.' lxix. 424.
45 46 47	M. F. F.	31 58 34	Unknown quantity of tinct. of stramonium seeds by mistake for brandy.	1½ hr. afterwards, the male and elder female were nearly comatose. Congestion of the face, and injection of conjunctiva; complete dilatation of the pupils, skin hot. Tongue and throat dry and parched. R. slow and laboured. P. 100-150, tense and full. The younger woman, who had taken a smaller quantity, was in a condition resembling *delirium tremens*, and unable to stand. P. 140. R. hurried.	With each, the treatment began 1½ hr. after the ingestion of the poison. The male patient had vomited. The stomach-pump was used with the females, and the stomach washed out. To each 40 drops tinct. opii were given at once; and to the male ½ gr. morphiæ hydrochl. at intervals of ½ hr., until he had taken 3½ grs. of morphia. To the women, ¼ gr. of the salt of morphia every hr., until each had taken 2 grs.	At the end of 3¼ hrs. from the time of taking the poison, the pulse of the younger woman had fallen to 90, and she was now nearly well. The other female was at the same time in a fair way of recovery. The male had regained intelligence, and slept a little. At 4¼ hrs. the younger woman was quite well, and the other two patients were so much improved as to excite no further uneasiness. The women were well in the morning, and the man was able to walk, and read large type.	Dr. C. C. Lee, 'Amer. Jour. Med. Science,' 1862, vol. xliii. p. 54; 'Archiv. gén. de Méd.' 1864, p. 580.
48 49	M. M.	10 8	A quantity of nearly ripe stramonium seeds.	About 2 hrs. afterwards they were in a semi-comatose condition, with intermittent delirium; eyelids a little drooping. Pupils dilated and fixed. Eye and face suffused; slight dyspnœa; inability to articulate or walk. P. full and soft.	Sulphate of zinc and ipecacuanha caused free vomiting of undigested food and a quantity of stramonium seeds. Afterwards 7 drops of laudanum to each, and ½ hr. afterwards another emetic, and during the next 5 hrs. ♏xxx. more of laudanum.	7th hr. after the ingestion of the poison, they were able to sit up and articulate. P. 95, full and compressible; pupils less dilated. The poisoning took place at about 2.30 P.M. They went on well through the night, and next morning were almost quite well.	Dr. Paul Turner, 'Amer. Jour. Med. Science,' 1864, p. 552.

TABLE IV.—SYNOPSIS OF CASES IN WHICH OPIUM WAS SIMULTANEOUSLY ADMINISTERED WITH EITHER BELLADONNA OR HENBANE.

No.	Sex.	Age.	Case.	Reference to Author.
50	M.	55	ʒss. tinct. belladonna and ʒjss. Sydenham's laudanum. Somnolency, flushing of the face, a little injection of the eyes, dilated pupils followed. R. easy. P. 78, full and soft. The patient remained in a comfortable condition for 7 hrs., and awoke relieved of his pain.	Cazin, *op. cit.* p. 36.
51	M.	55	½ grain muriate of morphia, with ʒij. tinct. henbane, failed to produce sleep, which on previous occasions invariably followed the injection of half the quantity of morphia.	Mr. B. Bell, 'Ed. Month. Jour.' 1854; 'L'Union Méd.' 1859, p. 313.
52	F.	adult	Used three vaginal injections, containing the active matter of ʒj. of opium, and ʒjss. of belladonna leaves. After 1 hr. in a deep sleep. 3 hrs. after the last injection was completely insensible and motionless, face pale, pupils dilated and fixed. P. frequent and small. R. hurried. After the 5th hr. returning consciousness. Recovery. No injection was retained longer than 10 min.	Dr. Christison, 'On Poisons,' 4th ed. p. 973.
53	F.	9	4 grains each of opium and extract belladonna after a full meal. Sleep soon after. At 4th hr. awakened with difficulty, vomited freely. After 7 hrs. appeared fatigued and sleepy.	Dr. Coale, 'Amer. Jour. Med. Sc.' 1853, vol. xxvi. p. 69.
54	M.	5	Took ♏xxxvj. tinct. of opium and gr. vss. extract belladonna. Drowsiness and nausea followed. At 3rd hr. restlessness and slight delirium. After an emetic these symptoms increased, and the latter continued until the 7th hr., after which the symptoms passed off.	Mr. Wickham Legg, 'Med. Times and Gaz.' 1866, vol. ii. p. 473.

HYOSCYAMUS NIGER.

CHAPTER VIII.

Hyoscyamia is to henbane what atropia is to belladonna. It represents the whole activity of the plant. These two alkaloids are nearly allied, but each possesses distinctive characters, as will be seen from the following observations. The hyoscyamia used in every instance was prepared by myself from the recent seeds of the biennial plant, grown in the neighbourhood of London, in the season of 1867. In order to avoid the complex process of Geiger and Hesse, and the exposure to heat and other decomposing influences required in the other processes prescribed, I adopted the following method for the separation of the alkaloid.

Preparation of Sulphate of Hyoscyamia.—Having ground and sifted the seeds, place them in a percolator, and pour upon the powder ℥xvj. of water, holding fl. ʒiv. of sulphuric acid in solution. Macerate for 8 days, at a temperature of 60° Fahr.; then open the percolator, and by the occasional addition of hot water, thoroughly exhaust the seeds, until the filtrate has only a faint acid reaction, and a feeble dilating power on the pupil. About six pints of fluid will be thus obtained. Spread this in shallow layers on a number of flat dishes, and evaporate spontaneously in a dry warm room. About ℥vj. of light-brown syrupy fluid, containing a considerable quantity of white crystalline matter (sulphate of potash), will remain. Nearly neutralise with caustic ammonia (about fl. ʒxj. of a saturated aqueous solution will be needed), and then add a little powdered carbonate of ammonia, or chalk, to complete the saturation. Now wash with successive portions of a mixture of alcohol and æther (æther 9 parts, alcohol 1 part) until nothing more is removed. Distil the æther by means of warm water, and allow the alcoholic solution of hyoscyamine to evaporate spontaneously to dryness. Dissolve the residue in water, faintly supersaturate with sulphuric acid, and filter from resinous matter. Let the aqueous solution of sulphate of hyoscyamine evaporate spontaneously in a current of warm dry air, and dry over a dish of sulphuric acid.

One pound of the seed yielded about 20 grains of the salt.

Characters.—A bright sherry-brown gummy semicrystallized deliquescent mass, of a heavy tobacco-like odour, and of a bitter tobacco-like taste. A solution carefully evaporated, deposits a radiated mass of soft prisms. When kept over sulphuric acid, it remains as a brittle, gummy mass. Very readily soluble in water: a solution of 1 part in 100,000 dilates the human pupil from $\frac{1}{8}''$ to $\frac{1}{6}$, and maintains it so for five or six hours. My friend Dr. Guy thus describes its behaviour in the process of sublimation:—'Heated on porcelain, it darkens, smokes, and yields a bulky black ash, and a sublimate containing numerous delicate feathered crystals.'[1] The salt is easily preserved in a dish standing over sulphuric acid, under cover. I have thus kept it unimpaired for more than a year. The aqueous solution is as permanent as that of sulphate of atropia.

Physiological Action on Man.— Hyoscyamus or its active principle, when given in small doses, and such as are insufficient to produce positive dryness of the mouth, rapidly subdues ordinary excitement of the pulse, and reduces it, within an hour or two, to its slowest rate; that is to say, to the condition in which it is usually found after a long period of complete rest of mind and body. For example, the pulse of a man ordinarily engaged shall be 80. After a small dose of hyoscyamus ($\frac{1}{40}$ of a grain of sulphate of hyoscyamia, or 4 drachms of tincture of henbane), it will gradually fall to 60 or 50. In another person, whose pulse may be 72, we shall at the end of the same time find it steadily beating about 45. Schroff states that $\frac{1}{333}$ of a grain of hyoscyamine reduces the pulse from 79 to 18. In all my

[1] 'Principles of Forensic Medicine,' 3rd edition, p. 514.

experiments with hyoscyamus and its active principle I have never observed the pulse to fall lower than 40.

After doses ($\frac{1}{16}$ to $\frac{1}{12}$ of a grain) sufficient to produce complete dryness of the tongue and the hard and soft palates, the pulse will generally experience an acceleration of 10 or 20 beats, and be increased slightly in force and volume. This change in the pulse will be observed in from 10 to 20 minutes after the subcutaneous injection of hyoscyamine; the acceleration does not usually continue for longer than 20 or 30 minutes, and rarely lasts for an hour. Then the pulse slowly declines, and gains a little in force and volume. It usually decreases about 5 beats for every interval of 20 or 30 minutes, until, at the end of an hour and a half or two hours, it attains its minimum rate. After a small dose ($\frac{1}{40}$ of a grain) the pulse will usually fall without any previous acceleration. Apart from these accelerating or depressing effects on the pulse, the following symptoms will be observed after moderate doses ($\frac{1}{30}$ to $\frac{1}{24}$ of a grain):—In 10 to 20 minutes from the time of injection, the tongue more or less completely dry, rough, and brown, the hard and soft palates dry and glazed, excessive giddiness and a weight across the forehead, somnolency, the cheeks occasionally a little flushed, and the membranes of the eye sometimes slightly injected. After continuing for about an hour, these symptoms pass off; and the tongue and hard and soft palates become covered over with a sticky, acid, offensive secretion, agreeing in all respects with that which follows the action of belladonna. The pupils slowly dilate during the latter part of the action of the medicine, and at its close attain their maximum degree of dilatation.

If larger doses than $\frac{1}{12}$ of a grain be given, the above-mentioned effects will be increased, and prolonged for 2 or 3 hours; and they will be accompanied either by wakeful quiet and usually pleasing delirium, with illusions of the sight; or with such excessive somnolency that the patient cannot keep the eyelids raised for a few seconds, but, when aroused, lapses again into a dreamy sleep, broken by occasional mutterings and slight jerking of the limbs. In either case, the power of maintaining the erect posture will be lost, and at best the patient reels like a drunken man.

As it is my intention to show the effects of the combined action of opium and henbane on the same individuals in whom the combined effects of opium and belladonna have already been examined, I will complete my account of the physiological action of henbane on man by a series of observations on these patients. In each case the medicine was given subcutaneously dissolved in from ♏ v. to ♏ xv. of water:—

Obs. 106.—Subject, Samuel M. (see *Obs.* 106, p. 296, &c.) Pulse 74. Pupils $\frac{1}{9}$. Resp. 20.

(*a*). $\frac{1}{24}$ *grain hyoscyamiæ sulph.* by the subcutaneous tissue. *After* 23 *minutes*, pulse accelerated 15 beats. *After* $\frac{3}{4}$ *hour*, pulse accelerated 10 beats, regular, giddiness, tongue dry and hard. *After* 1 *hour*, pulse at the initial rate, 74, of natural volume and power; mouth clammy, throat dry. *After* 2 *hours*, pulse 72; pupils dilated to $\frac{1}{8}''$; mouth quite moist; throat felt very dry.

(*b*). $\frac{1}{10}$ *grain*. *After* 20 *minutes*, pulse increased 18 beats, regular; great somnolency; excessive giddiness, and staggered much in walking. *After* 1 *hour*, pulse accelerated 8 beats; pupils dilated to $\frac{1}{8}$; the legs felt too weak, and he was too giddy to walk, without assistance; continued very sleepy; no dryness of the mouth. *After* $1\frac{1}{2}$ *hours*, pulse accelerated 5 beats; pupils dilated to $\frac{1}{7}''$; continued too giddy to walk without laying hold of the furniture, or supporting himself by the wall, and there was great somnolency still; mouth and throat had been very dry and parched since the last date, and remained so. *After* 2 *hours*, pulse 72; reduced 2 beats, of the same volume and power as before the injection; pupils $\frac{1}{6}''$; tongue dry and brown at the centre; both palates dry and glazed; somnolency nearly passed off, but the giddiness continued, and he still required help in walking. *After* $2\frac{1}{2}$ *hours*, pulse still 72, regular, and of good volume and power; pupils remained $\frac{1}{6}''$; a portion of the hard palate dry; giddiness rapidly passing off, but he walked very cautiously. The remaining giddiness and dryness soon passed off.

Obs. 108. Subject, Charles V. (see *Obs.* 21, p. 127). Pulse 76; pupils $\frac{1}{10}$, sideways $\frac{1}{7}$.

(*a*). $\frac{1}{48}$ *grain hyoscyam. sulph. After* 20 *min.*, pulse decreased 18 beats. *After* $\frac{3}{4}$ *hr.*, pulse decreased 21 beats; pupils dilated to $\frac{1}{8}''$; sideways $\frac{1}{6}''$. *After* 1 *hr.*, pulse reduced 22 beats. *After* $1\frac{3}{4}$ *hrs.*, pulse reduced 25 beats, now numbering 51; inclined to yawn. *After* $2\frac{1}{4}$ *hrs.*, pulse reduced 26 beats; pupils as at $\frac{3}{4}$ hour; slight dryness of the mouth. At first the force and volume of the pulse were slightly increased, but afterwards it continued unchanged and regular throughout. There was neither somnolency nor giddiness.

(*b*). $\frac{1}{30}$ *grain. After* 40 *min.*, pulse decreased 10 beats. Giddiness came on 10 minutes after the injection, and he now staggered a little on rising from the chair; no somnolency; pupils slightly dilated; mouth generally dry. *After* $2\frac{1}{4}$ *hrs.*, pulse decreased 24 beats, regular, of unchanged volume and power; pupils at light $\frac{1}{7}''$; sideways $\frac{1}{5}''$.

(*c*). $\frac{1}{12}$ *grain. After* 20 *min.*, pulse increased 15 beats, regular, a little increased in volume; face a little flushed; felt heavy. *After* 1 *hr.*, pulse increased 4 beats; pupils $\frac{1}{8}''$; sideways $\frac{1}{7}$; sclerotic and conjunctivæ a little injected; the face —chiefly the cheeks—hot and flushed; tongue dry and brown down the centre, the rest of the mouth very clammy; much somnolency and giddiness, and had slept the last 10 minutes. *After* 2 *hrs.*, pulse 20 beats below the initial rate, of natural volume and power; pupils $\frac{1}{7}''$, sideways $\frac{1}{5}$; injection of the membranes of the eye and flushing of the face quite gone; mouth moist and clammy, and exhaling an offensive odour; had slept during the last hour, and still experienced somnolency.

(*d*). $\frac{1}{8}$ *grain. After* 30 *min.*, pulse accelerated 26 beats, of undiminished volume and power; pupils $\frac{1}{8}''$; conjunctivæ a little injected; tongue uniformly dry; both palates dry and glazed; dull, heavy, and sleepy, and very giddy; reeled much in walking. The giddiness and dryness came on 12 minutes after the injection. *After* 40 *min.*, pulse accelerated 24 beats. *After* 50 *min.*, pulse accelerated 12 beats; eyes suffused, eyeball restless; face hot and flushed; looked very heavy, had slept, and great somnolency continued; mouth still moist. *After* 1¼ *hrs.*, pulse fallen to 9 beats below the initial rate, regular, and of undiminished power and volume; pupils 1/7″, sideways 1/5; face less flushed; mouth quite moist, but the throat was very dry, and during the last hour a dry tracheal cough was present; very giddy and sleepy still, but he was able to walk with care and a little support; frequently yawned and sighed. *After* 2¼ *hrs.*, pulse decreased 18 beats, regular, softer, of good volume; pupils 1/6″, sideways ¼ nearly; injection of eye and flushing of the face disappeared; the somnolency was now passing off. *After* 3 *hrs.*, pulse decreased 21 beats, being now 53, regular, and of natural volume and power; pupils as at 2¼ hours; conjunctivæ and face natural; mouth moist; the giddiness and somnolency entirely passed off; no headache nor any other after-effects, excepting dryness of the mouth and throat. He now walked a distance of 4 miles.

The influence of the drug on the urinary function in the patient is shown in Table II. *Case* 2, p. 338.

Obs. 109.—Subject, Mr. J. W. (see *Obs.* 20, p. 126, &c.) Pulse 78; pupils 1/6.

1/12 *grain* produced on two occasions greater somnolency and giddiness than ¼ grain of acetate of morphia (used subcutaneously). The effects came on within five minutes, and he slept for two hours soundly and continuously. During the first 20 minutes, the pulse rose 18 beats, and then during the next 2 hrs. gradually fell to 20 beats below the initial rate. During the first half-hour the force and volume were slightly increased; the pulse afterwards assumed its usual character. There was some flushing of the face, and the mouth was parched. The effect on the vision is noted at p. 232; on the urine in Table II. *Case* 3, p. 338.

Obs. 110.—Mrs. F., æt. 47. Pulse 72; pupils at the light 1/8″.

1/24 *grain* caused great somnolency for 3 hours, with tottering and giddiness, so that she had to make two or three efforts before she could raise herself from the chair, and then began to reel and could not walk safely without support. Felt a great weight across the forehead. The pulse fell during the first 15 mins. 16 beats, and during the next 2¼ hrs. 8 beats more, when it attained the minimum, numbering 48. During the next hour it rose to 50. At the end of 30 mins., the tongue was dry and brown, and both palates dry and glazed, and remained so for the following 1½ hrs. 3½ hrs. after the injection, the pupils measured ⅛″, and the general symptoms had passed off completely.

Obs. 111.—Subject, Mrs. E. W. (see *Obs.* 25, p. 129). In much pain; pulse 88, after sitting an hour; pupils 1/10″. 1/24 *grain* injected over the back of the middle phalanx of the affected finger. *After* 15 *min.*, pulse decreased 6 beats; once or twice a minute a beat came very slowly, otherwise quite regular; much somnolency and giddiness. *After* 30 *min.*, pulse further decreased 2 beats, quite regular, slightly increased in volume and power; tip of the tongue and the hard palate quite dry; face hot and flushed; continued very sleepy. *After* ¾ *hr.*, and between the 44th and 46th minutes, the mouth suddenly and completely moistened; and in the same short interval the pulse fell 7 beats, and attained the minimum depression of 19 beats; the flushing of the face began to decrease; pupils ⅛″, sideways ⅙; somnolency and giddiness remained, and she continued to sleep during the next 1½ hour, when disturbed awaking with a start. At the end of this time, the pulse was 72, having increased 3 beats since the last mention; it was regular, and of slightly increased volume and force.

The 1/24 grain produced the same effects on two other occasions.

The foregoing cases illustrate the more common effects of the drug; the following will serve to show its deliriant action:—

Obs. 112.—Subject, John C. (see *Obs.* 55, p. 178). ⅛ *grain* was injected, for the relief of obstinate neuralgia, over the skin of the sixth cervical vertebra. *After* 5 *min.*, giddiness. *After* 20 *min.*, pulse increased 12 beats, full and regular; tongue and mouth generally dry; hard palate quite dry; pupils dilating; felt sleepy and giddy. On attempting to rise from the chair, he began to reel; the legs were very weak, and he could not walk without assistance. *After* 45 *min.*, pulse increased only 2 beats, of natural volume and force; pupils dilated to ⅛″; tongue, excepting the margin, quite dry and rough; both palates completely dry and glazed; articulation indistinct from the great dryness of the tongue and throat. Complained of giddiness across the forehead; no somnolency. Mind quite clear when engaged in conversation, but when left undisturbed, he relapsed into a dreamy condition, with the eyes wide open. Twice he reached out a hand to an object on the table, and began to look about on the floor, and when I asked him 'What for?' he said, 'Oh, I thought something had dropped off the table —the walls appear to move a little.' There was no flushing of the face, or injection of the eye. He could not rise from his chair, or walk without assistance; and as he sat, the extensors of the legs were slightly twitched now and then, so as to advance the foot with a little jerk. *After* 1 *hr.*, pulse fallen 1 beat, full and regular; pupils a little over ⅛″; excepting the gums, the mouth quite dry. Continued in the same condition, without manifesting the least tendency to sleep. He remained quiet, but inclined to be meddlesome when his attention was not engaged in conversation, grasping at objects on the table, or in the waste-basket on the floor, and attempting to remove them before the hand reached them, evidently misjudging the distance. The muscular system generally was flaccid, he did not sit erect in the chair, and a foot was occasionally advanced with a little jerk, and once or twice the hand was suddenly pronated and supinated with a jerk. *After* 2 *hrs.*, the pulse had decreased 6 beats, but still remained of good volume and force; pupils between ⅛″ and ¼; the mouth as dry as before; no flushing of the face or injection of the eye; no headache nor somnolency. In answer to my questions, he said that he only felt giddy. The limbs were fidgety, and occasionally affected with slight twitchings. Left alone, he lapsed into a state of forgetfulness and dreamy meddlesome delirium, picking at objects, and, having reached them after several unsuccessful efforts, fumbling them about until he dropped them, and then, in the attempt to pick them up, losing his balance, and, but for constant attention, falling over. When aroused, he rubbed his hands, gaped, and answered my questions readily. He was quite unable to walk. The eyeballs were unsteady. A remark which I made at this time excited a risibility which he was unable to restrain, and every now and then the suppressed chuckle burst out into a hearty laugh. During *the next* 1¼ *hr.*, he became somewhat restless, and if I left his side for a minute, he would attempt to get up, reel for a few paces, and then fall together like a drunken man upon the carpet. At the end of this time the giddiness diminished, and a slight inclination for sleep came on. *After* 3¼ *hrs.*, he was gapish and tired. He could now walk without assistance, but still reeled a little. The anterior part and margins of the tongue were now quite wet with an acid secretion. The eyeballs were still unsteady. *After* 3¾ *hrs.*, the pulse was 60, very regular, and of natural force and volume; it had continued so from the second hour. The pupils were nearly ¼″ at the light; the mouth everywhere moist; and the cerebral symptoms were rapidly subsiding. At this time he ate a plateful of cold meat; and half an hour afterwards, only slight giddiness and heaviness remained, and he now walked home.

When taken *by the mouth*, hyoscyamus, or its active principle, produces exactly the same effects. They are fully developed about an hour after the ingestion of the medicine.

Three fluid ounces of a succus hyoscyami (prepared for me according to the B. P. formula for the succus conii, by Mr. Buckle, of 77 Gray's-Inn Road, from some very fine biennial wild plants, grown in Essex) produced in an adult aged 40, effects exactly equivalent to those which followed the subcutaneous injection of ⅛ of a grain of sulphate of hyoscyamia in *Obs.* 112. There was the same inability to maintain the erect posture, and reeling when an attempt was made to stand, and an equal amount of muscular twitching. But instead of insomnia there was for three hours excessive somnolency, with dreaming and occasional muttering. When aroused, the condition of the patient was that described in *Obs.* 112, but so great was the tendency to sleep that he could not keep the eyelids open for many seconds, and he dozed off with a half-finished sentence on his lips.

Two ounces of good tincture of henbane produced equivalent effects, and ℥jss. of succus, or ℥j. of tincture, were always required to procure sleep. Smaller doses caused an insufficient inclination for sleep.

Modifying Influences.—The action of henbane is modified by the same conditions as those which affect the operation of belladonna (see p. 209). The most important of these practically is *age*. *Children* will usually bear a very large quantity of henbane. I have frequently given a fluid ounce of the succus or tincture to children under 12 years old, with no other effects than an acceleration of the pulse 30 or 40 beats (partly due to the alcohol), which after continuing for an hour has gradually declined; and towards the close of the action, a moderate dilatation of the pupil. The mouth has usually remained clean and wet throughout, and there has often been no trace of giddiness or sleepiness. In many *young adults and some children* hypnotic effects are soon developed, and in these ʒvj. of tincture, or ℥j. of succus, produce a grateful soporific effect. *Old persons* are very readily influenced by henbane, and generally delirium is the chief symptom. Five grains of extract, or ʒss. of tincture, will sometimes produce this effect, the intensity of which appears to be proportionate to the bodily weakness. It is therefore necessary to be cautious in administering henbane to those of advanced age, and especially when associated with great muscular weakness. The following will serve as an illustration. The gentleman whose case is referred to in *Obs.* 28, p. 123, was just able to walk, with assistance, from one room to another. He was unable to rest without taking from 1 to 2 grains of morphia in the 24 hours, and even then failed to obtain sound sleep. For some weeks previously and since the morphia had been increased, muscular twitchings, which were at first slight and partial, became severe and general, and in the fingers, arms, or legs constant. When dozing under the influence of the morphia the twitchings increased; and once or twice, as he sat in a chair, the whole body was jerked forwards and raised into the upright position. The paralysed limbs had for many months recovered their use, and both sides possessed equal power; the muscular weakness was general, and apart from the twitchings there was no indication of irritation of the motor centres. Fearing that the morphia, if it did not actually generate this excitation, greatly increased it, I prescribed on a certain occasion 8 grains of extract of henbane instead of the customary 1½ grain of acetate of morphia. After an hour, busy delirium, with constant muttering or talking and meddling with the hands, set in. These symptoms having continued with complete insomnia for 12 hours, I gave him ½ grain of morphia, and as this did not produce much effect, ½ a grain more 12 hours later on. Still less soporific effect followed this last dose, and during the next 24 hours, night and day, there was complete insomnia, great mental vivacity such as he had not exhibited for years, perpetual talking, and occasional catching at surrounding objects. During the whole of the night he engaged the nurses' attention and interest by recounting the adventures of a friend in the Peninsular campaigns. Throughout the next day he was intent upon taking a journey, but if any incident of his past life were suggested, he entered minutely into every particular, talking incessantly, with unwonted rapidity and emphasis. He used the simplest and most descriptive language, and he was quite independent of conversation, for it was necessary, in order to avoid any increase of excitement, to treat him with silence. Once a subject was named, no matter whether the attendant circumstances occurred the previous day or fifty years ago, it immediately engaged his attention until some incidental remark or an illusion suggested other ideas. If a subject with which he was not wholly familiar happened to be mentioned, he spoke of it rapidly and coherently as far as his knowledge extended, but then became confused, incoherent, and a little irritable and impatient. The connecting links in a particular train of thought were weakened and occasionally broken by illusions and delusions. The sight of a white napkin suggested, through milk, his former breakfasts in India, the milking of the cow at the door of the house, the appearance of the frothed milk in the silver basin—the tea freshly imported from China. His white handkerchief lying crumpled on the dark sofa cover recalled the ivory nut, and he entered into a minute and faithful description not only of this plant, its habits and fruit, but the characters of several other tropical vegetables. Then he wandered into the country, and suddenly pulling up a leg, exclaimed, 'Take care, give me your hand, that is a very deep step.' The next minute he introduced himself with a loud voice in a friend's house at Torquay, and, while engaged in imaginary conversation, suddenly raised the eyelids and looking across the empty space in the direction of the bare wall, said with much emphasis, 'That's a fine dahlia!' A few minutes afterwards he was engaged in Bristol. Several times he directed the carriage to be sent for, and supposing that it was at the door, made attempts to rise from his couch.

Thus 8 grains of ordinary extract of henbane aroused the mind of one remarkable for his calmness and tranquillity of thought, and who on account of general feebleness had been for months silent and averse from speech, into a condition of delirious excitement, and maintained it so for forty-eight hours. At the end of this time I gave the usual dose of 1½ grain of morphia; four hours afterwards he fell asleep, and slept tranquilly and almost continuously for the next twenty-six hours. During the action of the henbane the twitchings continued unabated, but after the refreshing sleep they began to subside. During the next fortnight he slept without any sedative. The twitchings meantime disappeared, and although it has been necessary to take as much as 1½ grain of powdered opium during the twenty-four hours, they have not returned after an absence of three months.

The action of henbane on the lower animals is essentially the same as on man.

On the Cat.—Under the influence of henbane the cat becomes dull and lethargic. When undisturbed she remains at first in a quiet dozy condition, the pupils widely dilated, and the mouth and nose rough and dry, the pulse accelerated. The mind apparently remains clear, and she begins to purr if caressed. Afterwards she becomes a little restless, and often changes her place, and in doing so makes constant use of the nose, as if her vision deceived her. She walks feebly, clumsily, and slowly, with the belly near the ground, and the power of springing is lost or impaired. If the animal wishes to attain an elevation such as a chair, she hesitates a long time, and then making a lazy ineffectual bound, fails to reach the seat, and falls to the ground. If placed on a table she soon discovers her situation, and walking to and fro along the edge, gets into a state of great distress to reach the floor, but hesitates to give the necessary spring; at last, after slipping about a long time, she loses her balance and tumbles awkwardly to the ground.

On the Dog.—In this animal precisely the same symptoms are observable; but the drug appears to have as much influence in accelerating the pulse as atropia itself; and, from my own experience, I should be unable to say whether, in a given case, the animal were under the influence of hyoscyamus or of belladonna. Hyoscyamus does not give so much force to the action of the heart as belladonna does; but this is a difference only to be appreciated by a comparative experiment.

The following observations illustrate the action of sulphate of hyoscyamia on the brown bitch (see *Obs.* 67, p. 198, &c.). Urine acid.

Obs. 113 (*a*).—$\frac{1}{24}$ *grain. After* 4 *min.*, pulse accelerated 120 beats, regular. *After* 6 *min.*, the maximum acceleration of 180 beats was attained—the systoles numbering 300—and was sustained for half an hour. *After* 10 *min.* the mouth and nose were dry, the pupils dilated, and the vision hypermetropic; on attempting to put her forepaws on my knees she fell short of them. *After* ½ *hr.* the trachea was dry, and there was slight coughing occasionally. *After* 1 *hr.* the pulse was accelerated 120 beats, and the pupils contracted in direct sunlight to $\frac{1}{8}''$. *After* 2 *hrs.*, pulse accelerated 60 beats; pupils contracted to ¼ in direct sunlight; moisture running down from the nostrils, and moisture returned to the mouth; the animal was quite frolicsome. *During the first* 1½ *hr.* she lay down and slept, awaking with a start when any one approached her. The respiration throughout varied from 16 to 12; the breathing was interrupted once every 2 or 3 minutes by a deep sighing inspiration, after which there was a pause for 15 seconds. Slight twitches occasionally affected the muscles of the shoulder and thigh. Between 1½ and the 4th hrs. she voided urine 6 times; both that voided at 1½ hr. and that at 2½ hrs. were acid, and dilated the pupil.

(*b*). ⅙ *grain* reproduced the above-mentioned effects. *After* 10 *min.* the pulse attained a maximum increase of about 250 beats, the systoles being full and strong, and numbering between 300 and 400; it was sustained for about ½ an hour. *After* ¾ *hr.* the acceleration was only 120; respiration 14, regular; pupils in direct sunlight measured ¼″. At the end of 3½ hrs. the effects had entirely passed off, and the animal was as frolicsome as usual. The hypnotic effect did not appear to be so great as after the $\frac{1}{24}$ grain. She remained sleeping lightly for 2½ hrs., but was very easily disturbed, and answered me with a wag of the tail at any time during the action. Urine, very rich in solid constituents and acid, was voided at 1¼ hr. and again at 3½ hrs. A non-crystalline film, composed (under a ¼-inch object-glass) of brown spherules, and smelling strongly of the original hyoscyamia, was obtained by means of chloroform from ℥ss. of the urine. The solution in ♏iv. of water was neutral, and one drop dilated my pupil within ½ an hour from ⅛″ to ⅙.

On the Mouse.—The action of hyoscyamia on this animal resembles that of atropia, but the soporific effect is greater. The following will serve as illustrations:—

Obs. 114.—Injected ⅙ *grain* hyoscyam. sulph. beneath the skin of a half-grown male mouse A. He became drowsy almost immediately, and continued to sleep with short and infrequent intervals of restlessness for the next 4 hours, when the observation was intermitted; the heart's action was excessively rapid; resp. 240. *After* 12 *hrs.* he was lying on the side, cold and torpid, and struggled feebly when handled, the belly and hind-legs were wet with urine, and a large quantity had been voided. Heart's action too feeble to be heard; resp. 50; pupils completely dilated. During the next 2 hours he improved, got up, and began to crawl a little. As he couched there was considerable muscular tremor, and slight irritation of the thick part of the tail or pads caused strong jerks. *After* 15 *hrs.* the heart's action was audible and regular, 300; resp. 68, forced; crawled well, and resisted when taken up; reflex action as before. *After* 24 *hrs.*: remained much the same up to this time, but the heart again became inaudible at the 20th hr., and remained so, and the respiration gradually fell to 60, and became very faint. The little animal appeared to be now urged along in a slow creep, and the hind-legs and tail were frequently twitched so as to throw him on the hip. After an hour he became quiet and torpid. *At the* 28*th hr.* I left him in the same state, and next morning he was dead, in a semiflexed posture; rigor mortis natural; lungs partially collapsed and rose-coloured. Right heart contained black fluid blood, and on puncturing the auricle it collapsed as the blood flowed out. Left heart pale, empty and firmly contracted; gall-bladder full of bile; stomach empty; small intestines pale, flabby, and full of bilio-mucous fluid; mucous membrane pale; kidneys free from congestion, but the suprarenal organs, which are always pale, were congested and as dark as the kidneys. Urinary bladder nearly empty. Brain and spinal cord dusky, but firm and free from congestion. Veins of the diploe dusky.

Obs. 115.—Injected ¼ *grain* beneath the skin of an adult female mouse B. After a short interval of excitement—running and jumping about—she closed the eyes and became very drowsy; and during *the next four hours* continued to sleep soundly. The heart's action, as in the former case, was strong and too rapid to count, being 600 at least. Resp. 160. *After* 12 *hours* the heart was inaudible, the resp. 30, and the torpor greater than A.; had passed much urine repeatedly. *After* 15 *hrs.* there was the same improvement; the systoles were 296, and distinct, and the resp. had risen to 120, crawled and struggled well when disturbed, otherwise remained couching, semi-torpid, and very tremulous. *After* 21 *hrs.*, was much recovered; systoles 320, and quite distinct;

resp. 100, good and regular; crawled feebly like a new-born kitten, and began to look about a little and sniff. *After* 22 *hrs.*, passed urine, the first time during the preceding 10 hours. *After* 24 *hrs.*, more lethargic, and rarely moved; systoles 320, regular; resp. 98. *At the* 28*th hr.*, in the same condition. *At the* 36*th hr.*, lying on the belly, with the legs sprawling, breathing faintly 60; heart inaudible, eyes milky white. On lightly touching her, she fell over on the side, and the limbs and body were fully extended in a feeble spasm —more like a slow rigid stretch, but identical in its nature with that produced by thebaia. The spasm relaxed after a few seconds, and I set her on the legs again. When handled she struggled faintly; the resp. was accelerated to 70; and the inspirations fuller. *After* 42 *hrs.*, she had experienced two such spasms when disturbed; the resp. had fallen to 16; on pricking the thick portion of the tail, or the pads, strong reflex actions of the legs and rump were immediately induced; after exciting these movements 4 times, the resp. fell to 12; after another minute the tail faintly quivered, the legs were slowly and gently drawn towards the abdomen, and the respiration ceased; from the moment of death, no further reflex action could be excited. *Two minutes* after the breathing ceased, the chest was opened; the lungs were collapsed, and of a red-lead colour; the right heart distended with very dark fluid blood, the coronary veins very turgid, the auricle contracted faintly and regularly 164 times, and the ventricle 36 times a minute. The left heart was empty, and contracted. The veins of the chest and abdominal walls were moderately injected with dark blood; gall-bladder full of good bile; the small intestines distended by air and yellow mucous fluid; the lungs contracted, and the other organs as those of A.

Elimination.—Like atropia, hyoscyamia is, when given in quantities mentioned in the preceding observations, wholly eliminated by the kidneys. Its presence may be detected in the urine 18 minutes after the subcutaneous injection of $\frac{1}{48}$ grain of the sulphate. (See also *Case* 6, Table II.) That it is wholly removed by the kidneys, and passes undiminished out of the body, may be inferred from the fact that the urine secreted during the action of a dose of the alkaloid dilates the pupil as readily as the same bulk of water to which the dose has been added. In *Case* 2, 10 drops of a solution of 1 part of sulphate of hyoscyamia in 100,000 parts of urine dilated the pupil from $\frac{1}{10}$ to $\frac{1}{7}$ for several hours. The diuretic action of henbane is often more marked than that of belladonna on account of the larger increase of the water. This is apparent in all the cases except 4 and 5, in one of which the action of the medicine was induced too soon after dinner to furnish satisfactory results. In the other, the medicine was given by the stomach, which would retard the diuretic action somewhat. It will be seen, from an examination of the accompanying table, that there is an increase of urea and the sulphates and phosphates, just as occurs during the action of belladonna. (See next page.)

The general action of henbane on the secretions and nervous system agrees in all respects with that of belladonna, and the results of its action are the same. (Compare *Obs.* 83, p. 227; and *Obs.* 114, p. 334.)

The difference between the two drugs may be summed up in these few words. Compared with belladonna, the influence of henbane on the cerebrum and motor centres is greater, while its stimulant action on the sympathetic is less.

Action on the Heart and Bloodvessels.—In moderate doses henbane decreases the irritability of the heart, and speedily reduces its action to that proper to a state of complete rest of mind and body; and at the same time it increases its force. These effects are very manifest in disease. It affects the healthily excited heart in exactly the same way as belladonna influences the organ when unduly excited in the morbid processes (p. 247). Both drugs directly stimulate the heart, but after moderate doses the action of henbane results in a sedative effect. Small doses of belladonna excite the heart, and large doses depress it. Small doses of henbane are sedative and tonic to the heart, large doses excite it; excessive doses depress it almost as readily as those of belladonna.

The influence on the minute arteries of the frog is the

TABLE II.—URINE BEFORE AND AFTER THE ACTION OF HYOSCYAMUS.

Name, Diet, Hyoscyamus.	Urine before the Hyoscyamus.	Urine after the Hyoscyamus.	
1. John C. *Obs.* 112. ⅛ gr. hyoscyam. sulph. at 3.48 P.M. 3¾ hrs. after tea and bread and butter.	A. Urine secreted between 2 and 3.48 P.M. ℥ij. sp. gr. 1016, acid, pale. 1000 gr. measures contained:— Grs. Chlorine . . . . 4·56 Urea . . . . . 14·80 Phosphates and Sulphates 7·01 Uric acid, no excess.	B. Urine voided 3¾ hrs. after the injection ℥ix. sp. gr. 1008, acid, nearly colourless. 1000 gr. measures contained:— Grs. Chlorine . . . . 2·42 Urea . . . . . 9·00 Phosphates and Sulphates 3.75 Uric acid, a trace.	3 hrs. after the injection, an urgent call to pass water; dysuria and discharge in a feeble dribbling stream; partial retention. 10 drops of B dilated my pupil from ⅛ to ⅙, and caused slight conjunctivitis with great dryness of the membrane, and aching of the eyeball.
2. Charles V. *Obs.* 108. 1/16 gr. hyoscyam. sulph. at 11.7 A.M. 2¼ hrs. after bread and butter and a pint of tea.	A. Urine secreted between 8.30 and 11.7 A.M. ℥v. sp. gr. 1015, acid, pale sherry coloured.	B. Urine voided 2¼ hrs. after the injection ℥xvijss. sp. gr. 1010, feebly acid, whey-coloured.	
1/10 gr. at 11.15 A.M. 3 hrs. after breakfast as above and a little fish.	A. Urine secreted between 8·30 and 11.15 A.M. ℥x. sp. gr. 1016, acid, much uric acid.	B. Urine voided 2 hrs. after the injection ℥xvss. sp. gr. 1008, alkaline, much less uric acid.	
1/16 gr. at 10.30 A.M. 2½ hrs. after breakfast as above, without fish.	A. Urine secreted between 8.30 and 10.30 A.M. ℥v. ʒvj. sp. gr. 1016, acid, much uric acid.	B. Urine voided 2¼ hrs. after the injection ℥xviij. sp. gr. 1008 neutral, a trace of uric acid.	
1/12 gr. at 11 A.M. 2½ hrs. after breakfast as above, no fish.	A. Urine secreted between 8.30 and 11 A.M. ℥xijss. sp. gr. 1010, faintly acid.	B. Urine voided 2 hrs. after the injection, ℥xvijss. sp. gr. 1005, very alkaline.	10 drops of B dilated my pupil from 1/10 to ⅛″ for several hrs.
⅛ gr. at 10.47 A.M. 2½ hrs. after bread and cheese and a pint of tea.	A. Urine secreted between 8.30 and 10.47 A.M. ℥viij. sp. gr. 1015·2, acid, bright on boiling. 1000 gr. measures contained:— Grs. Chlorine . . . 2·95 Urea . . . . 12·32 Sulphates and Phosphates 4·16	B. Urine voided 2¼ hrs. after the injection, ℥xiv. sp. gr. 1008, alkaline. Phosphatic opalescence on boiling. 1000 gr. meas. contained:— Grs. Chlorine . . . . 1·55 Urea . . . . . 5·33 Sulphates and Phosphates 3·24	C. Urine passed 3 hrs. after the injection ℥ix. sp. gr. 1006·5, more alkaline than B. 1000 gr. measures contained:— Grs. Chlorine . . . 1·85 Urea . . . . 5·66 Sulphates and Phosphates 3·78
3. John W. *Obs.* 109. 1/12 gr. hyoscyam. sulph. at 10.55 A.M.	A. When the bladder was emptied of ℥iij. of high-coloured acid urine.	B. Urine voided 3½ hrs. after the injection ℥xvss. sp. gr. 1007·2, glaucous green. Chloroform extracted from ℥ij. a brown varnish, smelling strongly of the original hyoscyamia, 2 drops of its solution in ♏x. of water dilated my pupil within an hour from ⅛ to ⅕″.	
4. George G. æt. 15. Epilepsy. ℥ij. succi hyoscy. with ℥j. aquæ at 7.5 P.M. 2 hrs. after tea and bread and butter.	A. Urine voided ¼ hr. after the henbane ℥ij. ʒvj. sp. gr. 1017, acid. 1000 gr. meas. contained:— Grs. Chlorine . . . 3·00 Urea . . . . 16·66 Phosphates and Sulphates 3·60	B. Urine voided 2½ hrs. after the henbane, ℥ij. ʒvj. sp. gr. 1023, acid. 1000 gr. meas. contained:— Grs. Chlorine . . . . 5·49 Urea . . . . . 21·66 Sulphates and Phosphates 3·80	A few drops of urine B. dilated my pupil from 1/10 to ⅕″, and the dilatation persisted for 8 hrs. Both urines contained a normal amount of uric acid.
5. Alfred L. æt. 21. 1/12 gr. hyosc. sulph. by skin at 7.35, 1½ hr. after dinner.	A. Urine secreted during the previous 2½ hrs. ℥xiij. ʒij. sp. gr. 1010, acid, bright on boiling.	B. Urine voided 2½ hrs. after the injection ℥iijss. sp. gr. 1025, alkaline, much phosphatic opalescence on boiling, and deposit of triple phosphate on standing. Dilated the pupil.	
6. Samuel M. *Obs.* 1/12 gr. hyos. sulph. at 7.30 P.M.	A. Urine voided at time of injection, sp. gr. 1034·4, acid.	B. Urine voided 22 minutes after the injection ℥iij. sp. gr. 1038, acid. A few drops dilated my pupil from ⅛ to ⅙, and maintained it so for 10 hrs.	

same as that of atropia. After the subcutaneous injection of $\frac{1}{80}$ grain hyoscyamiæ sulph. the circulation was increased in rapidity and tone, with slight contraction of the main artery of the web for two days; at the end of which time the pigment-cells became fully radiated, and the artery relaxed.

In its action on the cerebrum, henbane approaches very nearly to opium. And generally we may regard henbane as opium, *minus* the excitant action on the motor centres, which is so essential a part of the action of the latter. Taking another view, and in reference to the powerful action of henbane on the motor centres, its effects very closely resemble those produced by the combined action of hemlock and opium (see p. 95). The connection between alcoholic intoxication and the effects of henbane on the cerebrum and motor centres is undoubtedly very close. The delirium of henbane is generally of a quieter kind, while the depression of motor power is much greater in proportion to the delirium in henbane than in alcoholic intoxication.

The effects of belladonna and henbane respectively on the reflex function are identical. Muscular twitchings and restlessness of the eyeball are certainly more frequently seen during the action of the latter. In the last stages the effects appear to be equal (see pp. 240 and 335–6).

In the frog, the reflex action of the cord is greatly excited, and the slightest disturbance of an animal who has had a large dose of hyoscyamia a day or two before, will often produce a rigid spasm.

The effects on the respiration resemble those on the heart. When the patient sleeps or remains tranquil after a dose of henbane, the breathing quickly assumes the rate characteristic of complete repose. But if the brain remain active under the influence of a large dose, the respiration is liable to temporary acceleration from mental impressions, just as occurs in the nervous stage of enteric fever, or any other nervous condition.

The Medicinal Use of Henbane is so well understood, that I shall content myself with a very few general remarks on this topic:—

1. *As a general sedative to the heart*, it claims our first consideration; and in functional disturbance arising from emotion, the subcutaneous use of $\frac{1}{48}$ of a grain of sulphate of hyoscyamia exercises a most speedy and beneficial influence. —2. In *cardiac and pulmonary asthma*, it is the appropriate remedy, and when used subcutaneously will often bring immediate relief.—3. In *neuralgia* it is a powerful anodyne, but in affections of the nerves of common sensation it possesses no advantage over atropia.—4. In neuralgic affections of the internal viscera, especially of the genito-urinary organs, it is more efficacious than belladonna. Administered in any way, it is an invaluable remedy in *renal affections dependent on the oxalic or gouty diathesis*.—5. In *nephritis*, both recent and chronic, henbane may sometimes be substituted with advantage for belladonna.—6. In *spasmodic affections of the uterus, the bladder, and the urethra*, the antispasmodic and anodyne effects of henbane are very decided. The *enuresis* of young persons is very speedily ameliorated, and ultimately removed, by the judicious use of this plant.—7. In *hypochondriasis*, and in *epilepsy* arising from emotional disturbances, I have found henbane very serviceable. It must be given in full doses, and occasionally. I usually give from ʒiv. to ℥jss. of the succus, or from ʒiv. to ℥j. of the tincture. In other varieties of this disease, and in *convulsive affections generally*, it has proved useless in my practice. The plant undoubtedly exercises a considerable depressing influence on the corpora striata, but it fails to diminish the excitability of the spinal centres, if it does not actually exalt it. As the will loses much of its directing power under the influence of henbane, the conditions induced by the drug are generally favourable for the development of convulsive action. Its use in these affections is therefore contra-indicated, unless we know that they have their origin in cerebral or peripheral irritation, or in some derangement of the sympathetic. Even in cases of cerebral irritation, henbane may do harm, unless its action tends to produce hypnosis rather than delirium.

CHAPTER IX.

COMBINED ACTION OF OPIUM AND HENBANE.

After the full consideration which has been given to the combined action of opium and belladonna, but little need be said on this subject, and I shall content myself with a few illustrations. I am glad to be able to refer to observations on individuals in whom the effects of opium, belladonna, and henbane have been previously fully ascertained.

Obs. 116.—Subject, Samuel M. (see *Obs.* 106, p. 324, &c.). Pulse 74; pupils $\frac{1}{9}$; resp. 20.

(*a*). $\frac{1}{8}$ *grain morphiæ acet.* and $\frac{1}{24}$ *grain hyoscyamiæ sulph.* by a single injection. *After* 20 *min.*, pulse accelerated 12 beats; staggered much; intense somnolency; tongue dry and brown anteriorly. *After* 30 *min.*, pulse attained a maximum acceleration of 13 beats, regular, of unchanged volume and force; the anterior part of the tongue dry and brown; both palates dry and glazed. *After* $\frac{3}{4}$ *hr.*, the pulse was accelerated only 4 beats, and the tongue was completely dry from front to back. *After* 1 *hr.* 20 *min.*, the pulse had resumed its initial rate; *within three minutes the mouth moistened, and the pulse at the same time fell* 8 *beats. After* 2 *hours* 20 *min.*, the pulse had increased 2 beats, and the mouth was partially dry again.

(*b*). $\frac{1}{4}$ *grain morphiæ acet.* and $\frac{1}{12}$ *grain hyoscyam. sulph.* by a single puncture. *After* 10 *min.*, pulse accelerated 10 beats; resp. 19; great somnolency. *After* 1 *hr.*, pulse accelerated 14 beats; resp. 16; tongue, excepting the margin, completely dry and brown; both palates dry and glazed; throat dry, and voice husky. *After* 2 *hrs.*, pulse fallen 4 beats below the initial rate; resp. 15; pupils dilated to $\frac{1}{6}$; mouth still dry; no flushing of the face; a little injection of the sclerotic.

The general effect on both occasions was that of tranquil sleep; on the last occasion it was so deep that he could hardly maintain his position in a chair. He retired to bed, and slept soundly and continuously until a late hour in the morning. Drowsiness was the only after-effect. The pulse maintained its natural force and volume throughout.

(*c*). $\frac{1}{18}$ *grain hyoscyam. sulph.*, and *after* 1 *hr.* $\frac{1}{4}$ *grain morphiæ acet.* 1 *hr. after the hyoscyamus*, pulse accelerated 10 beats; resp. 16; mouth dry; very giddy and drowsy. 20 *min. after the morphia*, pulse increased 5 beats more, and a little stronger; resp. 15; sleeping soundly, and continued to do so, after one disturbance, for the next 12 hrs.

(*d*). $\frac{1}{12}$ *grain hyoscyam. sulph.*, and *after* 1 *hr.* $\frac{1}{4}$ *grain morphiæ acet.* 1 *hr. after the hyoscyamus*, pulse increased 10 beats; resp. 16; mouth dry and glazed; pupils dilated to $\frac{1}{6}$; had been sleeping ever since the injection, and was now scarcely able to keep the eyes open a few seconds; had to make 3 or 4 attempts before he could rise from the chair, and then could not walk without support. 15 *min. after the morphia*, the pulse had decreased 6 beats; resp. 15; sleeping soundly. The dryness of the mouth and dilatation of the pupils persisted, and the patient slept soundly and continuously for the following 10 hrs.

Obs. 117.—Subject, Charles V——(see *Obs.* 108, p. 325, &c.). Pulse 76; pupils $\frac{1}{10}''$.

(*a*). $\frac{1}{12}$ *grain hyoscyam. sulph.*, and *after* $\frac{3}{4}$ *hr.* $\frac{1}{4}$ *grain morphiæ acet.* $\frac{3}{4}$ *hr. after the hyoscyamia*, his condition was as described in *Obs.* 108 (*c*). 25 *min. after the morphia*, the pulse was decreased from its rate at the time of the morphia injection 10 beats, being now 68; pupils *contracted* from $\frac{1}{8}''$ to $\frac{1}{10}$. *After* 30 *min.*, no further change. *After* 1 *hr.*, pulse decreased 25 beats; sclerotic injected; eyeballs unsteady; mouth clammy. *After* 1 *hr.* 40 *min.*, pulse decreased 2 beats more, being now 50; pupils $\frac{1}{10}$ still. The morphia increased the somnolency within 10 minutes, and the patient slept soundly with interruptions for 2 hours, and then went home and slept for several hours.

(*b*). $\frac{1}{4}$ *grain morphiæ acet.*, and *after* 1 *hr.* 10 *min.*, $\frac{1}{24}$ *grain hyoscyam. sulph.* 1 *hr.* 10 *min. after the morphia*, pulse 72;

pupils $\frac{1}{12}$; mouth moist; no somnolency (see p. 127). 20 *min. after the hyoscyamia*, pulse accelerated 15 beats, and a little increased in volume and power; great somnolency. *After* 30 *min.*, pulse accelerated 26 beats, regular; eyeballs very unsteady; pupils $\frac{1}{12}''$, dilating freely; face flushed; conjunctivæ injected; very giddy. *After* $\frac{3}{4}$ *hr.*, pulse accelerated only 14 beats; pupils *dilated* to $\frac{1}{9}''$; mouth generally dry. *After* $1\frac{1}{2}$ *hr.*, pulse, as he slept, decreased since last mention 26 beats, being 60; after waking, and without stirring from his chair, it rose to 80, of good volume and power in both conditions; the eyeballs still unsteady; pupils further dilated to $\frac{1}{8}''$; mouth clammy. *After* 2 *hrs.* (3 hrs. 10 min. from the first injection), pulse 57. Sound comfortable sleep followed the injection of the hyoscyamia, and continued for 2 hrs., when he was disturbed, and went home. Somnolency continued for several hours. The following analysis illustrates, comparatively, the effects of opium and hyoscyamus on the kidneys:—

A. Urine passed at time of morphia injection—ʒiij. sp. gr. 1021·4, very acid, deposited lithates on cooling. 1000 gr. measures contained:— Grs.	B. Urine passed at time of hyoscyam. injection—℥v½. sp. gr. 1013·6, acid, bright. 1000 gr. measures contained:— Grs.	C. Urine passed 1½ hr. after hyosc. injection—℥viij½. sp. gr. 1012·8, acid, bright. 1000 gr. measures contained:— Grs.	D. Urine passed 2 hrs. after hyosc. ℥iij½ sp. gr. 1015·2, acid. 1000 gr. measures contained:— Grs.
Chlorine . . 3·76	Chlorine . . 1·8	Chlorine . . 1·41	Chlorine . . 3·03
Urea . . 23·0	Urea . . . 9·33	Urea . . 7·0	Urea . . . 13·33
Sulphates and Phosphates . 9·36	Sulphates and Phosphates . 3·0	Sulphates and Phosphates 2·91	Sulphates and Phosphates . 7·38

12 drops of urine (C) caused dilation of my pupil from $\frac{1}{8}$ to $\frac{1}{6}''$, and a smart attack of ophthalmia, consisting in dusky injection of the conjunctivæ up to the margin of the cornea, and uniform dusky redness of the inner surfaces of the lids, with great dryness, stiffness, and dull aching of the eye. It came on 3 hours after using the urine, and, having continued for about 6 hours, terminated with a free mucopurulent discharge. The urine itself was quite free from irritant action, and, as in the other case (see Table II. *Case* 1), the effect was due to the mydriatic.

Obs. 118.—Subject, Mrs. E. W. (see *Obs.* 111, p. 327, &c.). Pulse 80; pupils $\frac{1}{10}''$.

$\frac{1}{30}$ *grain hyoscyam. sulph.* and $\frac{1}{18}$ *grain morphiæ acet.*, by a single puncture, produced a slight acceleration of the pulse, with a little increase of volume and force, dryness of the mouth, slight dilatation of the pupil, and comfortable sleep for 3 hours.

It thus appears that in man, henbane, like atropia, not only increases the hypnotic effect of opium, but (as in *Obs.* 117) determines it when opium alone is unable to produce sleep. Opium and henbane in various proportions, according to the nervous peculiarity of the individual, constitute, I believe, the most powerful hypnotic or narcotising combination that can be formed. The following observations on the dog still further strengthen this conclusion :—

Obs. 120.—(*a*) Injected beneath the skin of the bitch (see *Obs.* 113, p. 333, &c.) ½ *grain morphiæ acet.* and $\frac{1}{12}$ *grain hyoscyamiæ sulph.* by one puncture. *At the 4th min.* she retched violently twice, but did not vomit. *After* 15 *min.* the henbane effects were fully declared; pulse 240, regular and strong; resp. panting, and regular, about 240; mouth, lips, and nose completely dry; pupils fully dilated, and the animal lay motionless on the side completely narcotised. *After* 30 *min.* the panting ceased, and the resp. was 40 one minute, and the next 36, with a long-drawn sigh ; pulse 210, regular and strong; made no resistance or motion when I pushed my finger beyond the fauces. The breathing fell, and *after* 1½ *hr.* the resp. was 17, the pulse 140. *After* 2 *hrs.* the pulse was 104, and began to assume the respiratory character. Up to the end of 3¼ *hrs.* she lay in a most complete state of narcotism, and during the 2nd and 3rd hour I could not obtain a reflex movement on pushing my finger down to the glottis, or on touching the eye. At the end of this time I shook her roughly, and she got up, crawled two paces, and then fell down on the belly again, and remained for the next 5 hours in the condition of mixed stupor and restlessness usual after the above-mentioned dose of morphia. Excepting the dilatation of the pupil, the effects of the hyoscyamus had passed off at the end of 3¾ hrs., when the mouth was quite moist, the pulse 108, and regular; the resp. 18; urine was voided at the 6th, 7th, 8th and 9th hours. The first caused dilatation of the pupil; the last failed to do so.

The effect of henbane at a time when the influence of opium is declining is well seen in the following, which is a continuation of *Obs.* 9, p. 110 :—

(*b*). After 3¼ hours, when the somnolency had nearly passed off, ⅙ *grain sulphate of hyoscyamia* was injected. *After* 25 *min.* the pulse had risen to 120, was stronger, and unaffected by the breathing; resp. 14; mouth and pupils unchanged; no return of somnolency. *After* 35 *min.*, pulse 200, strong, and regular; resp. 14, nose dry; now threw herself on the side, with the head on the rug. *After* 1 *hr.* she was completely narcotised; made no resistance to my examinations, nor gave any response to my calls; pulse 230, resp. 15, both regular; mouth generally dry, but not completely so. *During the next two hours* she lay motionless, and insensible to external impressions. *At the 3rd hour* the pulse was 164, regular; resp. 13, regular and equal; she made no movement when I tickled the fauces, or pricked the nose. Shortly afterwards I aroused her by shaking, but after a few seconds she relapsed into her former condition, and for the *next* 1½ *hr.* she lay motionless on the table in a deep sleep, and took no notice of any disturbances. 5 *hrs. after* the hyoscyamia she still slept, awaking occasionally and walking slowly round the room. The pulse was 120, the resp. 13, and the mouth still dryish.

The conclusions to be derived from the foregoing observations are clearly those which have been already formed respecting the combined action of opium and belladonna, and they are briefly as follows :—1. Opium prolongs and intensifies the effects of hyoscyamus, even to producing an acceleration of the pulse some 15 or 20 beats for an hour or more. 2. Hyoscyamus increases the hypnotic action of opium, and to a certain extent is able to prevent the derangement of the vagus nerve, which is frequently the first effect of opium. 3. Opium, given in combination with hyoscyamus, does not prevent the elimination of hyoscyamia by the kidneys.

In reference to the particular question of antagonism, the preceding observations show that in the horse, in the dog, and in man, no antagonism of action on the brain and spinal cord, excepting purely local effects, exists between opium on the one hand, and the mydriatics on the other.

Dr. Bois[1] had arrived at the same conclusion with regard to cats; and Dr. Camus concludes, from his careful 'experimental researches on the antagonism of opium and belladonna,'[2] 'that the antagonism which has been suspected between opium and belladonna appears to be the result "de faits mal interprétés"—that this antagonism does not exist in the rabbit, nor in the sparrow—and 'we think,' he says, 'it does not exist to a greater extent in man' (p. 503).

[1] 'Gazette des Hôpitaux,' June 1865.
[2] 'Gazette hebdomadaire de Méd. et de Chir.' tome ii. 1865.

APPENDIX.

NOTE 1.

On the Excitant Action of Meconine (see p. 183).—$\frac{3}{4}$ grain of meconine, dissolved in 50 minims of hot water, was injected at intervals beneath the skin of a vigorous mouse. The same effects followed as are described under *Obs.* 37. *After* 2 *hrs.* the animal was in a state of torpor, approaching complete narcotism; respiration 140 to 200; pulse 200 to 300. At this time he fell over on the side in a slight convulsion, but soon recovered the feet again; slight twitchings of the limbs and general tremor of the body when disturbed, continued to the time of death at the 9th hr. *After* $8\frac{1}{2}$ *hrs.* the pulse was 240, distinct and regular; the respiration 120, very shallow; but it now gradually became imperceptible, and ceased at the 9th hr. During the last few minutes the limbs were alternately twitched, and this continued after the breathing had finally ceased. Surprised by the activity of these post-mortem movements, after some minutes I completely obstructed the respiratory apertures, thinking that there must be an imperceptible interchange of air in the lungs. But this was not the case, for the movements continued as before, and the pupils, instead of dilating as usual at the minute of death (see p. 138), remained contracted. The movements ceased 10 minutes after death. The chest was opened a minute afterwards. The lungs were collapsed and of a salmon colour; the left heart was empty, contracted, and motionless; the right auricle was contracting vigorously 70, and the ventricle 60 times a minute. Fifteen minutes later on both right cavities were contracting 30 times a minute. The kidneys were congested, the urinary bladder distended, and the urine, as did that passed two hours after the dose, contained many blood-corpuscles. The time was insufficient for the development of uræmic symptoms; the convulsive movements must therefore be referred to direct excitement of the cranio-spinal axis, which continued after the death of the animal. I have never witnessed such movements in any other animal, but in a lady aged 62 I observed slight twitching and retractile movements of the wrist half an hour after death.

NOTE 2.

In reference to the excessive excitability of the spinal cord produced by atropia in the frog (alluded to at p. 240), it is most satisfactory to me to find that Dr. Thomas R. Fraser, of the University of Edinburgh, has independently observed the same fact. In a letter to me, dated December 5, 1868, he writes: 'I am engaged in preparing a communication on the production of tetanus in frogs by sulphate of atropia, and am anxious to know if you have observed such symptoms in your experiments. It is a well-marked and very extraordinary effect. It always follows the paralysing effects of atropia, and these may exist for three or four days, when, as the animal begins to recover, it enters a stage in which it appears as if it were suffering from strychnia.' My own observations entirely accord with these statements; and I need not add more, but refer to the 'Proceedings of the Royal Society of Edinburgh,' and the 'Journal of Anatomy and Physiology' subsequent to the date above given, in which the reader will find a full account of Dr. Fraser's observations on this interesting feature.

NOTE 3.

North American Hemlock.—Dr. E. R. Squibb, of Brooklyn, New York, informs me (October 22, 1868) that *Conium maculatum* has become so common throughout that section of the United States as to be a great nuisance to agriculturists, and that it is now considered as indigenous.

In reference to the fruit described and examined at p. 94, Dr. Manlius Smith has favoured me with the following information:—'It was grown at Oneida, a village some 17 miles from Manlius. The average temperature of this place during the season of 1868, when the fruit was grown, was during June 65° Fahr., July 77·3°, and August 69°. The fruit was collected between the 22nd and 30th of July. The weather was quite dry, and for the most part unusually hot. The umbels were collected and dried in thin layers in a shaded room, and afterwards stored for a month in a darkened room warmed by a stove-pipe.'

NOTE 4.

Succus Conii.—I have stated, at p. 70, that no appreciable variation was observed in the activity of the several specimens of succus examined. Mr. Daniel Hanbury has recently called my attention to a preparation which I find to be at least *twice* the strength of the ordinary succus. Mr. Hanbury has kindly furnished me with the following particulars :—'It was made in the latter part of June 1865; the quantity of leaf used was 21 lbs., and the yield by hydraulic pressure 6 pints; the plant was in full flower.' The succus has the high sp. gr. of 1016·8, a strong hemlock taste and odour, and a dark reddish-brown colour.

Now I have already shown, with respect both to fruit reared in a very dry and hot situation at the back of my house, and that grown at Oneida during an unusually dry and hot season, that these are conditions most favourable for the development of the active properties of hemlock. The ancients, however, appear to have thought otherwise. My friend, the Rev. T. O. Cockayne, who has kindly supplied me with extracts from their writings, remarks, 'they picked their plants in special places.' My old edition of Theofrastos says, 'at Susa, in ground cold and shaded.' Dioskorides remarks, 'the most potent sort grows in Krete, Megaris, Attika, Khios, and Kilikia.'

The juice in question was evidently prepared from plants grown in a dry situation, the quantity obtained being only 35 per cent., which is less than half that which may be obtained from more succulent plants (see p. 68). On turning to the Registrar-General's Reports, we shall be reminded that the summer of 1865 was unusually hot; the *average* temperature for the whole year was greater than that of either of the previous 17 years at least; and the average temperature (56·20° Fahr.) for the months of April, May, and June, exceeded that of any other year by 2°, and of the general average of 17 years by 4°. Further, the summer of 1865 was preceded by an unusually dry season, the year 1864 being by far the driest of all the 18 years.

It thus appears that the activity of the *Succus conii* is dependent on season, and consequently on situation—not, probably, that a succulent plant contains less conia, but because it possesses more water. In prescribing the '*Succus*' this liability to variation must be borne in mind. Generally a dark colour and a high specific gravity will indicate increased activity, and of such a preparation the initial dose should not exceed a fluid drachm.

NOTE 5.

Continuation of Case XVI. (p. 267). Taking belladonna in full doses :—

1869. Jan. 19. Night urine, sp. gr. 1016·0, contained in 1000 gr. measures 0·28 gr. albumen.
Afternoon " " 1024·8 " " " " 1·08 "

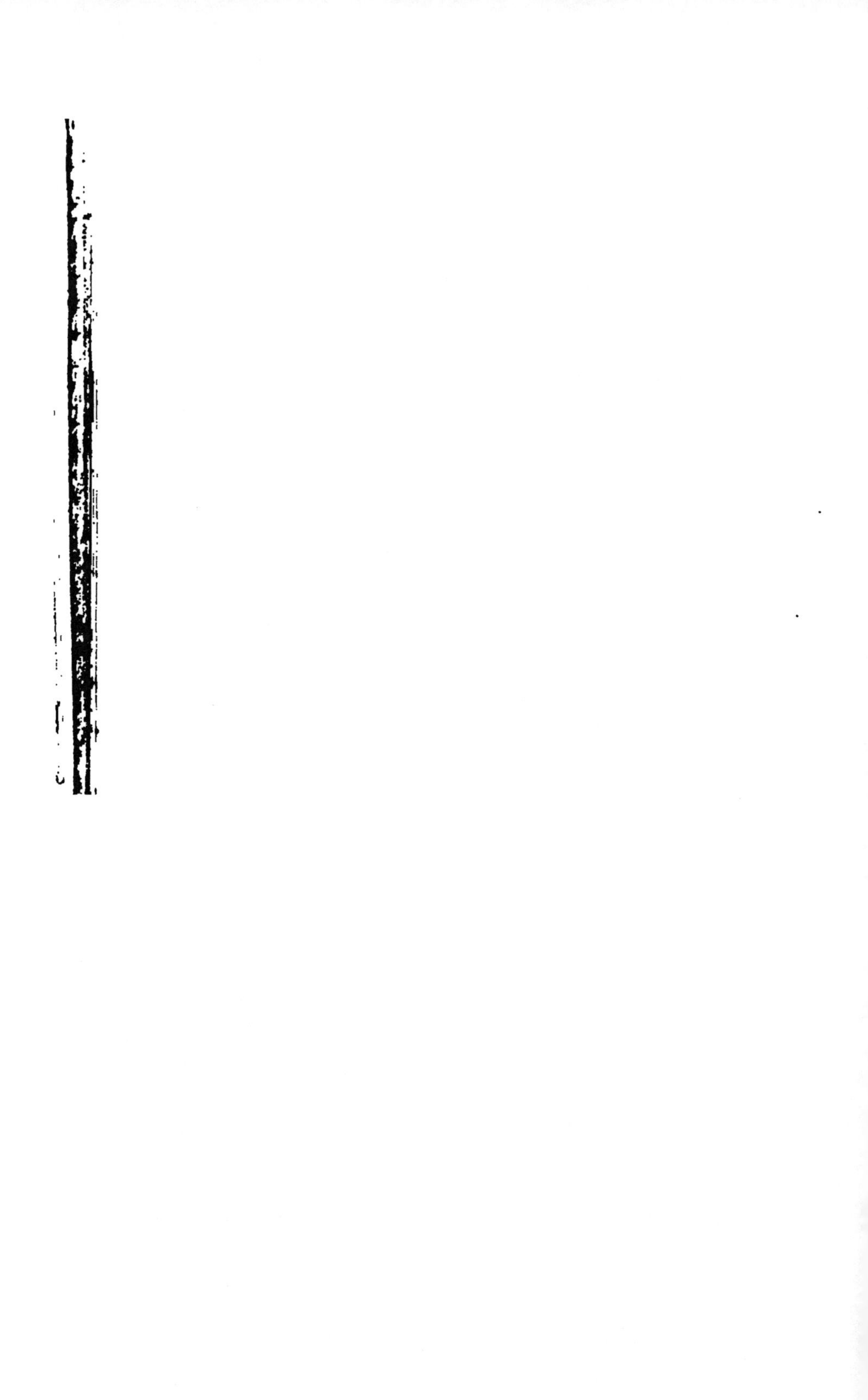

INDEX.

LONDON: PRINTED BY
SPOTTISWOODE AND CO., NEW-STREET SQUARE
AND PARLIAMENT STREET

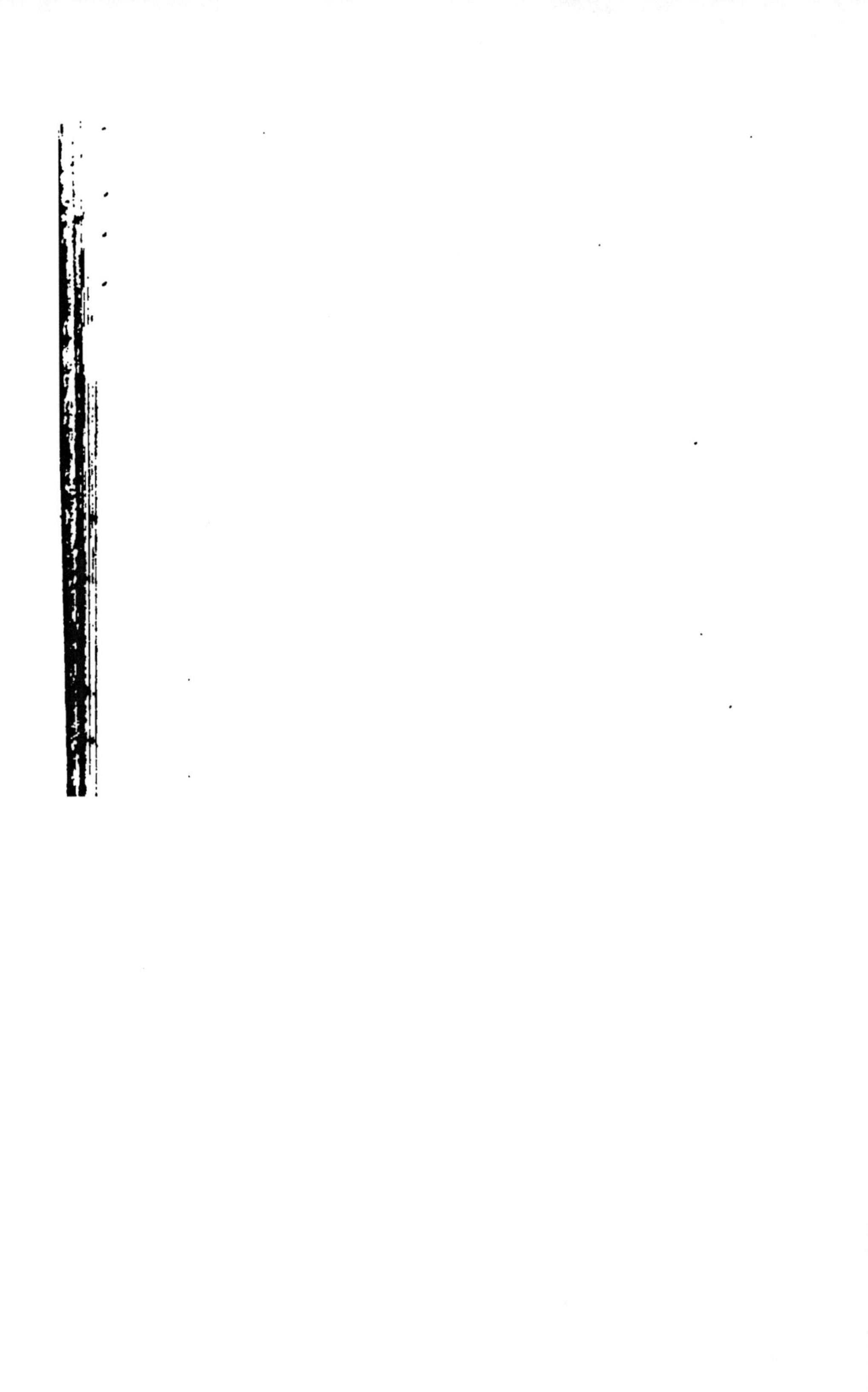

CPSIA information can be obtained at www.ICGtesting.com
Printed in the USA
BVOW061428250612

293604BV00003B/5/P